Tourism and the Emergence of Nation-States in the Arab Eastern Mediterranean, 1920s-1930s

Critical, Connected Histories

Critical, Connected Histories is a series of works that explore unfamiliar social, cultural, and political issues that connect people of Asia, the Middle-East, Africa, America and Europe in the modern age. Building on trends in historiography that look at transnational flows and networks and foreground themes of circulation and connection, the series aims to break down artificial boundaries between regions that have long dominated traditional area studies and the discipline of history. It aims to build new ways of mapping these networks and journeys of people, ideas, and goods without omitting dynamics of power and the resilience of the state. The works in this series seek to sharpen the edges of such inquiries and challenge legal and imagined boundaries while examining the durability of their structures of power. In addition, the series cautions against the potential universalization of histories, and situates queries within different vantage points.

Multi-disciplinary in their approach, the books in this series draw from sources in multiple languages and media. They seek to expand our understanding and historical knowledge of events, processes, and movements. In denaturalizing the practices of historical writing, this series offers a forum for scholarship that is both theorised and empirically rich, and one that in particular addresses the violence and inequality of the modern ages as well as its promises.

Series Editors
Nira Wickramasinghe, Leiden University
Tsolin Nalbantian, Leiden University

Editorial Board
Fred Cooper, New York University
Engseng Ho, Duke University
Ilham Khuri-Makdissi, Northeastern University
Susan Pennybacker, University of North Carolina, Chapel Hill
Bhavani Raman, Toronto University
Willem van Schendel, Amsterdam University

Other titles in this series:

Myra Ann Houser, *Bureaucrats of Liberation. Southern African and American Lawyers and Clients During the Apartheid Era*, 2020

Nadine Willems, *Ishikawa Sanshirō's Geographical Imagination. Transnational Anarchism and the Reconfiguration of Everyday Life in Early Twentieth-Century Japan*, 2020

Alicia Schrikker and Nira Wickramasinghe (eds), *Being a Slave. Histories and Legacies of European Slavery in the Indian Ocean*, 2020

David Henley and Nira Wickramasinghe (eds), *Monsoon Asia. A reader on South and Southeast Asia*, 2022

TOURISM AND THE EMERGENCE OF NATION-STATES IN THE ARAB EASTERN MEDITERRANEAN, 1920s-1930s

Jasmin Daam

LEIDEN UNIVERSITY PRESS

The Open Access edition was made possible by a grant from Kassel University Library.

This book is based on the dissertation *Tourist Spaces: The Arab East as a Tourist Destination in the 1920s and 1930s* submitted by Jasmin Daam for the degree of Doctor of Philosophy at the Faculty of Social Sciences, University of Kassel, and defended on 30 October 2020.

Critical, Connected Histories, volume 5

Cover design: Andre Klijsen
Cover illustration: The swimming pool of the Saint Georges Hotel, Beirut, in the mid-1930s (Photographer: René Zuber. Reproduced by kind permission of the Fonds photographique René Zuber and the Bibliothèque Nationale de France)
Lay-out: Crius Group

ISBN 978 90 8728 391 9
e-ISBN 978 94 0060 437 7
https://doi.org/10.10.24415/9789087283919
NUR 692

Printed and bound by CPI Group (UK) Ltd, Croydon, CR0 4YY

To Jinan, Assem, Shadi, El Paco, Ali and Hussein

Table of contents

List of illustrations

Acknowledgements

As this book is nearing completion, I look back on seven years of precious experiences, during which so many persons supported me and my research that it is impossible to name them all. The project took shape at the University of Konstanz. As a student assistant, I benefited from the inspiring environment of the Leibniz Prize Research Group *Global Processes: Eighteenth to Twentieth Centuries*, directed by Jürgen Osterhammel, to whom I owe my greatest thanks. From the outset, he encouraged me to venture into this project and all of our conversations guided me towards new levels of reflection. It was a fortunate privilege to be able to learn from him and from the members of staff at the Chair of Modern History.

Julia Hauser granted me the opportunity to realise the project at her Chair of Global History and the History of Globalisation Processes at the University of Kassel. I am deeply grateful for her encouragement and support, not least with regard to the numerous research leaves the project required, as well as for the atmosphere of critical debate, mutual trust and appreciation she created at the university. I also owe my positive memories at both universities to Heidi Engelmann, Petra Klein and Silke Stoklossa-Metz.

The project could not have been realised without the generous help of a number of additional institutions. Scholarships from the *Studienstiftung des deutschen Volkes*, the German Historical Institute in London and the German Historical Institute in Paris, as well as the graduate centre of the University of Kassel (KIGG) funded the essential research leaves during which the project took shape. The library of the University of Kassel generously funded the major part of the open access publication.

Moreover, I profited immensely from membership in the research network *The Modern Mediterranean: Dynamics of a World Region, 1800–2000* funded by the German Research Foundation DFG and directed by Manuel Borutta. The regular meetings and the inspiring exchange since 2016 have raised stimulating questions and confronted me with new lines of thought.

During those past years and many travels, I presented and discussed parts of the book with colleagues and audiences in various contexts and countries that are too numerous to be named individually. I am grateful for the tips, advice, questions and thoughts they generously shared. Particularly precious, at various stages of the project, were conversations with Alexandra Bechtum, Malte Fuhrmann, Agnes Gehbald, Jens Hanssen, Elke Shogig Hartmann, Waleed Hazbun, Valeska Huber, Andreas Hübner, Sarah Irving, Nora Lafi, Simon Jackson, Jan C. Jansen, Meike

Knittel, Judith Madgalena Kolmeigner, Lisa Korge, Esther Möller, Christine Pflüger, Jörg Requate, Karène Sanchez-Summerer, Cyrus Schayegh and Nadya Sbaiti. With Janina Santer, I shared the most inspiring lunchbreaks during these years, and I am greatly indebted to her for the knowledge she shared, archive materials she pointed out to me, doors she opened and for many discussions that helped me clarify my thoughts.

Agnes Gehbald, Meike Knittel, and Lisa Korge as well as Janina Santer read chapters and drafts at various stages of the writing process, and Judith Magdalena Kolmeigner thoroughly read the entire work. Hussein Al Sultan proofread my translations of Arabic sources. Rosemary Maxton carefully corrected and polished the language and style of a non-native English speaker. Obviously, all remaining mistakes, misunderstandings, mistranslations and shortcomings are my own responsibility.

Searching for documents on tourism has been a challenge as the often ephemeral sources are not necessarily catalogued in the major archives and libraries. The fact that I found many more needles in the haystack than I ever expected is due to the generous assistance and precious advice of colleagues, friends, librarians and archivists who shared their knowledge and admitted me to their collections. Ibrahim Assi and his staff welcomed me to the Lebanese National Archives, where I worked in the most pleasurable atmosphere. Monika Borgmann and Lokman Slim from the *Umam Foundation* in Beirut shared a brochure with me that proved to be particularly important. Rowina Bou Harb and Yasmine Chemali allowed me to delve into the treasures of the *Fouad Debbas Collection* at Sursock Museum and provided me with a charming workplace, especially on the late-opening Thursdays. Isabelle Chassin helped me in my research at the archives of the *Fédération française des clubs alpins et de montagne* in Paris. Clémence Ducroix and Hélène ten Hove guided me through the collections of the French Lines Association in Le Havre. Wolf Dieter Lemke introduced me to his stunning postcard collection and shared his vast inside knowledge. Magda Nammour unearthed treasures from the uncatalogued holdings of the *Bibliothèque Orientale* at the Université Saint-Joseph. Rachel Rowe, in charge of the *Royal Commonwealth Society Archives* at the University Library Cambridge, Stefan Seeger at the OIB Library in Beirut and Debbie Usher at the *Middle East Centre Archive* in Oxford assisted me in working with their collections. Charbel Saad from the *Arab Image Foundation* opened the library and archives for me. The library of the *Institute for Palestine Studies* in Beirut hosted me at the very beginning and at the very end of my research in Lebanon, where I was warmly welcomed and kindly assisted by Jeanette Sarouphim and the staff.

The exceptionally kind assistance of the staff at the *Centre des Archives Diplomatiques* in Nantes has to be highlighted, and the librarians at the *Jafet Library* (American University of Beirut) admitted me as a permanent resident in the microfilm reading room during the summer of 2017. In addition, I profited greatly from my research in the *Archives nationales* in Pierrefitte-sur-Seine, the *British Library* in London, the *Institut français du Proche-Orient* (Ifpo) in Beirut, the National Archives in Kew, the National Library of Scotland in Edinburgh, the National Maritime Museum in Greenwich as well as the SOAS Archives and Special Collections.

I am grateful to Tsolin Nalbantian and Nira Wickramasinghe that this book has found its place in the *Critical, Connected Histories* series. Romy Uijen and Saskia Gieling from Leiden University Press kindly assisted me during the process of preparing the manuscript for publication. I owe infinite thanks to the two anonymous reviewers and their attentive and critical reading of the initial version of this manuscript. Their comments and questions were incredibly helpful and helped me clarify my thinking. Of course, any remaining shortcomings are my own responsibility.

During these past years, travels and library retreats were indispensable to find answers to my questions, and they offered me insights I never dared to ask for. Friends and family, in particular Siegfried and Ines Daam as well as Siegfried and Edith Leinmüller, understood what this journey meant to me, accepting occasional phone calls and skype conversations as replacements for personal interaction. I am sure they are aware of my gratitude.

This book is dedicated to Shadi Ayyoubi, Ali Bassal, Hassan Nouri, Jinan Slim, Hussein Al Sultan and Assem Tarhini for thanks to them, Beirut has become home.

Note on Transliteration

I transliterated Arabic terms according to a simplified version of the system suggested by the *International Journal of Middle East Studies* (IJMES), omitting all diacritics except *ayn* (ʿ) and *hamza* (ʾ, except at the beginning of the word). My guiding principle in transliterating was to make places and names identifiable for the non-specialist, therefore I used common English spellings for proper nouns and geographic terms as far as they existed (e.g. Cairo instead of al-Qahira; Aleppo instead of Halab). If authors had published works in European languages, I opted for the transliteration they chose (e.g. George Antonius, rather than Jurj Antuniyus; Alexander Khoori, rather than Iskandar Khuri). If place names are well-established in French transliteration, which is the case with the Lebanese summer resorts in particular, I opted for this spelling (e.g. Dhour el-Choueir instead of Dhuhur al-Shuwayr), accepting inconsistencies where necessary. I tried to make sure that the interested reader can identify names and places easily by looking up these terms online.

All translations, unless otherwise stated, are my own.

Abbreviations

ANF	Archives Nationales (de France), Pierrefitte-sur-Seine
ANL	Archives Nationales du Liban, Beirut
AUB	Jafet Library, American University of Beirut
BL	British Library, London
BNF	Bibliothèque nationale de France, Paris
BO	Bibliothèque Orientale, Université Saint-Joseph, Beirut
CADC	Centre des Archives diplomatiques de La Courneuve
CADN	Centre des Archives diplomatiques de Nantes
FDC	Fouad Debbas Collection, Sursock Museum, Beirut
FFCAM	Fédération française des clubs alpins et de montagne, Paris
French Lines	Association French Lines, Le Havre
IfP	Institute for Palestine Studies, Beirut
Ifpo	Institut français du Proche-Orient, Beirut
LoC	Photograph Collection, Library of Congress, Washington, DC
MECA	Middle East Centre Archive, St Antony's College, University of Oxford
NA Kew	National Archives, Kew
NLI	National Library of Israel, Jerusalem
NLS	National Library of Scotland, Edinburgh
NMM	National Maritime Museum, Greenwich
RCSA	Royal Commonwealth Society Archives, Cambridge University Library
SOAS	Archives of the School of Oriental and African Studies, London

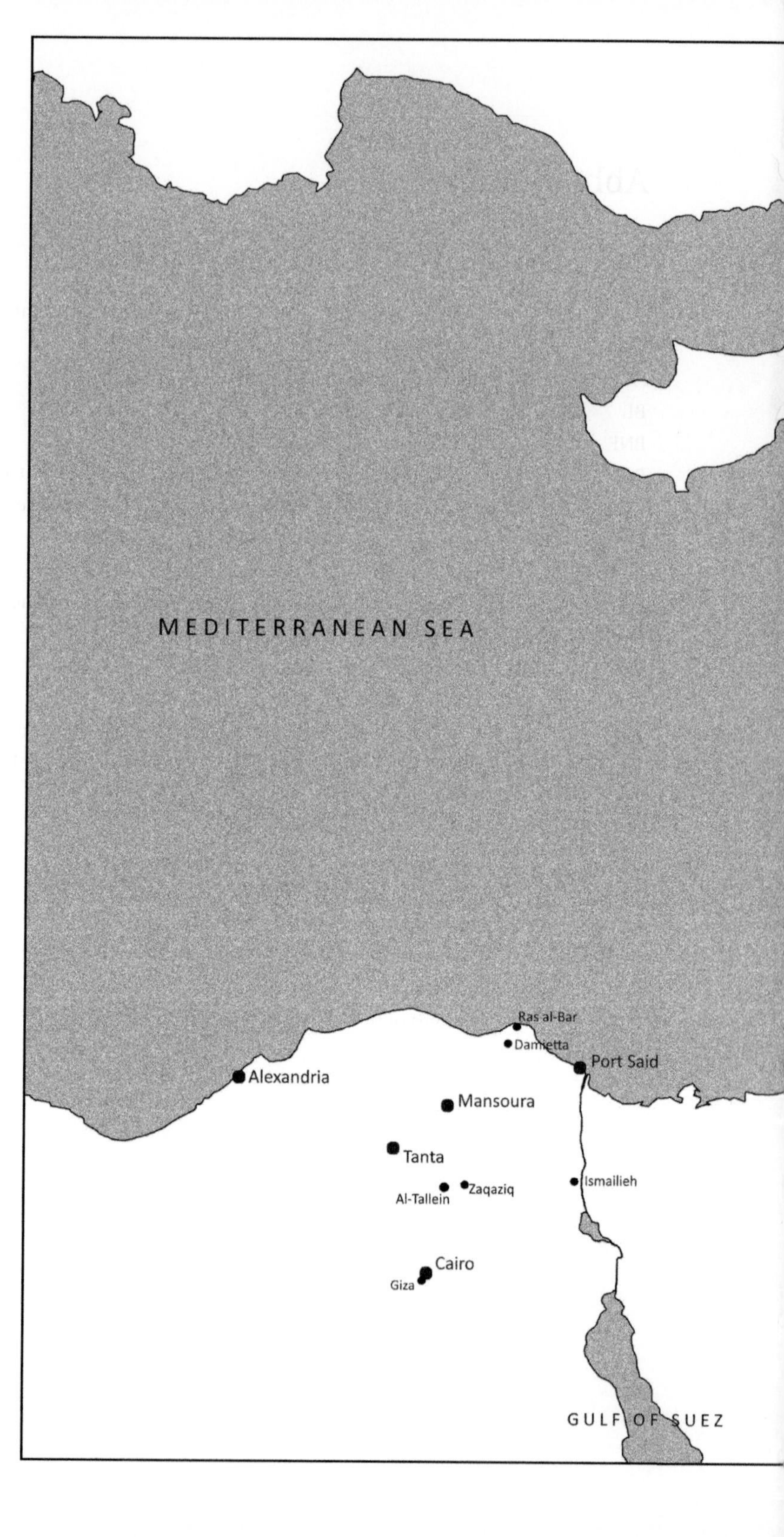

MEDITERRANEAN SEA
Ras al-Bar
Damietta
Port Said
Alexandria
Mansoura
Tanta
Al-Tallein
Zaqaziq
Ismailieh
Cairo
Giza
GULF OF SUEZ

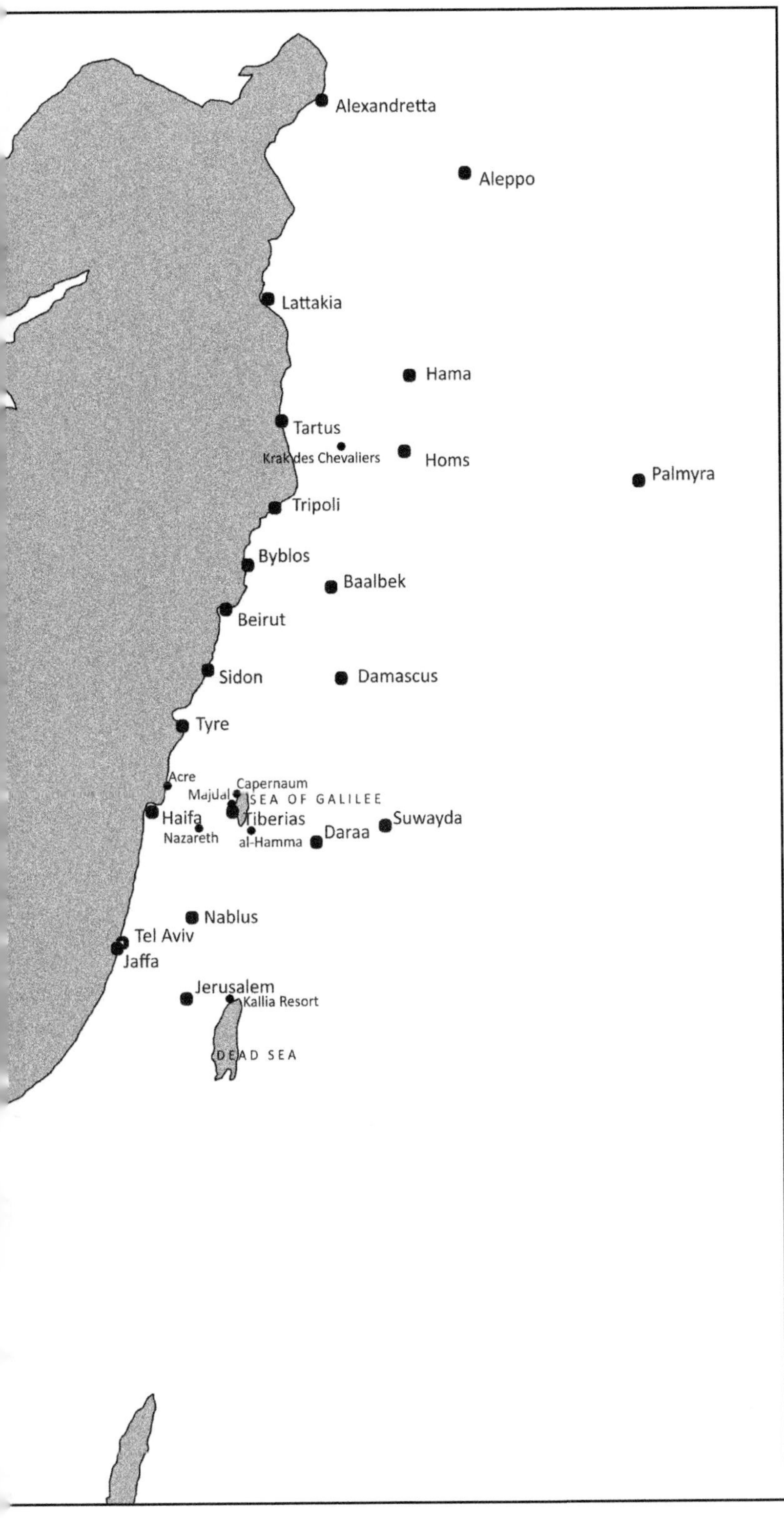

Map 1: Major places mentioned in the book

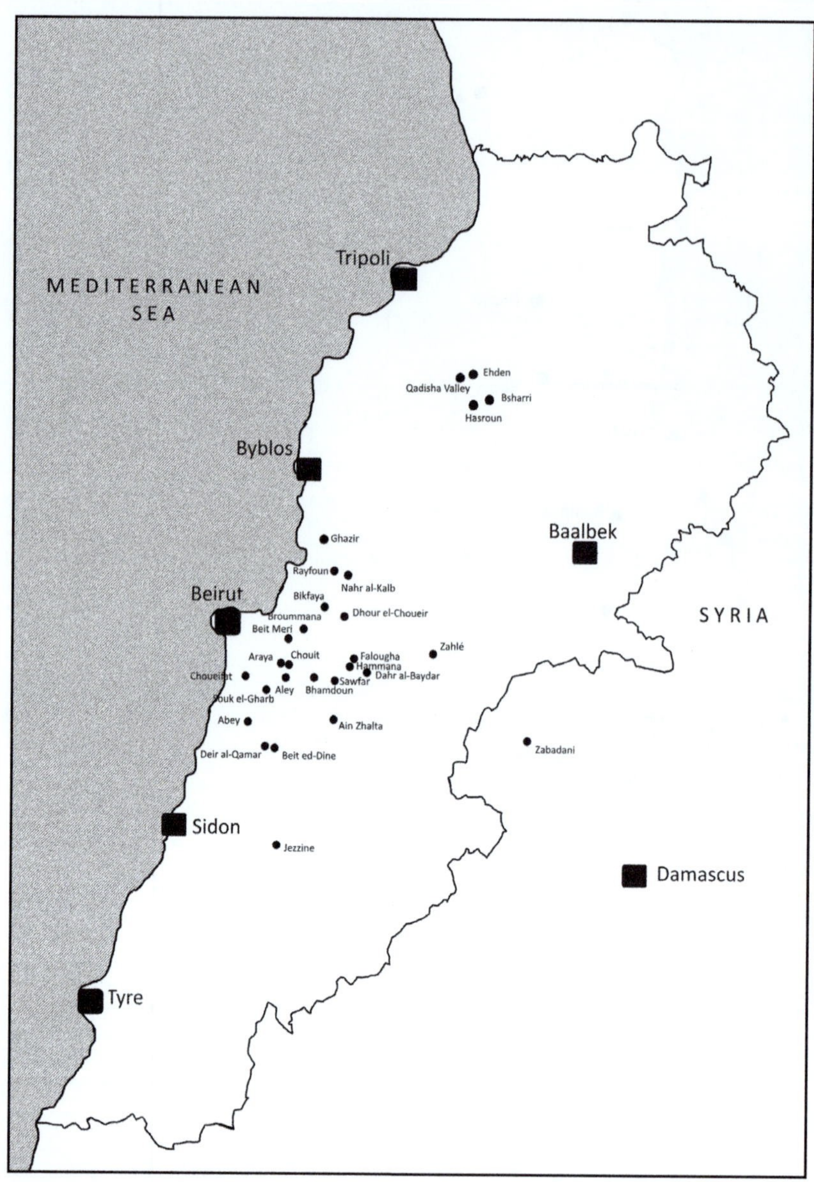

Map 2: Lebanese *estivage* villages mentioned in the book

CHAPTER 1

Introduction

On 1 April 1931, the Lebanese engineer Albert Naccache guided a group of tourists to the Qadisha Valley in Northern Lebanon. Nowadays, the Qadisha Valley figures in the World Heritage List of UNESCO, owing to its age-old cedars and monasteries dating back to the early days of Christianity.[1] Naccache, however, drew the attention of his visitors to a more recent attraction: the hydro-electrical power plant on the Abu ʿAli River.

Albert Naccache's guided tour to the power plant was depicted in the photograph album of one such French tourist (fig. 1). On the album pages, the black-and-white memories of this excursion appeared between photographs of pharaonic statues, the Umayyad Mosque in Damascus, the Church of the Holy Sepulchre in Jerusalem, and the Roman ruins of Baalbek. Since these renowned sights show that the tourist had followed the beaten tracks of numerous European travellers to the Arab East, it can be assumed that the visit to the electrical power plant did not indicate some eccentric hobby of hers. Instead, I suggest that the visit was initiated by Albert Naccache, the man wearing a Western suit and necktie in photograph no. 97.

In this way, the engineer-cum-tour guide Naccache questions our intuitive understanding of tourist attractions. If sights are commonly understood to be of historical or cultural relevance, remarkable due to their curious difference or seeming authenticity, Naccache's choice of presenting the local power plant to the group of tourists requires an explanation.[2] This excursion suggests that, for actors during the 1920s and 1930s, tourism was not a clearly defined business model they implemented, but rather a resource, the deployment of which remained to be identified.

It seems that in the Qadisha Valley, Naccache was on a political mission. The owner of the *Société de Kadisha* advocated the industrial development of Lebanon, and he perceived tourism as an opportunity to develop the Lebanese hinterland, an idea for which he had already lobbied in 1919. According to Naccache, a better road network and the electrification of Lebanon were necessary to achieve the economic independence of Lebanon, and only a viable economy would guarantee its political independence.[3] The scene at the Abu ʿAli power plant thus suggests that Naccache wished to showcase Lebanese technological progress and economic potential – but also its political ambitions – to the tourist party.

Figure 1: Album page depicting the excursion to the hydro-electrical power plant, 1931
(Reproduced by kind permission of the Bibliothèque Nationale de France)

Such an approach to tourism must be understood in the context of the international post-war order. At the end of World War I, claims for sovereignty gained ground in the territories under foreign rule and among those who feared being colonised.[4] After the defeat of the Ottoman Empire, the European colonial powers, at the height of their spatial expansion, intensified their presence in the Arab East in the framework of a modified colonial order: the mandates system. Although the system first and foremost served the legitimisation of prolonged foreign rule, according to Susan Pedersen, it irreversibly changed the modalities of colonial governance.[5] The mandates system suggested that the capacity of populations for self-government would be assessed according to internationally applicable criteria. Assuming that the inhabitants of the conquered territories were "not yet able to stand by themselves under the strenuous conditions of the modern world", the mandate agreement classified these territories according to their "stage of development" and stipulated that the formerly Ottoman Arab provinces should be assisted "by a Mandatory until such time as they are able to stand alone". This implied that the 'modernity' and the civilisational progress of those territories would be monitored and evaluated – once a sufficient degree of 'civilisation' was obtained, they would achieve full sovereignty.[6] Under these conditions, the presence of tourists from "advanced nations"[7] was a chance to negotiate the achieved rank of the states striving for independence. When Albert Naccache led the group of French tourists to the hydro-electrical power plant, he demonstrated the advanced level of progress and civilisation achieved by the Lebanese, justifying their claims for sovereignty.

Tourism therefore entered into the process of defining a new spatio-political order in the post-Ottoman Arab East. The question as to whether the formerly Ottoman territories would be administered by European colonial empires, gain independence as nation-states, or as a sovereign pan-Arab entity, stirred conflicts that were not only fought at the peace conferences and by means of violent protest and uprisings, but also by cultural means. One of these arenas was tourism. Actors in tourism presented historical bonds and mapped sites of symbolic relevance, thereby negotiating spheres of belonging. Moreover, as tourist movements produced centres and peripheries, created presence and justified the expansion of infrastructures, directing these movements became a method of implementing spatial visions. Analytically, tourism may thus serve as a lens through which we can trace the transition from a post-Ottoman spatial order to the emergence of nation-states. The combination of several spatial levels of analysis allows us to grasp both Ottoman continuities and new connections linked to imperial penetration and processes of globalisation as suggested by Cyrus Schayegh.[8] In this way, the analysis of tourism might complement our understanding of the formation of nation-states in the Arab East.

The book approaches tourism during the interwar period from a new angle as it examines tourism as a political resource in both its national and its imperial contexts. Whereas historical research on tourism analysed the phenomenon mainly in terms of mobilities, leisure culture, and perceptions of the Other – and thus from the perspective of travellers – in addition I address perspectives from societies in the tourist destinations, the active participation of which has been largely omitted in existing studies.[9] Even if the expectations and world views of tourists had been shaped in the imperial metropoles, they could be approached by local populations in different ways than colonial officials or military staff. Tourism in imperial contexts was more than an opportunity to disseminate imperial propaganda.[10] The presence of French and British tourists, informal representatives of the respective imperial power, created a forum for debate that potentially included the imperial metropole. While imperial power relations were not suspended, tourists depended on local assistants and interpreters due to their temporally limited presence and their lack of local knowledge. This book therefore focuses on tourists from Great Britain and France, even if tourists from other nationalities will occasionally cross our paths.

The attempts of colonial administrators, as well as associations and political representatives in the Arab East to communicate visions, shape perceptions, and define access and movements – creating spaces, that is – testified to their aims of shaping the future political order of the region.[11] Since these techniques were accessible to different actors, the analysis has to take into account processes of communication, negotiation, and shaping movements beyond the binary paradigm of coloniser and colonised, visitor and visited. In the 'visited' societies, tourism was a promising opportunity for some – both in economic and political terms – while it was a menace to others. Regarding the intercultural encounter, imperial propaganda, racial stereotypes and Orientalist thinking undoubtedly characterised Eastern Mediterranean tourism, but the relocation of tourists to the overseas possessions created an arena accessible to actors rejecting those narratives and presenting alternative views.[12] Such encounters allowed local tour guides, intellectuals, authors, politicians and entrepreneurs to address their messages directly to 'imperial amateurs' from the colonial metropoles, and their local expertise, on which tourists depended, questioned colonial configurations of power. The focus on these actors, their ambitions, achievements and failures in tourism development not only sheds light on an often-underestimated connection, but also demonstrates the agency of nationalist actors, who actively grasped tourism as a resource to shape sovereign nation-states.

Tourism

The relevance of tourists to the formation of spaces stemmed from their collective appearance and their organised, foreseeable itineraries. Naccache cared about the French tourist *because* she was a tourist like many others. 'Tourism' often has negative connotations and is contrasted with the more positive term 'travel', both by researchers and in popular imagination. Whereas travellers are reputed to be well-informed and open-minded explorers of foreign places and cultures, ignorance characterises the tourist: tourists allegedly follow the guide of the party rather than exploring unknown terrain; they are supposed to expect Western amenities, misbehave out of a lack of understanding, and mistake staged exotic performances for authentic expressions of local culture. Such attributions of ignorance and inadequate behaviour turned into clichés, and 'tourist' became a derogatory attribute mostly used by other tourists for self-distinction.[13]

To grasp the allegedly neutral core of the term, sociologists and historians have described tourism as travelling for the purpose of leisure. Travelling "for reasons basically unconnected with work"[14] defined tourism as a distinct category according to these authors, and as a distinctly modern phenomenon in comparison with varieties of travel motivated by trade or business, professional mobility (soldiers and sailors), or religious pilgrimage.[15] Yet authors referring to an inherent modernity of tourism often had to downplay evidence for pre-modern leisure travel.[16] In addition, the motivation behind a journey is difficult to assess. Reasons for travel overlapped more often than not and claims of travellers as to the purpose of their journey tell us much more about socially acceptable forms of mobility than about travellers' intentions. Empirical findings show that the practices associated with different types of travel often intersected. I came across soldiers based in Syria who travelled to Baalbek and Jerusalem for sightseeing, and businessmen who profited from the occasion to undertake excursions of a recreational nature. Religious Christians spending their Easter holidays in Jerusalem often embedded this pilgrimage in a larger journey, including visits to the Umayyad Mosque in Damascus or the Pyramids of Giza. In many cases sources such as photographs or postcards did not reveal the intention of travellers regarding their journey.

I therefore adopt an approach that focuses on structures of organised travel rather than aiming to define the individual intentions of travellers. In his paper on tourism to the Dutch Indies, Robert Cribb added a second feature to the definition of tourism that circumvents the problem of focusing on the self-perception of tourists, avoiding both the negative moral connotations of the term and the notoriously ambiguous and unstable individual statements of travel purposes. Instead, Cribb argued that a peculiarity of tourism was the high degree of organisation:

> Tourism [...] is perhaps best characterized not just by the element of pleasure, but also by a loss of independence. In exchange for access, the tourist surrenders autonomy whether by joining a tour group or simply by following a guidebook which maps out one or more beaten paths. If we can take guidebooks and tour groups as the key signs of modern tourism, then we can date international tourism in the Indies to around the beginning of the twentieth century.[17]

While the Dutch Indies were a relative latecomer in international tourism, Cribb's definition of tourism as a highly organised form of travelling is instructive as it shifts attention from the mere tourists to infrastructures and services related to tourism. The modernity of tourism is ascribed to the material conditions of travelling: innovations in transport and the organisation of labour contributed to the emergence of the organised 'package' tour as a new form of travel, which made excursions and voyages more broadly accessible. Simultaneously, these new forms of group travel implied a scheduling of the journey and necessitated careful planning before departure. In these processes, tourists were assisted not only by travel agents, but also by guidebooks issued by various publishing houses such as *Baedeker, Hachette* or *Murray*.[18] These developments of the late nineteenth century fundamentally altered the way of travelling for everybody – regardless of the intended travel purpose. As a result, 'tourism' was no longer a conscious choice but the default option. Based on Cribb's definition, the terms 'traveller' and 'tourist' (resp. 'travel' and 'tourism') will be used synonymously throughout this book.

By the 1920s, organised travel was well-established in several countries around the Mediterranean. The British tour operator *Thomas Cook & Son* led a first tour to Egypt in 1869 and the business quickly expanded.[19] In French North Africa, the *Compagnie Générale Transatlantique* (CGT) offered tickets for circular tours from 1882 onwards.[20] When European tourists on packaged tours started to visit the Eastern Mediterranean, they merged with other leisure-seekers. Since the 1860s, 'wintering' stays had become popular among wealthy Europeans who escaped the cold European winter season, seeking refuge in the mild Mediterranean climate of the French Riviera, Algeria and Egypt.[21] Around the same time, Lebanese and Syrian mountain villages started to attract summer guests from the Arab East. Among them were emigrant families visiting their places of origin as well as summer guests from the administrative centres of the Ottoman Empire. Quite similar to the *hiverneurs* seeking refuge from the European winter, the latter escaped the heat and humidity of Beirut, Baghdad, Cairo and Damascus and spent the summer months in the cooler Mount Lebanon area.[22] *Estivage* (*al-istiyaf*) or *villégiature* were the terms used both in British and French sources to describe this practice.[23] It was characterised mainly by practices of sociability, but photographs suggest that it also included sightseeing excursions, for example to the ruins of Baalbek.[24]

The proponents of tourism development had to face several setbacks caused by the political and economic crises of the interwar period, which will be discussed in the following chapters. Protests and revolts against the imperial overlords such as in Egypt in 1925 or in Syria in 1925–27 caused downturns, and in Palestine violent uprisings against the effects of the Jewish immigration movement hampered tourism development and led to a noticeable decline in the number of visitors in 1929 and 1936–39. On a global level, the Great Depression from 1929 onwards led to a cancellation of bookings and brought several hotels in the region into turmoil. As a reaction to these difficulties, advocates of tourism development attempted to reach out to new customers.

The general expansion of organised tourism attracted new social groups to the Mediterranean shores. In the 1930s, proponents of tourism development advertised excursions to the seaside among the local urban working classes. Moreover, an increasing number of members of the European middle classes, unmarried women, and even workers were able to afford journeys to the Arab East at least "once in a lifetime".[25] Some tour operators consciously targeted travellers on a low budget. The shipping line *Fabre Lines* or associations such as the British *Workers' Travel Association*, for example, offered comparatively inexpensive tickets or tour packages.[26]

The tour operators also targeted female travellers like the aforementioned author of the photograph album. By 1925, she had already undertaken a long voyage to East Asia, where she visited Japan, China, Cambodia and Indonesia, among other places. Both on this voyage and on her journey to the Arab East in the spring of 1931, she appears to have travelled alone.[27] In particular *Cook & Son* had advertised organised tours as a possibility for 'respectable women' to travel without their husbands or other relatives, and contributed to rendering the solitary travel of bourgeois women socially acceptable by the interwar period.[28] Not all tour agents would go as far as the 1934 guidebook *"Sur les routes du Levant"* though, which recommended a journey to the Eastern Mediterranean to solitary female travellers as a (last) adventure before marriage.[29]

Whereas organised tourism facilitated travelling for some social groups, other travellers faced new restrictions. Valeska Huber argued that we should consider 'mobility' in the plural. Having demonstrated that the general growth in mobility triggered increasing controls and restrictions for certain groups, she coined the term "other mobilities" to grasp the restrictions and sometimes refusals of mobilities which lower-class travellers in particular were confronted with.[30] While we will encounter a number of solitary female tourists, for example, women intending to work in the tourism sector, often as performers, experienced far greater difficulties in obtaining visas.[31] Moreover, British and French authorities critically observed and sometimes impeded the movements of politically or morally suspect travellers.

Such restrictions notwithstanding, the expansion of organised travel to the Eastern Mediterranean was the condition for the political implications that characterised tourism after World War I, and the overall number of tourists reached peaks in the late 1920s and the mid-1930s. The growing numbers of European and American tourists in Egypt and the mandate states turned these visitors not only into a target group for entrepreneurs in the transport and accommodation sectors, but also into an audience accessible to politicians, entrepreneurs, and intellectuals. Viewed from such an angle, tourism has to be understood as an essentially transnational phenomenon, not because European travellers crossed national borders, but because it established occasions that allowed, for example, an Arab entrepreneur like Naccache to communicate with a French tourist.[32] Since Great Britain and France were the imperial powers in control in the countries under scrutiny, I focus on the British and French tourists to whom these messages were mainly addressed.

That said, the connection between tourism and empire is, according to Eric G. E. Zuelow, still among the major understudied aspects in the history of tourism.[33] Only in recent years have historians begun analysing tourism in this context, often as an element of imperial culture. Ellen Furlough was among the pioneers, convincingly arguing that French imperial administrators exploited tourism in order to disseminate propagandistic messages. This was consolidated by subsequent studies of tourism in the French and Italian imperial contexts, for example by Colette Zytnicki in her research on tourism to French Algeria, Brian McLaren's work on Italian colonial Libya, and Aline Demay's study on French Indochina.[34] As far as British imperial tourism is concerned, research often focused on travel literature and self-conceptions of travellers. Few studies analysed the role of the British Empire in the expansion of travel infrastructures or the institu-tionalisation of tourist practices.[35] Studies on nineteenth-century travel to Egypt often demonstrated in a Saidian tradition of research how the exhibition of local culture and the "tourist gaze" at the Other were linked to imperial ambitions and a colonial mindset.[36]

In the Mediterranean, the propagandistic side of tourism was particularly pro-nounced in the French imperial possessions, presumably due to the important role the Mediterranean played in French imperial self-conceptions. The claim to restore the grandeur of the ancient Roman Empire resulted in policies of archaeological discovery, preservation and exhibition of monuments that were promoted as tour-ist attractions. The reference to ancient Roman history implied that imperial rule was the 'natural' condition of the Mediterranean and its proponents stipulated a parallel between the French colonisers and the ancient Romans, who had allegedly brought knowledge, science, and culture to the region.[37] A second strand of French preservationist policies was dedicated to 'traditional' local culture, prominently developed under Hubert Lyautey in the French protectorate of Morocco, and later applied in Mandate Syria. Its proponents argued that local traditions, conceived of as expressions of an allegedly authentic local culture, had to be protected against the spoiling influence of contemporary inventions and technologies. Accordingly, efforts to maintain the production of local crafts paralleled measures of conservation and exhibition, while preserving an allegedly 'primitive state' and demonstrating that these peoples were not ready for self-rule.[38]

Forms of creating heritage, as well as the implications of tourism for both travelling and accommodating societies, have been studied by sociologists, anthro-pologists, and geographers who dominated research on tourism until recently.[39] In these contexts, tourism as a variety of cultural contact has been judged rather critically. Authors argued that most tourist practices at best confirmed pre-travel stereotypes of tourists and at worst destroyed local culture. The authenticity tour-ists requested, sought "in other historical periods and other cultures",[40] seemed to impede encounters on equal terms. In the last decade, however, authors have nuanced such negative views of tourism.[41] Among others, historical perspectives contributed to a new understanding of the phenomenon. Shelley Baranowski argued in her work on the German national socialist travel organisation *Kraft durch Freude* that tourists should not be mistaken for victims of political propaganda. She reminded her readers that even if travel might have been intended as an occasion to spread propagandistic messages, such intentions could be subverted by tourists, who often pursued their own interests and agendas.[42]

Moreover, contrary to what is generally assumed, tourism was not a one-way movement of Westerners to the colonised world. French tourism advertisements in Egypt, for example, aimed at attracting Egyptian visitors to France, rather than vice-versa. The Egyptian bourgeoisie was such an important target group that in 1935, the French commercial attaché in Alexandria, Grandguillot, did not have any reservations about a tax-free exchange of tourist brochures, stating that "in return for one French tourist who spends a week in Egypt in winter without spending much money, we receive in summer one hundred Egyptian tourists, who stay for three months and spend at least 25,000 francs per person."[43] Tourism was thus a shared practice among a "global bourgeoisie".[44] The social and geographical range of such middle-class tourists widened considerably in the interwar period and we have to assume that European tourists encountered the inhabitants of their destinations not only in subordinate positions such as cameleers or waiters, but also as middle-class Arab travellers on ships or in grand hotels. At the same time, European tourists were a socially heterogeneous group. Elite circles existed, such as the touring clubs and automobile associations spreading across the globe. They constituted transnational elitist networks for the exchange of advice, recommendations and experiences.[45] The more significant development in terms of numbers, however, was the emergence of package tours and organised excursions attracting travellers from a broader social range.

Indeed, members of the middle classes appeared to be the dominant actors shaping tourist practices during the 1920s and 1930s. Regardless of their national backgrounds, they used the same media to document their journeys, they founded similar institutions and businesses catering to the needs of travellers, and they seemed to have a vital interest in drafting narratives to market their hometowns or countries as tourist destinations. In referring to these actors as members of a global middle class, I am drawing on recent reflections of Christof Dejung, David Motadel and Jürgen Osterhammel. They analysed the "global bourgeoisie" from a global historical perspective as a social group having emerged across the globe at a similar historical moment. Its often interconnected members were identified as the "most effective proponents" of processes of globalisation.[46] At the same time, the authors described the ambitions of the middle classes to shape their states and societies.[47] These ambitions were intertwined with claims for modernity, thoroughly analysed by Keith Watenpaugh, which were not perceived as an adoption of 'Western' modernity but rather understood as the participation in a universal standard.[48] I argue in this book that these claims for modernity, globalisation, mobility and access to shaping politics were intertwined with their ambitions to promote tourism in the Arab East.

The hopes members of the middle class placed in tourism were thus of political rather than economic character, and they were related to their aims to foster the emergence of viable nation-states. Tourism addressed two major requirements of nation-building: it contributed to the attribution of sense to an imagined national community as described by Benedict Anderson, and it contributed to the integration of a given, bounded, territorial entity as described by Charles Maier.[49] As a specific asset, tourism made a global audience participate in these processes.

Spaces

Tourism was an asset to actors aiming to shape the future nation-states in the region, and this potential explains the relevance middle-class actors attributed to tourism. 'Space' was defined as "to discern an order" ("*das Erkennen einer Ordnung*") by the geographer Judith Miggelbrink.[50] I adopt her definition because her emphasis on the verb 'to discern' has two advantages. First, the wording excludes the idea that 'space' was some sort of a quasi-natural container. Instead, it frames space as a process, which implies that spatial formations are subject to change. Second, the verb underscores the active participation of individuals in producing space. Spaces result from activities: persons identify an order in their surroundings and attribute meaning to it. These properties imply that "a plurality of different spaces" of various scales coexist, overlap, and are interconnected – and that these connections are similarly subject to change.[51] From such a spatial perspective, tourism, that is the organised movement of people, matters in two regards: First, in the minds of tourists and their guides emerge imagined abstract world orders, a hierarchy of sites and places, based on experiences made and explanations given, that remain even after the travellers have left the destination. Second, regular tourist movements create a concrete order on the ground because the mobilities tourism generates are intertwined with the creation of infrastructural connections.[52]

With regard to imaginaries, two major arguments about the relationship of tourism and nationalism have been put forward, both of which draw on the notion of journeys as a "meaning-creating experience".[53] First, it has been argued that the experience of encountering the Other when travelling abroad allowed tourists to grasp the specific properties of their 'own' communities.[54] Second, national governments promoted domestic tourism with the intention that tourists would get to know 'their' country.[55] Such experiences of travellers visiting national museums or outstanding cultural or natural sights of their respective nation may be likened to the experience of educational or administrative "pilgrimages" described by Benedict Anderson. He argued that the emergence of educational or bureaucratic centres attracting students or functionaries from all corners of the state allowed them to discover their belonging to a shared, imagined community.[56] My main interest in this book, however, lies in the perspective of those who organised travel rather than those who travelled. I wonder how the mobilities of others, namely mobilities of distinction rather than mobilities of belonging, contributed to the creation of meaning from the point of view of the 'visited' societies, in particular in imperial contexts.

I suggest that transnational tourism allowed nationalist middle-class actors to present their nations as distinct yet equal to those of their visitors. The imaginaries of the nation were based on established categories. For South East Asia, Anderson argued that the nationalist imaginaries drew on colonial predecessors, defining the nation in terms of "the nature of the human beings it ruled, the geography of its domain, and the legitimacy of its ancestry."[57] The concrete narratives nationalist tour guides presented often differed from imperial descriptions of the territory though, sometimes actively contesting the interpretations applied by European researchers and colonial officials. The idea of what tourists considered an attraction was not yet canonised as the example of Albert Naccache reminds us; apparently, he imagined a different tourist space than that of the authors of the UNESCO World Heritage List in the second half of the twentieth century.

Moreover, tourism helps us to understand that after the Arab nationalisms of the late-nineteenth century, which were framed mainly in historical terms, a second, territorial nationalism attributed sense to both an Egyptian nation-state and the newly shaped political entities of the mandates.[58] Beyond representing the nation on tourist maps, actors in tourism development aimed to exhibit – and create – a coherent territory to be experienced by citizens and foreign visitors alike. Yet, we will see that in Arab Palestine and Syria, such a territorial nationalism did not emerge.

Although spaces are tied to cognitive processes on an individual and a collective level, they cannot be reduced to mere imagination as the established order features material elements. Geographical places and their specific material properties shape the discernible order.[59] In addition, movements, access and the denial of access contribute to the perception of space. A 'tourist space' is a mental construct in the sense that actual and potential travellers identify the space as a possible destination, yet its recognition presupposes material elements such as infrastructures and sights enabling and stimulating organised travel.[60]

The contemporaneous formation of tourist spaces and a new spatio-political order led to an intertwinement of tourism development policies and the spheres of local and emerging national, as well as imperial, politics. Actors and activities overlapped, and contributions to shaping tourist spaces were often attempts to shape political entities. This connection was no coincidence, as directing tourist movements helped create a spatial order. The importance of space in historical analysis has been pointed out by the proponents of the 'spatial turn', who identified time and space as interconnected principles of ordering society.[61] The connection between space and rule, space and politics, and space and society established by these thinkers explains why tourism may serve as a lens in tracing the emergence of nation-states as the dominant political order in the Arab Eastern Mediterranean.[62]

I aim to assess the role of tourism in creating an order of nation-states in the Eastern Arab Mediterranean by examining which actors used tourism, and to what extent they pursued political agendas by means of tourism. The spatial approach to state-formation facilitates the inclusion of various groups of actors and their different, often contradictory, spatial visions. It leaves room for the blurriness, the mutual dependence and the reshaping of spaces, taking into account that the outcome of the process was not obvious to contemporary actors.[63] It allows us to consider the circulation of competing spatial visions for the region, and the range of actors advancing these notions. This study thus analyses the contribution of tourism to the process of the formation of Middle Eastern nation-states without taking the outcome (the nation-state) for granted.

Tourism in the 1920s and 1930s

In the Arab Eastern Mediterranean during the 1920s and 1930s, the three elements with which I am concerned – space, tourism and politics – collided in a historical moment that is interesting for three reasons. First, after the defeat of the Ottoman Empire, different actors in both the Arab East and in Europe advanced a multiplicity of spatial visions for a future order. These decades witnessed another extension of European colonial empires, while subjects contested imperial rule on the ground, in international diplomacy and in the imperial metropoles.[64] Second, the interwar period witnessed a significant growth in tourist movements to the area as well as the emergence of new forms of tourism. While in Egypt organised tourism was already set in motion in the nineteenth century, in neighbouring *Bilad al-Sham*[65] European tourism became a prominent feature in the 1920s. In the 1930s, attempts to foster domestic tourism were discernible in several parts of the region. Hence, tourism exemplifies Sönke Kunkel's and Christoph Meyer's description of the interwar period as an "experimental stage". In this time, they argued, major elements that would come to characterise the twentieth century were discernible, but not yet fully implemented.[66] Third, the emergence of tourism took place at a time of heated debates about the political and economic future of the various states. The mandates in Palestine, Syria, and Lebanon, but also the not-quite-concealed imperial rule in Egypt, rendered cultural policies a particularly relevant tool from both an imperial and a nationalist point of view. Urban notables, who had dominated the public administration in late-Ottoman times, but also members of the mercantile bourgeoisies, had to realise that their hopes of taking over the state administration would not immediately come into being. At the same time, they prepared themselves for the time when they would definitely take over. In the context of quasi-colonial rule, the domain of cultural policies provided nationalist actors with a greater room for manoeuvre than others.[67] Tourism was thus inter-linked with debates about how the emerging states would position themselves in an international and regional context.

Approach of the book

These observations reveal that in order to fully grasp the historical relevance of tourism, it is essential to consider its transnational dimensions. National histories of tourism and its socio-economic implications, or histories of national cultures of leisure, overlook the significance of tourism in the early twentieth century: it pro-vided local actors with the capacity to address an international audience.[68] From the point of view of its proponents, this capacity turned tourism into an important political resource – even if their attempts to place their messages were not always successful.

The relevance of the continuous transnational dialogue is discernible from a global historical perspective, which focuses – in the words of Sebastian Conrad – on transnational and transcultural interactions in the framework of a politically, economically and culturally integrated world.[69] Such an integrated world persisted from a tourist perspective, despite tendencies of disentanglement during the 1920s and 1930s.[70] Politically, the mandates system established a mechanism for the international oversight of governance in the post-Ottoman Arab provinces, creat-ing dialogue (and dispute) between Arab representatives, the Yishuv,[71] European mandate powers and the Council of the League of Nations. In Egypt, the British had secured imperial influence after the nominal Egyptian independence by the so-called 'Four Reserved Points' that limited Egyptian sovereignty and allowed Great Britain to interfere in the Egyptian economy, jurisdiction, foreign and military affairs.[72] Economically, visa controls and toll barriers between the mandate states were introduced only successively during the period.[73] Capital from all over the world financed economic projects in the Yishuv, among them ventures in tourism. Culturally, national culture was defined in the same categories, a global standard for the structure of guidebooks had emerged, hotels offered a standardised set of amenities, and leisure practices in the Middle East reminded visitors of the French Riviera.

The category of the 'nation' (and thus the 'transnational') is both important and problematic in this context. It is important because for contemporary observ-ers, the nation denoted the modern, sovereign state, independent from imperial rule. The mandate treaties stipulated that such sovereignty would be granted to nations. Whereas in the former provinces of the Ottoman Empire, national senses of belonging were far from established, tourism and sightseeing contributed to the emergence of national orders. As Dean MacCannell explained, sightseeing was a way of ordering the world, a process by which "entire cities and regions, decades and cultures have become aware of themselves as tourist attractions." Hence, presentations of the "national self" in national symbols, national histories, or national culture and traditions intended to attract tourists, had repercussions for local society and often justified its restructuring.[74] The international framework of the mandates and the targeting of European tourists thus enhanced the relevance of the national idea in tourism to the region.

At the same time, the application of the category is problematic because we cannot take its existence for granted from the outset.[75] During recent years, researchers of Middle Eastern nation-building turned away from histories of nationalism understood as histories of independence movements. Instead, they focused on the larger socio-political transformations of Arab societies in the nineteenth and twentieth centuries. Often, a focus on selected cities allowed them to trace the spatial shifts occurring in the wider region in which the cities were embedded. In these works, the constant shifting of notions of belonging and the instability in perceiving such notions have come to the fore.[76] Inspired by these works, I adopt a flexible framework. Both inner rifts and borders, as well as outer connections, are considered, in line with the processual understanding of space described above.

Another indispensable yet problematic category is 'modernity'. The term has been criticised for its Eurocentric implications, yet studies on the transformation processes in the Arab East during the nineteenth century demonstrated its relevance to local self-conceptions, reflection, and debate since the *Nahda* period.[77] The claim for modernity characterised the ambitions of certain Arab actors from the largely urban middle classes to participate in the economic, scientific, and cultural transformations at the basis of Europe's global dominance.[78] Despite local particularities in the social composition and ambitions of these advocates of modernity, they were united by their striving for what they perceived as a modern society. In all countries under scrutiny, their claims addressed both their own societies, advocating sociopolitical transformations, and an international audience, justifying demands for sovereignty. Therefore, I understand 'modernity' as an ambition, a claim and a self-conception, rather than as a heuristic category.[79]

In the selection of the case studies, I followed the trajectories of many British and French tourists on the Arab part of their round trips from Egypt via Palestine and the French Mandate of Syria and Lebanon to Turkey, Italy and then back to France or Britain. Thereby I attempted to trace structures of organised tourism that existed or were about to emerge. While Egypt differed from the three mandate states in its institutional and political history, tourism in Egypt was an important reference for both travellers and advocates of tourism development. Already in 1912, a *Tourist Development Association* was founded in Egypt, probably the model of several similar associations emerging across the region in the following years.[80] Moreover, actors from Syria to the Yishuv mentioned Egypt as a model in tourism development.[81] While the business of tourism in Egypt was largely in the hands of European tour operators, in political terms, the Egyptian Government and deputies had greater room for manoeuvre in shaping tourism policies and the case of Egypt thus offers an interesting example in contrast to the mandates.[82]

However, Iraq and Transjordan, two mandates under British tutelage, are not included in this study. Although Petra became a major attraction for tourists during the interwar period, and regular bus services between Beirut, Damascus and Baghdad facilitated travelling for both European tourists and Iraqi summer guests, hints at and references to visits to these places were so scarce in the diaries, notes and photograph albums of my sample of travellers that I decided to stay on the "beaten tracks" around the Mediterranean coast.

In order to assess the relevance of tourism in the context of shifting spatial orders, a broad range of sources had to be consulted. Documents from state archives and libraries were combined with sources obtained by what Lucie Ryzova termed the *"Ezbekiyya* methodology" (referring to Cairo's famous book market), i.e. looking for "what there *is*" beyond curated collections of historical documents and objects.[83] From central state archives to those of sporting clubs as well as beyond archival structures, I searched for sources that would reflect the perspectives of different actors within tourism – such as tourists, entrepreneurs and guides – but also of members of interest groups and of relevant political and administrative bodies. I consulted, for example, diplomatic correspondence as well as state legislation, official reports and development plans, published and unpublished travelogues and diaries, tourist brochures, guidebooks and newspaper articles, and postcards and photographs. In order to avoid a lengthy reflection on these very heterogeneous documents, reflections on some sources particular to the history of tourism will be part of the introductory sections in chapters 2–5: photographs, travel diaries, postcards and tourist brochures.

The structure of this book reflects its spatial approach to tourism. Each of the four chapters focuses on a national entity – Egypt, Palestine, Syria and Lebanon[84] – and their order mirrors the trajectories of many travellers on their tours around the Eastern Mediterranean. Although the nation-state as an integrated spatial entity only emerged during the period under investigation, legislation was organised on a national level, justifying a national approach. Already in the 1920s, both opponents and adversaries of tourism development addressed their desires and complaints to national governments or to the mandate powers. Differences between the four spatio-political entities, in terms of both their socio-structural properties and their approaches to tourism, justify their separate examination. In order to circumvent the trap of circular reasoning, however, the study pays attention to connections and exchanges between the countries, transnational activities and interest groups, cross-border movements, as well as alternative spatial visions, resistances, and counter-movements to the formation of nation-states. At times, the book rather narrates the stories of failed projects of nation-building, or argues that processes of transformation occurred unevenly within a national entity.

The individual chapters are organised according to a similar structure: the opening section of each chapter draws on a specific category of sources to approach tourism in the respective country, like photographs, travel diaries, postcards and tourist brochures. These often ephemeral and fragmented sources offer a glimpse at some topics of the chapter and shed light on often-overlooked actors shaping tourism. The second sections of the chapters suggest a periodisation of tourism in the respective context and introduce the main groups of actors promoting tourism. The third sections consist of case studies from selected places on the tourist map. These 16 miniatures add up to a kaleidoscopic picture of tourism in the Arab East, allowing us to grasp the implications of tourism development for different groups of actors. Notwithstanding its parallel structure, this is not a comparative study. The transformative processes of the spatio-political entities under scrutiny, the manifold entanglements between the regions and the great variety in sources do not provide the conditions for an insightful comparison.[85] The chapters analyse different facets of tourism development, aiming to grasp the particularities of each case, thereby adding up to a more comprehensive picture of tourism in the Arab Eastern Mediterranean.

Chapter 2 shows how Egyptian actors struggled with the legacy of organised European tourism, which had turned Egypt into the first large-scale destination in the region to attract organised tours in the late nineteenth century. The early emergence of travel on the Nile implied that during the 1920s and 1930s, tourists built on well-established routes, patterns and imaginations when visiting the country. After the country obtained partial independence in 1922, new ambitions, actors, and aims in tourism competed with existing structures. To the Egyptian *efendiyya*, tourism was a means of contesting established Orientalist visions of Egypt and presenting alternative national self-conceptions to an international audience. In the 1930s, they addressed Egyptians as a second group of potential tourists: excursions to the Egyptian sites of leisure should contribute to turning both lower middle classes and the allegedly cosmopolitan upper classes into better Egyptian citizens.

Chapter 3 studies the competition of Yishuvi and Palestinian actors in tourism to the British Mandate of Palestine. Although the Bible remained an important reference for most European tourists, the idea of visiting the 'Holy Land' faded into the background when, from the late 1920s, conflict narratives began to define tourists' perceptions of mandatory Palestine. Both Arab Palestinian proponents of tourism development and Yishuvi, in particular Zionist, actors, aimed to present their vision of Palestine to visitors. Other than Arab intellectuals presenting Jerusalem as a symbol of a historically rooted Arab Palestine, Zionist institutions pursued a territorial approach to tourism development. The Zionist's access to processes of political decision-making, capital, British support as well as a large reservoir of potential Jewish tourists allowed them not only to shape a coherent Yishuvi tourist space, but also to use it as a foundation for the envisioned future state.

Chapter 4 focuses on the French mandate administration in Syria as a major actor in tourism development. To the French High Commission, tourism development was part of their strategies of 'pacification', as it was considered a contribution to both economic development and territorial control. The implementation of tourism development plans remained limited though, among other reasons because cooperation with advocates of tourism development among the Syrian urban middle classes did not occur. The latter, by contrast, lacked the political support to realise their ambitions, and neither a coherent national narrative nor an integrated tourist space emerged in Syria.

Chapter 5 examines the broad consensus about the importance of tourism that existed in Lebanon since the early days of the mandate. For the supporters of an independent Lebanese state, tourism was a vital resource. Although the borders of 'Greater Lebanon' were by no means uncontested in the early 1920s, tourism contributed to both an imaginary and a material integration of the nation, while achieving the acknowledgement of Lebanese sovereignty by international observers and its regional neighbours. The access of the Lebanese middle-class nationalists to processes of political decision-making was vital to the successful establishment of both tourism and a nation-state.

The conclusion traces the spatio-political transformations of the region from the perspective of tourism. The case studies show that the members of the middle classes were the driving forces in shaping both tourism and the emerging nation-states. The Arab middle classes in Palestine and Syria, largely excluded from political processes, were not able to implement strategies of reaching out to tourists. In Lebanon, Egypt and in the Yishuv by contrast, where these middle classes had access to political decision-making, either by a sufficient degree of autonomy or by cooperation with the colonial state, they put well-thought-out tourism development policies into practice. The desire of tourists for information and guidance, as well as the steadiness of tourists' mobilities allowed them to use tourism as a resource of attributing meaning to the national entities and of consolidating territory, thereby creating viable nation-states.

Notes

[1]   United Nations Educational, Scientific and Cultural Organization (UNESCO), World Heritage Centre, 'Ouadi Qadisha'.

[2]   John Urry stipulates the variability of the "tourist gaze", yet stresses that its core is a perceived "difference": Urry and Larsen, *The Tourist Gaze 3.0*, pp. 1–3. Zuelow mentions that also in the Netherlands and the United States in the late nineteenth century, innovative technologies or institutions such as modern prisons and hospitals were promoted as tourist attractions: Zuelow, *A History of Modern Tourism*, pp. 36, 106.

[3]   Naccache, 'L'industrie de la villégiature'. Naccache, *Un autre Liban*, pp. 61–63, 243–245. He also guided journalists of the newspaper *L'Orient* around the power plant, explaining his ambitions to a Lebanese audience: Naccache, *Un autre Liban*, p. 245.

[4]   Manela, 'Dawn of a New Era'.

[5]   Pedersen, 'Meaning of the Mandates System', pp. 578–579, 581. Pedersen, *The Guardians*, pp. 402–406.

[6]   Yale Law School, 'The Covenant of the League', Art. 22. Pedersen, *The Guardians*, pp. 29, 148–149.

[7]   Yale Law School, 'The Covenant of the League', Art. 22.

[8]   Schayegh, *The Middle East and the Making of the Modern World*, pp. 15–21.

[9]   Wonderful exceptions are Mairs and Muratov, *Archaeologists, Tourists, Interpreters*, and Mairs, *From Khartoum to Jerusalem*. I am indebted to Jens Hanssen who pointed out these works to me.

[10]   Tourism has been interpreted as a form of colonial propaganda by: Cohen-Hattab, 'Zionism, Tourism, and the Battle for Palestine', pp. 63–65. McLaren, *Architecture and Tourism in Italian Colonial Libya*, pp. 43–45. Furlough, 'Une leçon des choses', pp. 472–473. Zytnicki considered it a tool of colonial domination: Zytnicki, '"Faire l'Algérie agréable"', p. 9.

[11]   Rau, *Räume*, pp. 135–142, esp. pp. 140–141.

[12]   In similar contexts, authors have referred to the term "contact zone" coined by Pratt, *Imperial Eyes*, p. 8. As Pratt defines the "contact zone" as a space of encounter for cultures between the members of which there had not been interaction before, I do not use the term in this book.

[13]   Buzard, *The Beaten Track*, pp. 1–5. Löfgren, *On Holiday*, pp. 260–267. Enzensberger, 'Eine Theorie des Tourismus'.

[14]   Urry and Larsen, *The Tourist Gaze 3.0*, p. 6.

[15]   Urry and Larsen, *The Tourist Gaze 3.0*, pp. 4–6.

[16]   E.g. Zuelow, *A History of Modern Tourism*, pp. 5–9. On overlaps between pilgrimage and tourism cf. also: Osterhammel, *Die Verwandlung der Welt*, p. 1277.

[17]   Cribb, 'International Tourism in Java', p. 193.

[18]   Rosenberg, *Weltmärkte und Weltkriege*, Geschichte der Welt, pp. 985–987. Chabaud et al., *Les guides imprimés*. Rauch, 'Du Joanne au Routard'. Morlier, *Les Guides-Joanne*. On the difference between travel reports and guidebooks: Behdad, 'Orientalist Tourism'.

[19]   Hunter, 'Tourism and Empire'. Hazbun, 'The East as an Exhibit', pp. 5–6.

[20]   Zytnicki, *L'Algérie, terre de tourisme*, p. 19. On the particular role of the CGT in North African Tourism: Kazdaghli, 'L'entrée du Maghreb'. Perkins, 'The Compagnie Générale Transatlantique'.

[21]   Nash, 'The Rise and Fall'. Zytnicki, *L'Algérie, terre de tourisme*, pp. 36–41. Zuelow, *A History of Modern Tourism*, p. 71.

[22]   Although Dewailly and Ouvazza, 'Le tourisme au Liban', dated estivage practices to the mandate period, there is evidence that estivage movements were taken up already in the late Ottoman period: for Bikfaya: Chébli, 'Evolution d'un centre d'estivage', p. 23. For Dhour el-Choueir: Sawaya, 'Un centre d'estivage libanais', pp. 45–46. Cf. also Barakat-Buccianti, 'Beyrouth sous le Mandat

français', p. 75. On the visits of émigrés cf. Naccache, 'L'industrie de la villégiature', pp. 210–211. As the visits of Lebanese émigrés presumably had their own, specific dynamics, revolving around familiarity rather than Otherness, I did not include these journeys in the analysis.

[23] I will stick to this terminology and use the French terms of *"estivage"*, and *"estiveur* (pl. *estiveurs)"*.

[24] AIF, *Norma Jabbour Collection.* AIF, *Azar-Chouceir Collection.* AIF, *Mohsen Yammine Collection.* AIF, *Hamdan Chafic Collection.*

[25] BL, Workers' Travel Association, *Notes on Summer Programme.*

[26] On the *Workers' Travel Association* cf. Barton, *Working-Class Organisations and Popular Tourism.* On the prices of Fabre Lines cf. FRENCH LINES, Compagnie des Messageries Maritimes, *Agence de Beyrouth, Exercice 1921,* Trafic, p. 5. FRENCH LINES, Compagnie des Messageries Maritimes, *Agence de Beyrouth, Exercice 1925,* Trafic, p. 4.

[27] Two photographs in the album from 1925 show a young woman labeled "adopted niece", yet it seems to have been rather a nickname for a travel acquaintance.

[28] Zuelow, *A History of Modern Tourism,* pp. 65–66. Rosenberg, 'Transnationale Strömungen', pp. 986–987.

[29] BNF TOLBIAC, Roux-Servine, *Sur les routes du Levant.*

[30] Huber, 'Multiple Mobilities'. Huber, *Channelling Mobilities,* pp. 304–307.

[31] NA, *Theatrical artists in M. E.* CADN, *Le Ministre des Affaires étrangères, 02/04/1928.*

[32] Conrad, *Globalgeschichte,* pp. 16–18. Osterhammel, 'Transnationale Gesellschaftsgeschichte'.

[33] Zuelow, *A History of Modern Tourism,* pp. 95–96.

[34] Furlough, 'Une leçon des choses'. Zytnicki, '«Faire l'Algérie agréable»'. McLaren, *Architecture and Tourism in Italian Colonial Libya.* Zytnicki, *L' Algérie, terre de tourisme.* Zytnicki and Kazdaghli, *Le Tourisme dans l'Empire français.* Demay, *Tourism and Colonization in Indochina.* Baranowski et al., 'Tourism and Empire'.

[35] Hunter, 'Tourism and Empire'. Kennedy, *The Magic Mountains.* Sacareau, 'Tourisme et colonisation'.

[36] Urry differentiates "gazing" from "looking at", stating that the gaze "orders, shapes, and classifies" the world, rather than merely reflecting it: Urry and Larsen, *The Tourist Gaze 3.0,* pp. 1–2. Mitchell, *Colonising Egypt,* p. 24. Gregory, 'Scripting Egypt', p. 146. Gregory, 'Emperors of the Gaze', pp. 224–225.

[37] Jansen, *Erobern und Erinnern,* pp. 284–297. On archaeology and imperial culture in British contexts: Cobbing, 'Thomas Cook and the Palestine Exploration Fund'. Colla, *Conflicted Antiquities.*

[38] Burke III, 'French Native Policy'. Arrif, 'Le paradoxe de la construction du fait patrimonial'. Notions of „traditional" and „authentic", as I understand them throughout this book, have been fabricated in the context and for the purpose of tourism: Groebner, *Retroland,* pp. 180–183.

[39] Lew, Hall and Williams, *A Companion to Tourism.* Smith, Waterton and Watson, *The Cultural Moment in Tourism.*

[40] MacCannell, *The Tourist,* p. 3. MacCannell, 'Staged Authenticity'. Cf. also footnote 38.

[41] Eg Watson, Waterton and Smith, 'Moments, Instances and Experiences'.

[42] Baranowski, 'Strength through Joy', pp. 225–226. From an anthropological perspective, Laurajane Smith has argued in favour of a new understanding of tourism: Smith, 'The Cultural "Work" of Tourism'.

[43] CADN, *G. Grandguillot, 04/12/1935.*

[44] Dejung, Motadel and Osterhammel, 'Worlds of the Bourgeoisie', pp. 2–4.

[45] In 1920, the reports of the Royal Automobile Club on "touring abroad" referred to France, Switzerland, Belgium. In 1924, the RAC had established cooperations with clubs in the Maghreb, South America, and other places: BL, Royal Automobile Club, *Royal Automobile Club Year Book.*

[46] Dejung, Motadel and Osterhammel, 'Worlds of the Bourgeoisie', p. 2.

47   Dejung, Motadel and Osterhammel, 'Worlds of the Bourgeoisie', pp. 20–21.

48   Watenpaugh, *Being Modern*. Dejung, Motadel and Osterhammel, 'Worlds of the Bourgeoisie', p. 16. For Palestine, Sherene Seikaly has described this social group: Seikaly, *Men of Capital*. For Egypt: Ryzova, *The Age of the Efendiyya*.

49   Anderson, *Imagined Communities*, pp. 6–7. Maier, *Once within Borders*, pp. 238–243. Maier, 'Consigning the Twentieth Century', p. 808.

50   Miggelbrink, 'Räume und Regionen der Geographie', p. 92. This is a main argument of Doreen Massey, yet in her definition of space as a "product of interrelations, constituted through interactions" the term "product" suggests a stability Massey does not intend. Therefore I referred to Miggelbrink's definition. Massey, *For Space*, p. 9. Rau, *Räume*, pp. 143, 170.

51   Schlögel, *Im Raume lesen wir die Zeit*, pp. 68–69. Rau, *Räume*, pp. 164–171.

52   Rau, *Räume*, p. 143.

53   Anderson, *Imagined Communities*, p. 53.

54   Zuelow, *A History of Modern Tourism*, p. 92. Anderson made the point that nations were imagined as inherently limited: Anderson, *Imagined Communities*, pp. 6–7.

55   Löfgren, 'Know Your Country'. Shaffer, *See America First*. Cf. Zuelow on the Niagara falls, drawing on John F. Sears: Zuelow, *A History of Modern Tourism*, pp. 103–107.

56   Anderson, *Imagined Communities*, pp. 53–58, 114–140.

57   Anderson, *Imagined Communities*, pp. 163–164, 175, 184–185.

58   Hourani, *Arabic Thought in the Liberal Age*.

59   Doreen Massey: "By 'place' we mean any part of the earth's surface, however large or small." Massey and Jess, 'Introduction', p. 3. On the materiality of space cf. Osterhammel, *Die Verwandlung der Welt*, p. 130.

60   Yet, in contrast to Remy Knafou, I do not assume that tourist spaces were identified as tourist spaces due to inherent qualities but rather because of attributions overlapping with other interpretations and uses. Knafou, 'L'invention du lieu touristique', p. 17.

61   Already in 1998: Osterhammel, 'Die Wiederkehr des Raumes'. Schlögel, *Im Raume lesen wir die Zeit*, pp. 11–12, 61–71.

62   Massey, 'Politics and Space/Time', pp. 154–159. Massey, *For Space*, pp. 11–12.

63   Schlögel, *Im Raume lesen wir die Zeit*, p. 64.

64   Kunkel and Meyer, 'Dimensionen des Aufbruchs', pp. 20–26. Osterhammel and Jansen, *Dekolonisation*, pp. 28–32. Schayegh, *The Middle East and the Making of the Modern World*, pp. 330–334.

65   *Bilad al-Sham*: usually translated as 'Greater Syria', the common term for the region of today's Syria, Lebanon, Israel/Palestine, and Jordan that were considered a cultural entity by most inhabitants of the region until at least the mandate period. The adjective is *Shami*.

66   Kunkel and Meyer, 'Dimensionen des Aufbruchs', pp. 8–9.

67   Dueck, *Claims of Culture*, pp. 229–233.

68   Eg Tissot, *Construction d'une industrie touristique*. Cross and Walton, *The Playful Crowd*. Barakat-Buccianti, 'Beyrouth sous le Mandat français'.

69   Conrad, *Globalgeschichte*, pp. 11–12.

70   Christof Dejung's balanced article on the world economy demonstrates that a characterisation of the 1920s and 1930s as a phase of 'deglobalisation' has to be nuanced: Dejung, 'Deglobalisierung?'.

71   *Yishuv*: The Jewish settler community in Palestine. Reich and Goldberg, 's.v. Yishuv'.

72   Goldschmidt Jr, *Modern Egypt*, pp. 71–77.

73   Schayegh, 'The Many Worlds of 'Abud Yasin', pp. 278–280.

74   MacCannell, *The Tourist*, p. 16.

75 Alternative spatial frameworks persisted long into the 1920s: Schayegh, *The Middle East and the Making of the Modern World*, pp. 8–14. Provence, 'Ottoman Modernity, Colonialism, and Insurgency', pp. 21–22.

76 E.g. Hanssen, *Fin de Siècle Beirut*. Watenpaugh, *Being Modern*. Doumani, *Rediscovering Palestine*.

77 *Nahda*: Initially coined to describe mainly literary reform movements, the term has come to describe more generally the cultural and political period in the Arab world. Cf. Hanssen and Weiss, 'Introduction', p. 1.

78 Watenpaugh, *Being Modern*, pp. 8–9, 28–30. Such an approch to modernity was not specific to the Arab world: Cooper, 'Modernity', p. 149.

79 Cooper, 'Modernity', pp. 131–132. Osterhammel, *Die Verwandlung der Welt*, pp. 1093–1095.

80 The Lebanese *Société de Villégiature au Mont Liban*, for example, was founded in Egypt: Santer, 'Imagining Lebanon', pp. 45–46.

81 CADN, al-Ghazzi, *al-Muqaddima [Preface]*. NLI, Settel, *Selling Egypt, 04/1937*.

82 Hunter, 'Tourism and Empire'. Hazbun, 'The East as an Exhibit'.

83 Ryzova, *The Age of the Efendiyya*, pp. 29–31.

84 Throughout this study, the term 'Syria' refers to the territory under the government of Damascus the French named '*Etat de Syrie*', as well as the temporarily detached entities of the Governorate of the ʿAlawites (of Lattaquieh), the Sanjak of Alexandretta and the Jabal al-Druze. The concept of 'Greater Syria' including the Mandates of Palestine, Transjordan, Syria and Lebanon is distinguished from this term by referring to the contemporary Arabic designation of *Bilad al-Sham*. The entity of the French Mandate (Syria and Lebanon) contemporary authors sometimes referred to as 'Syria' or 'Syrian Mandate', cf. Burns, *The Tariff of Syria*, p. 1, will not be treated as an entity because the patterns of governance as well as cultural representations of Syria and Lebanon differed to an extent that the separate analysis of both cases seemed necessary.

85 Osterhammel, 'Transferanalyse und Vergleich im Fernverhältnis', pp. 465–466.

References

Sources

Arab Image Foundation, Beirut
Azar-Chouceir Collection, AIF/0215az.
Hamdan Chafic Collection, AIF/0241ha.
Mohsen Yammine Collection, AIF/0009ya.
Norma Jabbour Collection, AIF/0083ja.

Bibliothèque nationale de France, Paris
ROUX-SERVINE, LOUIS, *Sur les routes du Levant: Petit manuel du voyage expérimental*, Paris: J. Barreau, 1934, BNF Tolbiac/8-G-13749.

British Library, London
ROYAL AUTOMOBILE CLUB, *Royal Automobile Club Year Book*, London, 1920–1924, BL/P.P.2498.kg.
WORKERS' TRAVEL ASSOCIATION, *Notes on Summer Programme*, The Travel Log 3.15 (1934), p. 11, BL.

Centre des Archives diplomatiques de Nantes

AL-GHAZZI, KAMIL, *al-Muqaddima [Preface]*, in: Majallat al-ʿAdiyat, tusdaru Shahriyan bi-ʿAnayat Jamʿiyya ʿAdiyat Halab, al-ʿAdad al-Awal Shahr Ayar 1931 [Société Archéologique d'Alep (ed.), Revue Archéologique, Mai 1931], Fonds Beyrouth, Cabinet politique, CADN/1 SL/1/V/626.

G. Grandguillot, Attaché commercial près de la Légation de France en Egypte, Alexandrie, à M. le Ministre du Commerce et de l'Industrie, Paris, 04/12/1935, Archives des Postes, Consulat du Caire. Dossier Exemptions douanières, 1935–1936, CADN/353 PO/2/230.

Le Ministre des Affaires étrangères à Ponsot, 02/04/1928, Bureau diplomatique, Règlementation du travail. Dossier: Prestige de la femme française en Syrie, CADN/1 SL/1/V/1406.

Association French Lines, Le Havre

COMPAGNIE DES MESSAGERIES MARITIMES, *Rapports d'Agence. Agence de Beyrouth, Exercice 1921*, French Lines.

COMPAGNIE DES MESSAGERIES MARITIMES, *Rapports d'Agence. Agence de Beyrouth, Exercice 1925*, French Lines.

Institut français du Proche-Orient, Beirut

CHÉBLI, HEND, 'Evolution d'un centre d'estivage au Liban (Bikfaya)', MA Thesis (Beirut, Université libanaise, 1971).

SAWAYA, MARIE, 'Un centre d'estivage libanais: Dhour Choueir', MA Thesis (Beirut, Université libanaise, 1963).

National Archives, Kew

Theatrical artists in M. E. territories: Restrictions on European female performers, 1927, Colonial Office, Palestine Original Correspondence, NA/CO 732/18/14.

National Library of Israel, Jerusalem

SETTEL, ARTHUR, *Selling Egypt: How Tourists Are Attracted to the Land of the Nile. Setting an Example for the Holy Land*, The Palestine Post, 25/04/1937, p. 5. Historical Jewish Press, NLI.

Other published sources

BURNS, NORMAN, *The Tariff of Syria: 1919–1932* (Beirut: American Press, 1933).

NACCACHE, ALBERT, 'L'industrie de la villégiature au Liban', *Revue Phénicienne*, 1/4–6 (1919), pp. 208–213.

NACCACHE, PIERRE, *Un autre Liban: Les écrits d'Albert Naccache* (Paris: Geuthner, 2018).

YALE LAW SCHOOL, 'The Covenant of the League of Nations: signed 28/06/1919, effective 10/01/1920, The Avalon Project, Lillian Goldman Law Library', 2008 <https://avalon.law.yale.edu/20th_century/leagcov.asp>, updated 2008, accessed 9 Jun 2020.

Secondary Literature

ANDERSON, BENEDICT R., *Imagined Communities: Reflections on the Origin and Spread of Nationalism* (2nd edn., London, New York: Verso, 2016).

ARRIF, ABDELMAJID, 'Le paradoxe de la construction du fait patrimonial en situation coloniale: Le cas du Maroc', *Revue du monde musulman et de la Méditerranée*, 73/74 (1994), pp. 153–166.

BARAKAT-BUCCIANTI, LILIANE, 'Beyrouth sous le Mandat français ou la "Nice de l'Orient"', in Colette Zytnicki and Habib Kazdaghli (eds.), *Le Tourisme dans l'Empire français. Un outil de la domination coloniale? Politiques, pratiques et imaginaires (XIXe–XXe siècles)* (Saint-Denis: Publications de la Société française d'histoire d'outre-mer, 2009), pp. 73–78.

BARANOWSKI, SHELLEY, 'Strength through Joy: Tourism and National Integration in the Third Reich', in Shelley Baranowski and Ellen Furlough (eds.), *Being Elsewhere. Tourism, Consumer Culture, and Identity in Modern Europe and North America* (Ann Arbor: University of Michigan Press, 2001), pp. 213–236.

BARANOWSKI, SHELLEY, ENDY, CHRISTOPHER, HAZBUN, WALEED et al., 'Tourism and Empire', *Journal of Tourism History*, 7/1–2 (2015), pp. 100–130.

BARTON, SUSAN, *Working-Class Organisations and Popular Tourism, 1840–1970* (Manchester, New York: Manchester University Press, 2005).

BEHDAD, ALI, 'Orientalist Tourism', *Peuples méditerranéens*, 50 (1990), pp. 59–73.

BURKE III, EDMUND, 'A Comparative View of French Native Policy in Morocco and Syria, 1912–1925', *Middle Eastern Studies*, 9/2 (1973), pp. 175–186.

BUZARD, JAMES, *The Beaten Track: European Tourism, Literature, and the Ways to Culture, 1800–1918* (Oxford: Clarendon Press, 1993).

CHABAUD, GILLES, COHEN, ÉVELYNE, COQUERY, NATACHA et al. (eds.), *Les guides imprimés du XVIe au XXe siècle: Villes, paysages, voyages* (Paris: Belin, 2000).

COBBING, FELICITY, 'Thomas Cook and the Palestine Exploration Fund', *Public Archaeology*, 11/4 (2012), pp. 179–194.

COHEN-HATTAB, KOBI, 'Zionism, Tourism, and the Battle for Palestine: Tourism as a Political Propaganda Tool', *Israel Studies*, 9/1 (2004), pp. 61–85.

COLLA, ELLIOTT, *Conflicted Antiquities: Egyptology, Egyptomania, Egyptian Modernity* (Durham: Duke University Press, 2008).

CONRAD, SEBASTIAN, *Globalgeschichte: Eine Einführung* (München: C.H. Beck, 2013).

COOPER, FREDERICK, 'Modernity', in Frederick Cooper (ed.), *Colonialism in Question. Theory, Knowledge, History* (Berkeley: University of California Press, 2010), pp. 113–149.

CRIBB, ROBERT, 'International Tourism in Java, 1900–1930', *South East Asia Research*, 3/2 (1995), pp. 193–204.

CROSS, GARY S., and WALTON, JOHN K. (eds.), *The Playful Crowd: Pleasure Places in the Twentieth Century* (New York: Columbia University Press, 2005).

DEJUNG, CHRISTOF, 'Deglobalisierung? Oder Enteuropäisierung des Globalen? Überlegungen zur Entwicklung der Weltwirtschaft in der Zwischenkriegszeit', in Sönke Kunkel and Christoph Meyer (eds.), *Aufbruch ins postkoloniale Zeitalter. Globalisierung und die außereuropäische Welt in den 1920er und 1930er Jahren* (Frankfurt a. M., New York: Campus, 2012), pp. 37–61.

DEJUNG, CHRISTOF, MOTADEL, DAVID, and OSTERHAMMEL, JÜRGEN, 'Worlds of the Bourgeoisie', in Christof Dejung, David Motadel, and Jürgen Osterhammel (eds.), *The Global Bourgeoisie. The Rise of the Middle Classes in the Age of Empire* (Princeton: Princeton University Press, 2019), pp. 1–39.

DEMAY, ALINE, *Tourism and Colonization in Indochina, 1898–1939* (Newcastle upon Tyne: Cambridge Scholars Publishing, 2014).

DEWAILLY, BRUNO, and OUVAZZA, JEAN-MARC, 'Le tourisme au Liban: Quand l'action ne fait plus système', in Mohamed Berriane (ed.), *Tourisme des nationaux, tourisme des étrangers. Quelles articulations en Méditerranée?* (Rabat: Université Mohammed V, Faculté des Lettres et des Sciences Humaines, 2009), pp. 83–114.

DOUMANI, BESHARA, *Rediscovering Palestine: Merchants and Peasants in Jabal Nablus, 1700–1900* (Berkeley: University of California Press, 1995).

DUECK, JENNIFER M., *The Claims of Culture at Empire's End: Syria and Lebanon under French Rule* (Oxford, New York: Oxford University Press, 2010).

ENZENSBERGER, HANS-MAGNUS, 'Eine Theorie des Tourismus', *Merkur*, 12 (1958), pp. 701–720.

FURLOUGH, ELLEN, 'Une leçon des choses: Tourism, Empire, and the Nation in Interwar France', *French Historical Studies*, 25/3 (2002), pp. 441–475.

GOLDSCHMIDT JR, ARTHUR, *Modern Egypt: The Formation of a Nation-State* (2nd edn., Boulder: Westview Press, 2004).

GREGORY, DEREK, 'Scripting Egypt: Orientalism and the Cultures of Travel', in James S. Duncan and Derek Gregory (eds.), *Writes of Passage. Reading Travel Writing* (London, New York: Routledge, 1999), pp. 114–150.

GREGORY, DEREK, 'Emperors of the Gaze: Photographic Practices and Productions of Space in Egypt, 1839–1914', in Joan M. Schwartz and James R. Ryan (eds.), *Picturing Place. Photography and the Geographical Imagination* (London: I.B. Tauris, 2009), pp. 195–225.

GROEBNER, VALENTIN, *Retroland: Geschichtstourismus und die Sehnsucht nach dem Authentischen* (Frankfurt a. M.: S. Fischer, 2018).

HANSSEN, JENS, *Fin de Siècle Beirut: The Making of an Ottoman Provincial Capital* (Oxford, New York: Oxford University Press, 2005).

HANSSEN, JENS, and WEISS, MAX, 'Introduction: Language, Mind, Freedom and Time: The Modern Arab Intellectual Tradition in Four Words', in Jens Hanssen and Max Weiss (eds.), *Arabic Thought beyond the Liberal Age. Towards an Intellectual History of the Nahda* (Cambridge, New York: Cambridge University Press, 2017), pp. 1–37.

HAZBUN, WALEED, 'The East as an Exhibit: Thomas Cook & Son and the Origins of the International Tourism Industry in Egypt', in Philip Scranton and Janet F. Davidson (eds.), *The Business of Tourism. Place, Faith, and History* (Philadelphia: University of Pennsylvania Press, 2007), pp. 3–33.

HOURANI, ALBERT, *Arabic Thought in the Liberal Age: 1798–1939* (2nd edn., Cambridge, New York: Cambridge University Press, 1983).

HUBER, VALESKA, 'Multiple Mobilities: Über den Umgang mit verschiedenen Mobilitätsformen um 1900', *Geschichte und Gesellschaft*, 36/2 (2010), pp. 317–341.

HUBER, VALESKA, *Channelling Mobilities: Migration and Globalisation in the Suez Canal Region and Beyond, 1869–1914* (Cambridge, New York: Cambridge University Press, 2013).

HUNTER, F. ROBERT, 'Tourism and Empire: The Thomas Cook & Son Enterprise on the Nile, 1868-1914', *Middle Eastern Studies*, 40/5 (2004), pp. 28–54.

JANSEN, JAN C., *Erobern und Erinnern: Symbolpolitik, öffentlicher Raum und französischer Kolonialismus in Algerien, 1830–1950* (München: Oldenbourg, 2013).

KAZDAGHLI, HABIB, 'L'entrée du Maghreb dans les circuits du tourisme international: Le rôle précurseur de la Compagnie Générale Transatlantique', in Colette Zytnicki and Habib Kazdaghli (eds.), *Le Tourisme dans l'Empire français. Un outil de la domination coloniale? Politiques, pratiques et imaginaires (XIXe–XXe siècles)* (Saint-Denis: Publications de la Société française d'histoire d'outre-mer, 2009), pp. 205–215.

KENNEDY, DANE, *The Magic Mountains: Hill Stations and the British Raj* (Berkeley: University of California Press, 1996).

KNAFOU, RÉMY, 'L'invention du lieu touristique: La passation d'un contrat et le surgissement simultané d'un nouveau territoire', *Revue de géographie alpine*, 79/4 (1991), pp. 11–19.

KUNKEL, SÖNKE, and MEYER, CHRISTOPH, 'Dimensionen des Aufbruchs: Die 1920er und 1930er Jahre in globaler Perspektive', in Sönke Kunkel and Christoph Meyer (eds.), *Aufbruch ins postkoloniale Zeitalter. Globalisierung und die außereuropäische Welt in den 1920er und 1930er Jahren* (Frankfurt a. M., New York: Campus, 2012), pp. 7–33.

Lew, Alan A., Hall, Colin Michael, and Williams, Allan M. (eds.), *A Companion to Tourism* (Malden: Blackwell, 2004).

Löfgren, Orvar, *On Holiday: A History of Vacationing* (Berkeley: University of California Press, 1999).

Löfgren, Orvar, 'Know Your Country: A Comparative Perspective on Tourism and Nation Building in Sweden', in Shelley Baranowski and Ellen Furlough (eds.), *Being Elsewhere. Tourism, Consumer Culture, and Identity in Modern Europe and North America* (Ann Arbor: University of Michigan Press, 2001), pp. 137–154.

MacCannell, Dean, 'Staged Authenticity: Arrangements of Social Space in Tourist Settings', *American Journal of Sociology*, 79 (1973), pp. 589–603.

MacCannell, Dean, *The Tourist: A New Theory of the Leisure Class* (Berkeley: University of California Press, 1999).

Maier, Charles S., 'Consigning the Twentieth Century to History: Alternative Narratives for the Modern Era', *American Historical Review*, 105/3 (2000), pp. 807–831.

Maier, Charles S., *Once within Borders: Territories of Power, Wealth, and Belonging since 1500* (Cambridge, Mass.: The Belknap Press of Harvard University Press, 2016).

Mairs, Rachel, *From Khartoum to Jerusalem: The Dragoman Solomon Negima and his Clients (1885–1933)* (London: Bloomsbury, 2016).

Mairs, Rachel, and Muratov, Maya, *Archaeologists, Tourists, Interpreters: Exploring Egypt and the Near East in the Late Nineteenth and Early Twentieth Centuries* (London: Bloomsbury, 2015).

Manela, Erez, 'Dawn of a New Era: The "Wilsonian Moment" in Colonial Contexts and the Transformation of World Order, 1917–1920', in Sebastian Conrad and Dominic Sachsenmaier (eds.), *Competing Visions of World Order. Global Historical Approaches* (Basingstoke: Palgrave Macmillan, 2007), pp. 121–149.

Massey, Doreen, 'Politics and Space/Time', in Michael Keith and Steve Pile (eds.), *Place and the Politics of Identity* (London, New York: Routledge, 1993), pp. 141–161.

Massey, Doreen, and Jess, Pat, 'Introduction', in Doreen Massey and Pat Jess (eds.), *A Place in the World? Places, Cultures and Globalization* (Oxford, New York: Oxford University Press, 2003), pp. 1–4.

Massey, Doreen B., *For Space* (Los Angeles: Sage, 2005).

McLaren, Brian, *Architecture and Tourism in Italian Colonial Libya: An Ambivalent Modernism* (Seattle: University of Washington Press, 2006).

Miggelbrink, Judith, 'Räume und Regionen der Geographie', in Paul-Gerhard Klumbies, Ingrid Baumgärtner, and Franziska Sick (eds.), *Raumkonzepte. Disziplinäre Zugänge* (Göttingen: Vandenhoeck & Ruprecht, 2010), pp. 71–94.

Mitchell, Timothy, *Colonising Egypt* (Berkeley: University of California Press, 1988).

Morlier, Hélène, *Les Guides-Joanne: Genèse des Guides-bleus: itinéraire bibliographique, historique et descriptif de la collection de guides de voyage, 1840–1920* (Paris: Les Sentiers débattus, 2007).

Nash, Dennison, 'The Rise and Fall of an Aristocratic Tourist Culture: Nice, 1763–1936', *Annals of Tourism Research*, 6/1 (1979), pp. 61–75.

Osterhammel, Jürgen, 'Die Wiederkehr des Raumes: Geopolitik, Geohistorie und historische Geographie', *Neue Politische Literatur*, 43/3 (1998), pp. 374–397.

Osterhammel, Jürgen, 'Transnationale Gesellschaftsgeschichte: Erweiterung oder Alternative?', *Geschichte und Gesellschaft*, 27/3 (2001), pp. 464–479.

Osterhammel, Jürgen, 'Transferanalyse und Vergleich im Fernverhältnis', in Hartmut Kaelble and Jürgen Schriewer (eds.), *Vergleich und Transfer. Komparatistik in den Sozial-, Geschichts- und Kulturwissenschaften* (Frankfurt a. M., New York: Campus, 2003), pp. 439–466.

OSTERHAMMEL, JÜRGEN, *Die Verwandlung der Welt: Eine Geschichte des 19. Jahrhunderts* (München: C.H. Beck, 2009).

OSTERHAMMEL, JÜRGEN, and JANSEN, JAN C., *Dekolonisation: Das Ende der Imperien* (München: C.H. Beck, 2013).

PEDERSEN, SUSAN, 'The Meaning of the Mandates System: An Argument', *Geschichte und Gesellschaft*, 32/4 (2006), pp. 560–582.

PEDERSEN, SUSAN, *The Guardians: The League of Nations and the Crisis of Empire* (Oxford, New York: Oxford University Press, 2015).

PERKINS, KENNETH J., 'The Compagnie Générale Transatlantique and the Development of Saharan Tourism in North Africa', in Philip Scranton and Janet F. Davidson (eds.), *The Business of Tourism. Place, Faith, and History* (Philadelphia: University of Pennsylvania Press, 2007), pp. 34–55.

PRATT, MARY LOUISE, *Imperial Eyes: Travel Writing and Transculturation* (2nd edn., London, New York: Routledge, 2008).

PROVENCE, MICHAEL, 'Ottoman Modernity, Colonialism, and Insurgency in the Interwar Arab East', *International Journal of Middle East Studies*, 43/2 (2011), pp. 205–225.

RAU, SUSANNE, *Räume: Konzepte, Wahrnehmungen, Nutzungen* (Frankfurt a. M., New York: Campus, 2013).

RAUCH, ANDRÉ, 'Du Joanne au Routard: Le style des guides touristiques', in Gilles Chabaud, Évelyne Cohen, Natacha Coquery et al. (eds.), *Les guides imprimés du XVIe au XXe siècle. Villes, paysages, voyages* (Paris: Belin, 2000), pp. 95–112.

REICH, BERNARD, and GOLDBERG, DAVID H., 's.v. Yishuv', in Bernard Reich and David H. Goldberg (eds.), *Historical Dictionary of Israel* (2nd edn., Lanham, Toronto, Plymouth: Scarecrow Press, 2008), p. 554.

ROSENBERG, EMILY S., 'Transnationale Strömungen in einer Welt, die zusammenrückt', in Emily S. Rosenberg (ed.), *Weltmärkte und Weltkriege. 1870–1945*, Akira Iriye and Jürgen Osterhammel (Geschichte der Welt, München: C.H. Beck, 2012), pp. 815–998.

ROSENBERG, EMILY S. (ed.), *Weltmärkte und Weltkriege: 1870–1945*, Akira Iriye and Jürgen Osterhammel (Geschichte der Welt, München: C.H. Beck, 2012).

RYZOVA, LUCIE, *The Age of the Efendiyya: Passages to Modernity in National-Colonial Egypt* (Oxford, New York: Oxford University Press, 2014).

SACAREAU, ISABELLE, 'Tourisme et colonisation: Les *hill stations* himalayennes de l'Empire britannique des Indes (Darjeeling, Simla Mussoorie, Nainital) (1820–1947)', in Amaury Lorin and Christelle Taraud (eds.), *Nouvelle histoire des colonisations européennes, XIXe-XXe siècles. Sociétés, cultures, politiques* (Paris: Presses universitaires de France, 2013), pp. 91–102.

SANTER, JANINA, 'Imagining Lebanon: Tourism and the Production of Space under French Mandate (1920–1943)', MA Thesis (Beirut, American University of Beirut, 2019).

SCHAYEGH, CYRUS, 'The Many Worlds of 'Abud Yasin: or: What Narcotics Trafficking in the Interwar Middle East Can Tell Us about Territorialization', *The American Historical Review*, 116/2 (2011), pp. 273–306.

SCHAYEGH, CYRUS, *The Middle East and the Making of the Modern World* (Cambridge, Mass.: Harvard University Press, 2017).

SCHLÖGEL, KARL, *Im Raume lesen wir die Zeit: Über Zivilisationsgeschichte und Geopolitik* (Frankfurt a. M.: Fischer, 2011).

SEIKALY, SHERENE, *Men of Capital: Scarcity and Economy in Mandate Palestine* (Stanford: Stanford University Press, 2016).

SHAFFER, MARGUERITE, *See America First: Tourism and National Identity, 1880–1940* (Herndon: Smithsonian, 2013).

SMITH, LAURAJANE, 'The Cultural "Work" of Tourism', in Laurajane Smith, Emma Waterton, and Steve Watson (eds.), *The Cultural Moment in Tourism* (London, New York: Routledge, 2012), pp. 210–234.

SMITH, LAURAJANE, WATERTON, EMMA, and WATSON, STEVE (eds.), *The Cultural Moment in Tourism* (London, New York: Routledge, 2012).

TISSOT, LAURENT (ed.), *Construction d'une industrie touristique aux 19e et 20e siècles: Perspectives internationales* (Neuchâtel: Éditions Alphil, 2003).

UNITED NATIONS EDUCATIONAL, SCIENTIFIC AND CULTURAL ORGANIZATION (UNESCO), WORLD HERITAGE CENTRE, 'Ouadi Qadisha (the Holy Valley) and the Forest of the Cedars of God (Horsh Arz el-Rab): World Heritage List', <https://whc.unesco.org/en/list/850/>, accessed 26 Mar 2020.

URRY, JOHN, and LARSEN, JONAS, *The Tourist Gaze 3.0* (3rd edn., Los Angeles, London: Sage, 2011).

WATENPAUGH, KEITH DAVID, *Being Modern in the Middle East: Revolution, Nationalism, Colonialism, and the Arab Middle Class* (Princeton: Princeton University Press, 2006).

WATSON, STEVE, WATERTON, EMMA, and SMITH, LAURAJANE, 'Moments, Instances and Experiences', in Laurajane Smith, Emma Waterton, and Steve Watson (eds.), *The Cultural Moment in Tourism* (London, New York: Routledge, 2012), pp. 1–16.

ZUELOW, ERIC G. E., *A History of Modern Tourism* (London, New York: Palgrave Macmillan, 2016).

ZYTNICKI, COLETTE, '"Faire l'Algérie agréable": Tourisme et colonisation en Algérie des années 1870 à 1962', *Le Mouvement Social*, 242 (2013), pp. 97–114.

ZYTNICKI, COLETTE, *L' Algérie, terre de tourisme: Histoire d'un loisir colonial* (Paris: Vendémiaire, 2016).

ZYTNICKI, COLETTE, and KAZDAGHLI, HABIB (eds.), *Le Tourisme dans l'Empire français: Un outil de la domination coloniale? Politiques, pratiques et imaginaires (XIXe–XXe siècles)* (Saint-Denis: Publications de la Société française d'histoire d'outre-mer, 2009).

CHAPTER 2

Space of sovereignty

Abolishing the colonial order of tourism in Egypt?

Abstract

Egyptian actors struggled with the legacy of organised European tourism, which had turned Egypt into the first large-scale tourist destination in the region as early as the nineteenth century. Whereas the economic relevance of tourism for Egypt remained limited, the members of the *efendiyya* aimed to use its political potential. They presented Egypt as a modern Mediterranean nation to tourists, aiming to modify the Orientalist visions that continued to be reproduced by travellers, and underlining their ambition to be acknowledged as a full member of the League of Nations. In addition, from the 1930s onwards they targeted a domestic Egyptian audience. These mobilities, they hoped, would enhance identification with the nation and contribute to the intellectual, moral and physical refinement of Egyptian citizens.

Keywords: tourism; Egyptian history; imperialism – history – 20th century; modernity; seaside resorts; photography – 20th century

On Saturday 26 January 1952, a group of protesters set the *Shepheard's Hotel* in Cairo on fire. After several days of fighting between British troops and Egyptian protesters, the latter destroyed several British institutions and symbols of the Western lifestyle such as banks, showrooms of automobile companies, clubhouses, *Groppi's* tea house and cinemas.[1] Along with these institutions, the burning *Shepheard's* and the *Thomas Cook* travel agency, which were situated in the same building, represented the British presence in Egypt resented by most Egyptians. For them, the institutions of the 'golden age' of Egyptian tourism symbolised Western imperialism.[2] Therefore, the burning hotel also stood for the failed ambitions of the Egyptian bourgeoisie to appropriate luxury tourism to Egypt.

Egypt had been the first and a particularly well-established destination for European and American tourists on the southern Mediterranean shore, serving as a model for tourism development in the neighbouring countries. Packaged tours along the Nile and an industry of grand hotels were already flourishing around the turn of the century. The attraction of Egypt surpassed the popularity of the neighbouring 'Holy Land': according to Robert F. Hunter, almost 11,000 tourists were counted in Cairo in the winter season of 1889/90, of which 1,300 joined the tours up the Nile, while during 1913/14, the record year for tourism to Palestine before the war, approximately 6,800 visitors were counted in Palestine.[3]

Organised tourism to Egypt had been established as a largely European business. Tourists travelled on the steamers of the French shipping company *Messageries Maritimes*, had their tour organised by *Thomas Cook & Son*, read *Baedeker*'s guidebooks, and stayed in hotels administered by the Swiss manager Charles Baehler. Photographers such as *Lehnert & Landrock*, Francis Frith or *Bonfils* shaped their visual expectations and the 'discoveries' of European archaeologists seemed to justify assumptions about European superiority. On the terraces of restaurants and coffee houses, only few Egyptian notables joined the European guests served by dark-skinned waiters in *galabiyyas*[4] and fezzes, supposed to add to the local touch (fig. 2). According to Timothy Mitchell's *Colonising Egypt*, such visual markers allowed tourists to identify the Orient they expected and were seeking for, to rediscover a place they were already familiar with upon arrival, as an exhibition, ordered in familiar terms:[5]

> The so-called real world 'outside' is something experienced and grasped only as a series of further representations, an extended exhibition. Visitors to the Orient conceived of themselves as travelling to 'the East itself in its vital actual reality'. But, as we saw, the reality they sought there was simply that which could be photographed or accurately represented, that which presented itself as a picture of something before an observer. [...] In the end the European tried to grasp the Orient as though it were an exhibition of itself.[6]

Such representations, Mitchell concluded, ultimately created a fundamental distinction not only between self and Other as identified by Said, but also between a material world and a conceptual framework, thus an ordering principle which, in turn, served as a strategy of power sustaining colonial rule.[7]

The colonial order of tourism will form the backdrop to the present chapter, which brings together both European visions of Egypt and Egyptian attempts to adapt or undermine these visions. While tourism had emerged in Egypt as an imperial practice, from the 1920s onwards Egyptian actors attempted to reshape tourism and its representations of Egypt. Their example shows that tourism policies in the interwar period were not a prerogative of European entrepreneurs, but that representatives of the middle and upper classes promoting an Egyptian modernity were well aware of the potential of tourism.

This potential was two-fold, from their point of view: tourism would allow them to present Egypt as a distinct, yet equal nation among others and justify Egyptian demands for membership in the League of Nations. In addition, from around the 1930s onwards, Egyptian deputies advertised domestic tourism as an opportunity

Figure 2: The terrace of
the *Mena House Hotel*
(The Tourist Development
Association of Egypt,
Egypt: A Travel Quarterly
2/3 (1939), jd private
collection)

to foster the integration of the nation and make its inhabitants identify with the
Egyptian state. The chapter builds on a range of recent publications that analysed
Egyptian attempts to create a modern society in the late nineteenth and early twentieth centuries.[8] The main advocates of this modernity are generally understood as
a new social group, the Egyptian *efendiyya*, studied in detail by Lucie Ryzova, who
identified distinct practices, motifs and imaginings characterising the members.
One of her observations seems particularly relevant in the context of tourism:
Ryzova observed that the rhetorical juxtaposition of 'old' and 'new', 'authentic'
and 'modern', served as a typical trope of an "*efendiyya* modernity". She explained
that this trope reconciled claims for modernity with a simultaneous distancing
from a 'Western' imperialist modernity. The combination of signifiers of modernity
and elements deemed authentic allowed *efendi* actors to counter accusations of
propagating an alienated Egyptian culture and to claim an indigenous Egyptian
modernity.[9]

Apparently, the claim for modernity stood in marked contrast to Orientalist
discourses influencing the views of most European travellers. The chapter
therefore asks how Egyptian actors intended to reframe the Egyptian tourist
space and analyses the implications of such a tourist modernity for Egyptian
nation-building.

Turning away: Photographing authenticity in times of transformation

> The only way to see the Sphinx, apparently, was to ride down on a camel (it must be nearly ¼ mile from Mena House, where their motor cars stoped[sic]). There must have been between 30-40 camels carrying those poor fools down to the Sphinx. Then they must be photographed in front of it (at 10/- a shot so our Dragoman said). Then, without going any further one rode back, clinging on to the back or front for dear life. One buys a small copy of the sphinx or a scarab, as a souvenir, it does not seem to matter if it is old or not. You are then an authority for the rest of your life on the Sphinx (which you have barely looked at) and the Pyramids, which you may have had jolted across [sic] your line of vision!!! Not a glance at the pyramids of Dashur or Sakkara and so a very interesting day in your tour of Egypt (total 4 days which includes the Mosques and bazzars[sic] comes to an end, and you go home and lecture your home town on "My tour in Egypt" and are looked on as a mighty authority on all places on Earth for the rest of your existence. He (or she) must be right because he has seen it!!![10]

This sarcastic comment on tourists was uttered by George H. Williams, a 30-year-old British entrepreneur in textiles, who visited Egypt in 1925 with his father, Howard. His remarks ridiculed the tourist as someone who travelled to distant places yet saw the world as a stage for his self-representation, photographing rather than contemplating the sights with his own eyes. The irony in Williams' account results

Figure 3: "You are then an authority for the rest of your life on the Sphinx"
(Photographer: George H. Williams, 1925. The Royal Commonwealth Society Archives.
Reproduced by kind permission of the Syndics of Cambridge University Library)

from the contrast between the superficiality of the tourist's visit and the expertise attributed to him after his return.

Williams' account demonstrates that the production of tourist photographs was embedded in a larger context of social relations. It suggests a triangular relationship between the photographer, the sitters, and the future audience – that is, the spectators 'at home'. This – imagined – audience was inherently involved in the production of the photograph, as the photographer took it (at least in part) to satisfy their expectations. Even though in the diary, Williams distanced himself ironically from this practice, he did not ignore the visual canon.[11] In his photograph album, Williams documented his own visit to the Sphinx in a similar fashion (fig. 3): the shot in sepia tones showed two tourists on camelback in front of the Sphinx, guided by a dragoman, the pyramids of Giza in the background. While Williams was aware of the cliché and ridiculed it, he nevertheless reproduced it.

In this section, I discuss three facets of Williams' travel photographs, trying to understand how an amateur photographer like him formed a picture of Egypt. First, I approach his photographs as reproductions of an Orientalist discourse addressing an audience 'back home'; second, I ask for the relation between Williams' impressions of Egypt and the motifs he chose to capture with his camera; third, I suggest that the persons sitting for his photographs had an active role in shaping their portrait and were well aware of how they liked to present themselves.

Orientalist landscapes: Stabilising the imperial order

The first axis of the triangular relationship – the connection between photographers and spectators in the imperial metropoles – has been studied thoroughly in research on Orientalist visions of Other societies, that is often contemptuous visual representations of seemingly backward societies.[12] Inspired by Edward Said's work, it has been argued that Orientalist photographs anticipated, prepared and sustained imperial rule, based on two assumptions: the denigrating perspective on local populations transported ideas of Oriental inferiority and, due to the serial production and dissemination of such views, these presumptions were mistaken as facts.[13]

According to Derek Gregory, the photographs produced by professional photographers and tourists in Egypt corresponded to an imaginary appropriation of the country, which was intertwined with its material occupation. He argued that this visual appropriation occurred in two stages, both of which contributed to an abstraction of the Egyptian space. In the first stage, historical monuments were photographed as solitary structures, ideally concealing the presence of other visitors. Thereby, they suggested an uninhabited 'empty space', seemingly awaiting appropriation and conquest. Sary Zananiri added that such 'empty spaces', apparently bereft of contemporary inhabitants and thereby any change, came to be equalled with an objective representation of space.[14]

Yet when armchair travellers turned into mobile visitors in the second half of the nineteenth century, the presence of local inhabitants could no longer be ignored and required an explanation. The idea of an empty, mythical and bygone Orient collapsed. As a result, Gregory argued that photographers replaced this notion with that of a present, yet 'backward', society, which, in turn, became an attraction in its own right. Such "ethnographic" photographs, as Gregory termed them, claimed to record instances of an 'authentic' Orient, doomed to perish, and thereby preserve its memory.[15]

The assumption that such representations sustained imperial power relations was advanced by Timothy Mitchell, who showed that the repetition of certain visual motifs turned them into clichés or stereotypes, ultimately mistaken for a truthful reality, which Mitchell termed a "reality effect".[16] More recent insights from visual studies confirmed that images do not represent realities, but constitute it, thus serving as agents.[17] Applied to tourist photographs, these strands of thought suggest that by attributing meaning to certain motifs and reproducing them, photographers contributed to creating an idea of the respective country.[18] Imaginings conceiving of the East as an unchangeable region had been firmly established in both French and British culture and, early in the nineteenth century, photographs had defined the tropes that came to represent 'the authentic' Oriental society. The contradiction between processes of 'modernisation' and the *"collection des clichés"* is expected to be of particular relevance in the context of Oriental tourism.[19]

Until the late nineteenth century, travel photography remained the domain of specifically trained photographers who were able to handle the necessary equipment.[20] The invention of the portable, easy-to-handle Kodak camera in 1888 made the journey more widely recordable, transportable, and shareable. As a result, the impressions of travellers and the expectations of those at home materialised in a multitude of photographs, lose or in albums, and the camera turned into the symbolic accessory of the ordinary tourist.[21] Private lantern slide shows and self-made photographic albums increasingly complemented and replaced the professionally produced stereoscopic views and photographs. Still, amateur pictures often reproduced the images defined by previous generations of explorers, travellers and photographers. The monumental and ethnographic motifs changed slowly, even though Egypt and the neighbouring countries underwent profound and visible transformations, for example in urban structures, professions and fashion.

The axis of photographer and audience generally explained the stability of representations in times of transformation. John Urry argued that from the tourists' perspective the idea of 'authenticity', namely seeking the experience of a local culture supposedly 'unspoilt' by modernity, was less significant than often assumed in research on tourism. Rather, he defined the "tourist gaze" as a cultural practice identifying cultural difference. The tourist gaze, according to Urry, attributed meaning to elements travellers considered extraordinary or simply different from their own everyday lives.[22] The identified difference did not have to be original: "the tourist gaze is largely *pre*formed by and within existing mediascapes."[23]

The audience, in turn, benefited from – and expected – a stable, static representation of the Orient. Ali Behdad argued that visual commonplaces, stereotypes and interpictorial references allowed the audience to decipher the represented foreign cultures.[24] To this assumption of better legibility, Anne-Gaëlle Weber added the notion of credibility. Her remarks on travel reports of the nineteenth century, which she characterised as a "factory of commonplaces", can be applied to photography as well: by reproducing topoi, the amateur photographer confirmed common expectations, thereby demonstrating that he did visit the well-known places.[25] Had Williams not photographed the Sphinx, his audience might have doubted that he properly visited Egypt. The topos thus had a mediating function between report and fiction, as it created the impression of an 'authentic experience', while being a consciously shaped element. As such, its "reality effect" was based on the seriality of its application.[26]

By providing new photographs that confirmed widespread views, amateur photographers participated in shaping spaces. They reproduced and popularised topoi coined in the nineteenth century, contributing to their longevity and stabilising views of the world order. The everyday production and consumption of photographic memories perpetuated views of 'backward' colonised societies and rendered the empire plausible for friends and relatives.[27] Exoticist commonplaces continued to characterise local societies in photographs even if they did no longer correspond to a widely experienced reality. Animals such as donkeys or camels, for example, were popular symbols expressing both 'authenticity' and 'backwardness' in a time of speedy modernisation and motorisation.[28] Hence, topoi persisted long after the realities they authenticated had changed, thereby creating the fantastic, imagined spaces still existent in the realm of tourism.

Searching for strangers: George H. Williams' views of Egypt

The reproduction of views that stood for an Other Egypt, 'unspoilt' by traces of modernity, required the conscious effort of the amateur photographer. Already upon his arrival, Williams carried a mental list of photographs he hoped to capture during his journey, 'preformed' by other pictures he had seen, guidebooks he had read, and by his own previous travels to North Africa.[29]

In particular, his preconceived ideas about how to photograph Egyptian monuments were very concrete. Often Williams had a precise perspective and angle already in mind, and he planned the monumental photographs carefully and far ahead. In his diary, he described the monuments in detail (often copying passages from the guidebook) and took notes as to how to proceed in the case of a second visit to Egypt. For such an occasion, Williams also noted the interesting sights for which sufficient film should be spared. In general, he was quite satisfied with the monumental photographs he was able to take. It seems that the knowledge he had acquired of which sights to expect in Egypt allowed him to sufficiently prepare for the photographs, and to avoid disappointment.[30]

Although the extensive descriptions of monumental photography dominated his diary, Williams' photograph album was characterised by 'exotic' photographs reflecting the Othering tourist gaze. Ali Behdad identified four types of Orientalist photographs: panoramic, monumental, exotic, and erotic photographs.[31] In Williams' photograph album, 9 pictures showed panoramic views, and 19 represented monuments or other antiquities, while erotic photographs were missing entirely. The 39 exotic photographs represented the allegedly picturesque daily life of the local population, corresponding to Gregory's category of ethnographic photography.[32]

Compared to monumental photographs, shooting exotic views of local society was more difficult to plan for Williams. While his monumental photographs were products of reflection (and reading), capturing persons required spontaneity in order to grasp occasions, as well as successful negotiations with the sitters. Moreover, Williams' vague ideas about the daily life of the Egyptian population left greater space for imagination – and disappointment.

His expectation that Egypt would resemble the Maghreb he knew from previous travels shows that, in terms of daily life and culture, he perceived Egypt as part of a North African continuum. For him, North Africa had come to represent the 'Orient'. Although his expectations about exotic motifs were rather vague ("street shots"), the encounter with different local realities in Egypt often caused frustration because Williams did not find what he was looking for. Mosques, for example, were often part of the dense urban structures of the inner cities, not allowing Williams to obtain the necessary distance for photographing, rendering them in his judgement "hopeless from a camera's point of view".[33]

Despite such vague expectations, the structure of the album indicates that these scenes and "street shots" were the photographs Williams was looking for. Whereas the photographs in the album roughly followed the chronological order of the journey from Cairo up the Nile, topics were the major ordering principle. Each page or double-page was dedicated to a topic. The fact that Williams left some empty frames on several double-pages (four pre-shaped frames on each double-page defined the position of the photographs) suggests that the pictures were intended to complement each other as variations of the topic. Occasional empty frames signalled that Williams was lacking additional views to illustrate the respective topic.

Most pages were dedicated to Egyptian life in the countryside. Workers on the Nile, souvenir sellers, women, children and young Egyptians from the villages were the main subjects in 56 out of 87 photographs. In nine other photographs, he portrayed Egyptians working with tourists, such as guides and cameleers. In contrast, Williams himself and his travel companions appeared in only 5 photographs, while 11 views indicated the presence of other tourists in the background. The marginality of travel companions in Williams' album differs from comparable albums of the time, in which tourists typically documented their own presence, their relatives, or other acquaintances at different tourist places.[34]

Not only the quantity, but also the framing of the photographs suggests that Williams attributed greater significance to portraits of Egyptians than to his European travel companions. While shots of his Egyptian sitters often presented them from close-up and with a focus on facial expression, Williams portrayed his travel companions in medium shots, thus from a greater distance and often while they were interacting with other travellers. Williams' own presence in the album may thus be likened to an omniscient narrative voice – as a narrator he created order, directed the narrative and suggested interpretations, without appearing to be personally involved.

The nostalgic representation of a romanticised rural population required the conscious omission of elements considered European, Western, or modern. The diary of George H. Williams demonstrates that he *did* perceive the transformations of the country, though without capturing them in his photographs. In contrast to his album, other photographs reveal which aspects of contemporary life in Egypt Williams did not document: there were, for example, no views of cars parking in front of cinemas, of the large squares and mundane department stores.[35] Rather than overlooking the urban modernity of Egypt, Williams consciously omitted it: in his notes, he described Egypt as the playground of European and American tourists; he mentioned the presence of cars in the streets, the electric lighting in the pharaonic tombs and the bicycles of Cairo's inhabitants, while regretting the disappearance of truly 'Oriental' bazaars in Cairo. Williams recorded the transformations induced by tourism, such as new houses built in Asyut "rather like those on the Rivera[sic], Cannes fashion", and three new hotels near Karnak.[36] Photographs documenting these changing architectural trends and the tourist expansion, however, were lacking in his album.

While the spectator of Williams' album had to assume that transport in Cairo was based on camels or donkey carriages, his diary reveals that in picturing animals, Williams in fact documented the end of an era. Like other travellers, he was

Figure 4: Egypt 'ancient and modern' (Photographer: George H. Williams, 1925. The Royal Commonwealth Society Archives. Reproduced by kind permission of the Syndics of Cambridge University Library)

searching for an Egyptian culture and society he conceived of as fundamentally different, a 'traditional' – in the sense of backward – society.[37] From his perspective, during a journey to the 'Orient' only these allegedly authentic Egyptian facets of lived reality were worth recording. In order to reproduce visions of the Orient as a topic, the photographer had to overlook elements connoting modernity.[38]

The effect, in Mitchell's sense, of an authentic Egyptian reality created by the topical ordering principle of Williams' album bore a sense of ambiguity though. The creation of visual icons in Orientalist tourism potentially converged with the essentialising imaginaries of nationalism. For example, Williams juxtaposed the portrait of a young woman outside a village on the Nile with the photograph of an Egyptian statue in similar posture (fig. 4). Such juxtapositions of ancient monuments and items that were identified as the modern version of the former were popular at the time.[39] According to Beth Baron, the representation of young Egyptians next to ancient sights claimed a genealogical connection between modern Egyptians and their alleged pharaonic ancestors, and stood for pride in the past and the future of the country. Drawing on Baron, the album page showed the allegory of Egypt as a woman in its modern and in its ancient version.[40] It thus offered two alternative interpretations to the spectator: as a contemporary Egypt which had preserved part of its ancient grandeur, or as a view of a romanticised unchanged society, which relegated the woman to a past distant from Western modernity.

The potential ambiguity stemmed from the inherent polysemy of images, and Williams did not provide the spectators with captions or similar textual explanations about what the photographs showed. The reliance on exclusively visual means of representation rendered his narrative more complex, as it required the spectator's ability to identify the visual topoi. Whereas it is likely that in private contexts the presentation of the album was accompanied by oral interpretations, the absence of written anchorage in principle created room for ambiguity.[41] Textual anchorage might have seemed dispensable to Williams because he referred to firmly established topoi and familiar motifs; yet by not doing so, the photographs maintained their potentially subversive polysemy.

Outside the frame: Encounters with amateur photographers

The polysemy inherent in photographs is at the core of recent debates about the interpretation of 'Orientalist' photographs. New studies on photography in colonial contexts questioned the efficiency of photographs in sustaining and reproducing a colonising discourse. Nancy Micklewright argued that the visions of an exotic, eroticised Orient considered typical by authors such as Alloula, Gregory and others were not representative of nineteenth-century photograph collections. Instead, she pointed out that the particularities of the collections revealed individual preferences, rather than an overarching discourse. She thus questioned the impact of Orientalist photography. Given that even collectors of similar socio-biographic backgrounds produced highly diverse collections, Micklewright concluded that a homogeneous stereotypical vision of the Orient did not exist.[42]

With regard to the production of photographs, three additional arguments questioned their unequivocal contribution to the stabilisation of an Orientalist discourse. First, Christopher Pinney argued that the analysis of visual discourses generally overlooked the contribution of the photographed subjects to their own image. This participation of the subjects could potentially subvert discourses of power.[43] Correspondingly, Urry and Larsen underscored that tourist photographs were not just serialised reproductions of given visual discourses, but that they had the potential of adding, changing, and subverting common views and visions. Therefore, they suggested replacing the static notion of the "tourist gaze" by grasping tourism as a performance. In visual analysis, this notion implies that the contexts of production, the actors involved and their agency impacted upon the image production.[44] Second, some authors criticised the fact that most assumptions about photography were based on images produced by European photographers, marginalising the contribution of non-European photographic traditions and appropriations of the technique.[45] Third, a growing number of researchers argue that a mere symbolic interpretation of images has to be complemented by an understanding of photographs as objects, in order to develop a contextualised understanding of their significance and implications.[46] These new calls to take both the materiality and non-European traditions, as well as the agency of sitters, into account invite us to question the implications photographs seemed to have at first sight and to focus on the conditions under which the photographs were produced.

In the context of tourism, this axis of the relationship between sitters and photographers seems particularly relevant, yet understudied. While the motifs did not differ greatly from earlier professional photography, as I argued above, amateur photographers captured their views under different conditions than the well-studied professional studio photographers, particularly as far as portraits of the local population were concerned. Williams and his fellow tourists neither hired models nor staged scenes in studios.[47] His documents constitute an especially valuable source in efforts to reconstruct what happened before the photograph was taken. The travel diary reveals that he was a passionate photographer who dedicated much energy, time and effort to capturing photographs. In particular, his reflections on photographic successes and failures are an invaluable source, for they provide us with insights into the production of photographs. His travel notes demonstrated that he depended on the collaboration of the local population, who self-confidently shaped the image they presented to the tourist.

While Gregory assumed that exotic photographs were "hunted" without obtaining permission from

Figure 5: A smile for the camera (Photographer: George H. Williams, 1925. The Royal Commonwealth Society Archives. Reproduced by kind permission of the Syndics of Cambridge University Library)

Figure 6: Mahmud guiding tourists (Photographer: George H. Williams, 1925. The Royal Commonwealth Society Archives. Reproduced by kind permission of the Syndics of Cambridge University Library)

the sitters, Williams' album suggests that in amateur photography, the sitters actively participated in the photograph.[48] The two portrait photographs of a young Egyptian woman on pages 31 and 44 exemplify this collaborative aspect. In one photograph, she was looking straight into the camera, while the second version showed her in the same pose with a smile (fig. 5). Standing on a stony slope, she was balancing an earthen jar on her head. The self-confidence in her face indicated that she was familiar with interacting with tourists and their cameras – she could handle the situation perfectly. Moreover, the change in her facial expression visualised that there had been some communication between the woman and Williams.[49] Either Williams had asked her for a smile or she had presented herself smiling of her own volition, apparently aware of photographic conventions.

This visible interaction differentiated amateur photographs from nineteenth-century studio photographs. Behdad observed that in order to create the illusion of an authentic situation, the 'exotic types' in studio photographs did not look into the camera.[50] Williams' portraits, in contrast, reveal the direct interaction that had preceded the photograph. By looking straight into the camera, the sitters of amateur photographers not only interacted with future spectators; they also made the photographer visible, hence subverting the idea of an objective, factual representation.

That the sitters knew how to handle the situation and how to shape their own portrait is also evident in the portrait of Williams' guide Mahmud, a man belonging to the Egyptian *efendiyya*. Williams dedicated two photographs to his guide, both of which suggested respect for the qualifications and assistance of his guide. A close-up portrait of Mahmud's face indicated a familiarity between Williams and Mahmud. The second portrait showed Mahmud in front of an ancient wall with hieroglyphic inscriptions (fig. 6). The guide had adopted a lecturing posture, revealing his self-perception as an expert. The portrait acknowledged this expertise, presenting him as a skilled and highly valued interpreter of the country and its antiquities. In the diary, Williams noted that Mahmud posed deliberately in front of the wall, offering the tourist group the opportunity to take his photograph.[51] Thus, it was the guide himself who defined the right moment, the right background and the right posture for his portrait. Not only did he participate in producing the photograph, but he also decided how he wanted to be seen by the tourists and their British audience.

For some inhabitants of the villages and towns along the tourist routes, the increasing touristification and the travellers' desire for 'authentic' portrait photographs offered an opportunity for income. Williams observed that while the persons posing in front of his camera in the Sudan, a less frequented tourist destination, did not ask for 'baksheesh', Egyptian sitters expected remuneration.[52] Assisted by their interpreters, travellers negotiated with the sitters about the price.[53] For some of them, acting as a model seemed to have become a remunerative activity. They waited on the shores for the tourist steamers to anchor in order to pose for photographs for money, actively searching for a way of profiting from the tourists' obsession with authentic scenes.[54] Notably, it was a way of contributing to the family income for children, and they often took the initiative in approaching tourists "for photograph and baksheesh purposes".[55]

In these situations of negotiation and eventual collaboration, local sitters were more than passive instruments for the realisation of the visual desires of foreign photographers. Mahmud defined the right moment for the photograph; other sitters negotiated over the price of their portrait; but none of the persons in Williams' album appeared insecure or uncomfortable in the situation and they were by no means at the mercy of the tourists.

In addition, they felt free to refuse the photograph. Williams reported on a discussion with a little girl who refused to sit for him. He had to accept that not even by offering money he was able to change her mind: "A little further on I saw a small girl dressed in a fringe and our guide tried to get her to stand for me, but she was bashful and the offer of Backshish was of no avail."[56] Thus, Williams realised that the girl felt uncomfortable in the situation – given her young age, it is plausible that she was insecure about how to behave with strangers – and he knew that negotiations with potential sitters were open. Combined with the photograph of the women in figures 4 and 5, the example demonstrates that it was a real option for the photographed subjects to set limits to the wishes of tourists. Amateur photographers like Williams required the cooperation of the local population, who actively shaped or refused the portraits that were taken of them – on equal terms.

These conditions of photograph production implied that the photographers recruited their sitters from mainly two sections of Egyptian society. As a first group, amateurs photographed their tour guides, cameleers, or street vendors working in the tourism sector. Second, the monetary incentives offered by tourists appealed predominantly to children, members of the lower classes and the rural population. As a result, (acquaintances from the tourist industry excepted) members of the Egyptian urban bourgeoisie remained invisible in the photographs of foreign visitors. The way the Egyptian population was represented photographically was not only defined by the Orientalist visions of the amateur photographers, but also by the limited incentives they were able to offer. The following sections demonstrate that the urban *efendiyya* did participate in the production of visual impressions of Egypt. However, they did so as a fourth actor, in addition to tourist photographers, sitters and the audience. As professional photographers and editors, rather than sitters, they aimed to profit from the seriality of photographs in order to shape and promote their visions of Egypt on a national and international level.

As far as Williams and his album are concerned, it seems that the polysemy of the photographs does not allow for a clear-cut interpretation. Williams repeated the prefabricated discourses on Egypt that were familiar to him and his future audience, thereby stabilising the idea of an authentic Egypt defined by nineteenth-century travel photographs. Whereas Egyptian sitters consciously shaped their individual portraits, we have to assume that, from the point of view of Williams' British audience, the seriality of the resulting photographs, as well as the ordering principle of his album, turned them into 'Egyptian types'. The discernible posing of sitters did not prevent them from being inserted into a preconceived overarching narrative of an allegedly backward society. The polysemy of photographs facilitated the coexistence of contradictory interpretations. The photographs of George H. Williams were neither 'impressions' nor 'snapshots' hunted illegitimately, but consciously staged compositions bringing together the interests of the photographer, Orientalist frames and Egyptian self-representations. Nonetheless, the audience continued to read them in familiar frames of Orientalist photography.

Negotiating tourism: Ambitions and limits of Egyptian tourism development

In 1933, Egypt hosted the International Congress of Tourism. The organisers expected 400 participants, for whom they drafted a large entertainment programme including receptions, tea parties and soirees in various Egyptian hotels, as well as excursions and visits to Egyptian antiquities.[57] The event demonstrated that Egypt was internationally recognised as one of the epicentres of the relatively recent tourism "industry".[58] While Egyptian governmental institutions took the lead in the organisation of the congress, the business of tourism was, however, a European one. The industrialists involved in the planning and preparation of the congress were mainly well-known international businesses: *Thomas Cook*, the *Société Internationale des Wagons-Lits*, *American Express* and others.

In order to organise and prepare for the congress, a local committee of approximately forty members was formed in July 1932. Tawfiq Doss Pasha, Minister of Communications, presided over the committee. It also included the general director of the national railway company, several under-secretaries of other state departments, the governors of Cairo and Alexandria, and representatives of numerous companies and organisations ranging from transport companies, hotels and banks to institutions such as the Arab Museum and the *Services des Antiquités Egyptiennes*. The presence of Faris Nimr and Gabriel Taqla on the committee signalled a certain public interest in tourism – they owned the two prominent Egyptian newspapers *al-Muqattam* and *al-Ahram*. The majority of the interested persons and companies, however, had a European background, such as Charles Baehler, the Swiss director of the famous *Shepheard's Hotel* in Cairo.[59]

The International Congress of Tourism took place in the midst of a new upturn in the Egyptian tourism business. After the end of World War I, European tourism to Egypt was slowly recovering. During the winter season of 1920 – at the time, tourists still wintered in the Mediterranean – hardly any foreign tourists visited Egypt.[60] Yet, although the repeated economic downturns during the 1920s negatively impacted the tourism sector, the number of tourists during most years of the 1920s and 1930s exceeded the number of travellers before the war, even those of the pre-war peak years.[61]

Tour operators in the Mediterranean also profited from Egyptian travellers. For the revenues of the French shipping company *Messageries Maritimes*, for example, summer tourism of Egyptians was much more lucrative throughout the period than the European traffic to Egypt. Egyptian vacationers either visited Europe or travelled to the mountain resorts in Lebanon and Syria. During the early 1920s, the *estivage* partly compensated losses from the European tourism business.[62] The regional economic crisis from 1925 onwards, in addition to political upheaval in Egypt, negatively affected the growth in this sector, but overall, the *estivage* traffic remained rather stable in contrast to the more volatile European tourism.

After the Great Depression, from around 1933, the number of European roundtrip tourists grew again and reached a peak in 1936, before the outbreak of the Palestinian Revolt again negatively affected the expansion.[63] This expansion was partly due to an increase in cheaper offers attracting new social groups to Egypt, reflected in growing numbers of second- and third-class travellers.[64] These travellers tended to book tours in summer during the low season, profiting from reduced prices during this time of the year: a decree from 1932, for example, stipulated that in Luxor, Korna and Karnak, the prices for donkey guides in winter were higher by 25% than in the summer season.[65]

Despite the expansion of the tourist sector, it continued to be dominated by European and American enterprises. As a result, Egyptian actors mainly appropriated tourism in terms of cultural politics. Associations and governmental institutions cooperated with international companies, rather than trying to oust them.

Cook's colony? Organised tourism in the early 1920s

The story of organised tourism to Egypt is often told as a one-man show starring John Mason Cook, who was responsible for the establishment of the Egyptian operations of *Thomas Cook & Son Ltd.* in the nineteenth century. The company was among the most famous tour operators of the time, not least because it was reputed to have invented the package tour, rendering travelling affordable by applying organisational principles of industrial production to the business of tourism.[66] Contemporary observers viewed *Cook* as a synonym for organised travel, and described its impact as having established 'mass tourism' on the Nile.[67] In addition to organising transportation, *Cook* diversified and expanded its entrepreneurial fields by taking over and running several luxurious grand hotels. The company purchased the *Luxor Hotel* in 1877 and held stakes in *Upper Egypt Hotels Ltd.*, which owned, in addition to the *Luxor Hotel,* the *Winter Palace* in Luxor and the *Cataract Hotel* at Aswan.[68]

While Martin Anderson questioned the pioneering role of *Thomas Cook & Son* in tourism in Egypt, Robert F. Hunter and Waleed Hazbun put forward more convincing arguments explaining the exceptional role of this particular British tour operator in the Egyptian market.[69] One of the keys to *Cook's* success in Egypt was its intertwinement with the late-nineteenth century British presence in the country. According to Hunter, British protection and the privileges Europeans enjoyed in the Ottoman Empire favoured the establishment of tourism along the Nile. Moreover, British geostrategic expansion in Egypt, especially after 1882, and the rise of the tour operator mutually benefited each other to the degree that Hunter considered the company vital in safeguarding British strategic interests in the region – and vice versa. The company backed the British quasi-colonial venture in Egypt, both materially and ideologically. The company's ships transported troops and supplies during the military campaign in 1884, and *Cook's* travel handbooks advertised the

British presence, which allegedly contributed to an improvement in living conditions in general and sanitary conditions in particular.[70]

British governmental support for the tour operator seems to have been complemented by John Mason Cook's personal skills in networking. Hunter described John Cook as a person able to win over his Egyptian employees by showing his appreciation, donating money and improving access to health care for the local population. Moreover, he maintained good relations with the Egyptian viceroys. Both Ismaʿil and his successor Tawfiq granted privileges to the company, notably concessions for passenger services and postal concessions on the Nile.[71]

Although *Cook & Son* was sold to the Belgian *Compagnie Internationale des Wagons-Lits* in the 1920s and re-established as a British firm only at the beginning of World War II, from the point of view of British travellers it remained a British company which they heavily relied upon.[72] They trusted in *Cook's* recommendations and services, as well as in the promise of reasonable prices that did not require negotiations. Most travellers to Egypt in my sample relied on *Cook's* services: they transferred money through them, had their cars and drivers organised by the company's agents, arranged pick-up services with them at harbours or train stations, and hired tour guides approved by *Cook's*.[73] The symbolic significance of the company should not be underestimated. From the perspective of contemporary observers, *Cook's* was more than a brand; it represented a trustworthy institution in an unknown country.

From an Egyptian perspective, *Cook's* and other mainly European companies represented foreign dominance in Egyptian political and economic affairs, resentment against which they expressed repeatedly.[74] Indeed, although Egypt was no longer officially under foreign rule since 1922, in the tourism sector, as in other spheres, foreign companies presided and often profited from privileges such as tax exemptions, which had not been abolished with Egypt's partial independence.[75]

European tour operators such as *Thomas Cook & Son* and the *Compagnie Internationale des Wagons-Lits* organised much of the transportation, and most grand hotels in Egypt were owned by internationally operating corporations. Egyptians profited from the tourism business at best as subordinate employees, as tour guides or bellboys, but even these professions were contested: the *Heliopolis House Hotel* advertised the discretion of their Swiss staff and a large number of employees in the agencies of the big shipping lines held European nationalities, notably those in middle and higher positions.[76] Due to the dominance of European companies in the tourist sector, the economic benefits of tourism were largely transferred outside the country. Although Egyptian actors offered an increasing range of services to tourists, the newly founded companies hardly threatened the dominant position of their well-established European competitors.

Tourism as a cultural ambition, mid-1920s to early 1930s

Around the mid-1920s, tourism entered Egyptian debates concerning the future of the country. It did so largely in the context of cultural policies, rather than as economic potential. In the debates and programmatic reflections of the Egyptian institutions concerned with economic development, tourism remained largely absent. This absence becomes visible particulary when compared to political speeches and debates from neighbouring countries. In the inauguration speeches of the Chamber of Deputies, the French High Commissioner of Lebanon and Syria repeatedly referred to tourism as having major economic potential for the Mandate. The Egyptian king, in contrast, did not mention this topic in any of the inauguration speeches.[77] Instead, King Fu'ad addressed mainly cotton production, agriculture and new technologies such as hydropower plants as Egyptian strategies for economic development.[78]

A similar absence characterised the Egyptian administration. Despite the prominence of Egypt as a tourist destination, neither a ministry nor a ministerial commission was designated for tourism. The Economic Council, a consultative institution reconstituted in February 1925, did not comprise any members involved in the tourism business or having expertise in the field.[79] Similarly, among the members of the Advisory Council at the Department of Commerce and Industry established in January 1928, expertise from within the tourism sector was lacking.[80] The Ministry of Commerce and Industry did not plan expenses in order to improve, expand, or sustain the sector, and the numerous commissions preparing legislation in the Chamber of Deputies hardly paid attention to tourism.[81] Only deputies whose constituencies attracted a great number of tourists occasionally asked for more support, but all in all, tourism remained a marginal topic within debates at Parliament and in the Senate.

Economic policies in Egypt generally focused on agriculture, rather than on services such as tourism, which contributed much less to the Egyptian economy.[82] Cotton production remained a central pillar of the Egyptian economy, and for the proponents of economic diversification, the industrialisation of Egypt was a more promising alternative to agriculture than tourism development. Their major instrument, *Banque Misr*, founded in 1920, invested capital mainly in companies linked to cotton production and trading.[83] Moreover, the economic historian Camilla Dawletschin-Linder argued that Egyptian politicians were by no means united in their ambition to diversify the economy. After all, the government and state institutions were largely dominated by prominent landowners, whose wealth was based on cotton production and who thus had an interest in directing state intervention and support to the agricultural business.[84] Government action therefore focused on the administration of water systems and canals, irrigation and the expansion of agricultural lands, as well as investments in infrastructural projects related to agriculture and industry. These political priorities mirrored the limited interest of the Egyptian population in the tourist sector, as the absence of tourism in petitions and proposals discussed by the Egyptian Senate indicates.[85]

The dominance of foreign companies in the tourism sector seems to have been another reason for the marginality of tourism in economic debates. From the point of view of several deputies, strengthening the business of tourism was tied to notions of transferring Egyptian funds to foreign corporations. During a heated debate in the Egyptian Chamber in 1927, half of the deputies rejected subsidies for a project of the Ministry of Public Works intending to enlarge the street connecting Cairo with the Pyramids. The other half backed such an incentive. In the end, the Chamber approved the necessary credit, but it was a narrow majority of 76 votes to 75.[86]

Opponents of the project voiced mainly financial concerns, doubting the beneficial effects of tourism. Arguing that mainly foreigners profited from such subsidies, several deputies attacked government expenditure on tourism in general. From their point of view, strengthening the business of tourism implied transferring Egyptian funds to foreign corporations.[87] Their position mirrored the extremely cautious policy of expenditure of the early Egyptian governments. Many deputies had the European financial control resulting from the bankruptcy of Egypt in 1876 in mind, a situation they wanted to avoid by any means for Egypt's future.[88] Among the critics, the deputy Dr Husayn Yusuf ͗Amir objected passionately to the project. He argued that enhancing the road to the Pyramids would benefit tourists, whereas roads used by Egyptians in other parts of the country desperately needed repairs. His position was backed by a large number of deputies who considered tourism a business organised by foreigners for foreigners on Egyptian ground, accusing the Government of neglecting the Egyptian population.[89]

Speaking in favour of tourism development, the Minister of Public Works argued that visitors spent large sums in Egypt, concluding that further investments in places of touristic interest would have positive economic effects. He feared that these financial benefits were at risk if tourists could not reach the Pyramids in a comfortable manner. Moreover, he pointed out that Cairo generated higher revenues than other Egyptian cities, which justified expenditure on the further embellishment of the capital. His main argument, however, was moral rather than economic, which he emphasised in a rhetorical climax at the end of his speech: the road had to "prove worthy of the greatest monument in the world: the Pyramids".[90] In order to convince the deputies of the necessity of the project, the minister highlighted that the international prestige of Egypt was at stake. Tourism was thus considered politically (rather than economically) relevant.

Accordingly, legislation in the context of tourism development aimed to improve the experiences of international travellers. Rather than setting incentives for the expansion of tourism, most measures introduced quality standards. Minor subsidies for the embellishment of tourist sites were included, but laws and decrees rather reacted to allegations voiced by tourists against their tour guides and service personnel. These ranged from lack of quality and of adequate knowledge to the opaque prices of services, often coupled with stereotypes about the greed, laziness and dishonesty of the local population.[91]

Regulation hardly targeted the large foreign companies dominating the tourism sector though. This was criticised by the deputy ꜥAbd al-Rahman ꜥAzzam, who lobbied for a direct involvement of Egyptian institutions in the tourism sector to protect Egypt's reputation as a tourist destination. He argued that state services would protect tourists from negative experiences with dishonest travel agents. Moreover, he stated that the measure would secure taxes for the Egyptian state. As a result of the Four Reserved Points, these taxes were not levied in the event that foreign agencies organised the tours. The Nile crossing of tourists aiming to visit the Valley of the Kings was one such example. At the time, the *Société des Hotels*, a Swiss corporation, organised the Nile crossing for tourists, charging them a fee of 100 piasters. ꜥAzzam argued that if the *Egyptian Railway Company* offered the transfer at a lower price, tourists would save money while the railway company and the Egyptian state would benefit from additional receipts and taxes.[92] Hence, the circumvention of foreign companies would be a bargain for both the state and tourists. However, his suggestion to break the agreement with the *Société des Hotels* was never put into effect. In regulating the business of tourism, it seems, the room for manoeuvre of the Egyptian parliament was restricted to regulating local entrepreneurs.

Already in the late nineteenth century, decrees had defined official prices for services provided by dragomans and guides.[93] In the later 1920s, the Egyptian directorates updated and systematised tariff regulations for such services, gradually extending the range of the regulations to towns all over the country.[94] In Cairo as in other places, donkey owners were obliged to publicly present the tariffs, their registration number and their official licence to customers.[95] The previously established prices defined by the regional directorate not only avoided competition between guides, but also guaranteed foreign tourists that the prices were fixed, reliable and not subject to chance or bartering skills.[96]

In addition to the regulation of tariffs, the authorities introduced quality standards in the early summer of 1929. The decrees stipulated that from then on, any person intending to serve as dragoman or guide had to be officially authorised. The authorities at the municipal or district level determined the number of licences available, which varied from fewer than ten guides in the towns of minor tourist attraction to one hundred in Giza and Luxor, even reaching as high as 350 in Cairo. In order to apply for a licence, the candidates had to provide identification documents and proof of their "good life and morals". Moreover, the applicants had to pass an exam testing them on foreign language skills (at least one foreign language was required), as well as their knowledge of monuments, antiquities and places of touristic interest in Egypt. After one year, the validity of the licence expired and had to be renewed. In addition, the possession of the licence enforced 'decent behaviour' on the guides, who were not allowed to actively approach potential customers, nor to enter hotels on their own in search of customers.[97] Few years later, similar regulations and quality standards were applied to the business of renting out camels to tourists.[98] As in the case of tour guides, the regulations aimed at creating transparency, safety and comfort for the tourists, rather than improving the working conditions of the guides.

During the 1920s, tourism development did not play a significant role in governmental economic policies. Active promotion of tourism development by means of subsidies remained contested because a significant portion of the deputies felt that foreign corporations stood to profit most from such an expansion. Governmental intervention in tourism was focused on regulating services. Measures aiming to regulate the behaviour of tour guides, cameleers and other professions in close contact with tourists seemed a response to complaints and clichés about Egyptian service staff. Since the tourists' experiences in the country would influence their judgement about the population, correct and decent behaviour mattered. In doing so, the Egyptian authorities shaped this intercultural encounter, rather than the tourism sector itself, which testifies to tourism being perceived as a quasi-diplomatic concern, as opposed to an economic potential.

Alternative visions? Transnational cooperation in the mid-1930s

In the 1930s, several organisations in Egypt launched new efforts to promote tourism. In particular, two interest groups achieved greater visibility: the *Tourist Development Association* and the state-run *Tourist Office*. After the Great Depression, the numbers of tourists on the steamers of the *Messageries Maritimes* started to resume slowly from 1932 onwards, while the crisis seems to have led to a decrease of tourists from Britain and the US, though still major countries of origin for Egyptian tourism. Yet around the mid-1930s, proponents of tourism development hoped that the political "instability" of European destinations in the wake of the rise of fascism would bring back a larger number of tourists to Egypt and launched a large advertisement campaign.[99]

The economic value of tourism continued to play a subordinate role in Egyptian political discussions. Local observers did not expect tourism to contribute significantly to Egyptian economic development, as the doctoral thesis of the economist Abdul Hamid Sidky showed, in which he analysed the economic potential of contemporary Egypt in 1931. Sidky did not even consider tourism as a relevant branch in this context.[100] Even in December 1934, when a new Ministry of Commerce and Industry was created under King Fu'ad I in order to foster the economic recovery of Egypt after the depression, its economic strategies did not comprise any reference to tourism.[101]

This perspective may have resulted from the structures of the Egyptian tourism sector. European travellers trusted in the reputation of the big tour operators, and Egyptian enterprises – even the state – could not rival the capital, knowledge and networks of these companies. Rather than establishing competitors, it seems, Egyptian initiatives, companies and associations intended to establish collaborative structures uniting Egyptian and European members. Among the investors in the Egyptian company *Pharos*, for example, which was founded in 1928 to offer services in transport and tourism, were both Egyptians and foreign nationals.[102] Foreigners from the Egyptian tourism and hotel business also dominated the lobbying group *Tourist Development Association of Egypt* (TDA). The TDA had already been founded in 1912, yet with the exception of a tourist map published in collaboration with the German photographers *Lehnert & Landrock*,[103] regular activities of the organisation are discernible only in the 1930s. By that time, it was running information offices in Cairo, Paris, London and New York. The TDA organised, for example, an important reception in Luxor in the context of the International Congress of Tourism, and published a number of guidebooks and brochures for European tourists.[104]

Egyptians held the representative positions in the upper echelons of the TDA's administration. The group was under the patronage of the King of Egypt and presided over by the General Manager of the Egyptian State Railways, Telegraphs and Telephones. In 1933–34, the Governor of Cairo (Mahmud Sidqi), the Governor of Alexandria (Husayn Sabri), as well as the Inspector General of the Department of Commerce and Industry (Muhammad Amin Yusuf), served as honorary members of its administrative committee. However, the majority of the elected members on the administrative committee were foreign nationals. Out of the 17 members, only two represented Egyptian companies: Talʿat Harb Pasha, the founder of *Banque Misr*, and Muhammad Beghat Chimy as representative of the Egyptian State Railways. The other representatives stood for the international tourism business, representing companies such as *American Express, Thomas Cook,* the *Egyptian Hotels Ltd., Grand Hotels d'Egypte, Lloyd Triestino,* the *Messageries Maritimes,* the *Anglo-American Nile and Tourist Co., Imperial Airways,* and the *American Export Lines.* Charles Baehler, the Swiss hotel magnate, served as vice president of the General Council.[105]

It appears that these members shaped the contents of the TDA's publications to a greater degree than the Egyptian representatives. The tourist magazines issued to attract visitors were edited by Europeans, for example by Philip Taylor, the Cairo-based correspondent of the *Daily Telegraph*. In the magazine, mostly British authors presented an exotic Egypt, describing the romantic charms of its Islamic architecture and the wilderness of the desert. The illustrated special section of the magazine published photographs selected by the industrialist and representative of *Lloyd Triestino* Guy U. Perera in collaboration with A. Otto.[106] Next to their own photographs, Perera and Otto reprinted older views distributed by famous European studios such as *Lehnert & Landrock*, who had already left Egypt by that point.[107] The narratives and views in the brochure characterised Egypt as an archaic society, in which visual symbols of modernity were attributed to European enclaves only.

The presence of Egyptian representatives on the board of the TDA did not, therefore, modify the image of Egypt circulating in tourist brochures. It was only around 1935 that a shift towards alternative visions of Egypt became discernible in touristic advertising, and this was connected with the establishment of a new, Egyptian institution concerned with tourism. In July 1935, the Egyptian Government created an official tourist office (*Office du Tourisme de l'Etat Egyptien*), modelled on similar institutions in other countries. Its first director, the former Director of the Municipality of Alexandria Ahmad Sadiq, had a budget of approximately 32,000 Egyptian pounds (£.E.) at his disposal and pursued four major strategies in advertising tourism to Egypt: (print) advertising; finding multipliers; the use of modern media; and practical improvements for tourists. First, the office intended to spend about 12,000 £.E. on commercial advertisement. Roughly half of this budget was reserved for advertisements in Great Britain and Ireland; the other 50 per cent was to be spent on advertising in the United States, Canada, France and other countries. Second, the strategists of the *Tourist Office* invited foreign students as well as scientific researchers, artists and journalists to Egypt, offering partial funding for their trip. The staff at the *Tourist Office* reasoned that such eminent and trustworthy personalities would serve as multipliers, disseminating information and spreading enthusiasm about Egypt as a destination after their return, and thereby attracting more tourists. Third, the office intended to make use of various media: radio programmes and documentary films on Egypt would arouse the curiosity of foreigners, and posters along the Suez Canal addressed transit passengers. The office would publish tourist brochures in Arabic, English, French and German, as well as a periodical review. Fourth, a reformed Office of Interpreters would facilitate the stay of tourists in Egypt. While French observers dismissed the ideas of Sadiq and his staff as neither new nor original, it was a new step for the Egyptian Government to assume responsibility for the advertisement of tourism to the country, and thereby for the presentation of Egypt on the international stage.[108]

The new office quickly resumed work. From the outset, a new presentation of Egypt was on the agenda. A review entitled *Egypt: A Tourism Quarterly* was published between 1936 and 1939. Lavishly illustrated and with bilingual (French and English) texts and captions, the magazine addressed an international, mainly European audience. In contrast to the earlier TDA publications, this magazine largely offered the vision of a modern, young, athletic, urban Egypt. An excerpt from the diplomatic correspondence suggests that this vision was the result of a conscious collection and selection of documents. In 1935, Sadiq's successor as general director, L. Hakim, asked the Minister of France in Egypt, Pierre de Witasse, for copies of some photographs de Witasse had exhibited shortly before at a salon at the French Legation in Cairo. Hakim stated that he intended to use these photographs in the publications of the *Tourist Office*. Unfortunately, the photographs have not been preserved in the correspondence, but the captions of the photographs Hakim listed indicate that they showed the Egyptian irrigation canals, characterising Egypt as an innovative country with a modern, flourishing agricultural production: *On the dyke* ("Sur la digue"), *Indentured labour at the waterways* ("La Corvée d'eau"), *Assuan*, and *High tide* ("La crûe"). Thus, Hakim and his collaborators at the Egyptian *Tourist Office* apparently replaced familiar photographic views of Oriental or Pharaonic Egypt by showcasing images of a productive, industrial country.[109]

Despite these activities, the *Tourist Office* lost its autonomy in designing and editing publications and advertising campaigns shortly after. The reasons behind the policy change are not documented, but in 1936, the Egyptian Council of Ministers granted a credit of 50,000 £.E. to *Thomas Cook & Son* to launch a large advertising campaign in the United States and Great Britain. The campaign comprised press campaigns, posters and publications demonstrating the beauty of Egypt's natural and historical sites. Whether British influence had led to this shifting of responsibility cannot be proven; a newspaper article suggested that the Council of Ministers attributed greater professionalism to the British company than to the *Tourist Office*. Its author stated that the collaboration would persist at least until the Egyptian *Tourist Office* would be able to meet the numerous demands concerning tourism advertisement in foreign countries.[110] If this reasoning indeed motivated the decision of the council, we may conclude that the experiences and long-established networks of the European companies, both in Europe and in Egypt,[111] provided these foreign actors with an advantage over the domestic, recently formed associations.

The 1930s saw a growing number of local Egyptian actors voicing interest in the tourism business, but they faced difficulties in circumventing the international corporations who maintained close relations with high-ranking Egyptian officials and offered their long-standing experience in the sector. The networks and experiences they built on rendered it difficult for newly established Egyptian competitors or even state-run agencies to join the field. Even though the claims and comments of Egyptian deputies demonstrated their awareness of the political relevance of the phenomenon, in competition with a long-established European business and ongoing privileges the established structures persisted.

Despite the protests in 1952 and the nationalisation of important economic sectors under Nasser, the implications of the coup for the tourism sector remained moderate. Matthew Gray showed that while the Egyptian Government claimed ownership of land and infrastructure of the grand hotels, the management was leased to international chains. Similarly, sea and air traffic to Egypt continued relatively undisturbed.[112] In the mid-1970s, investment in the Egyptian tourism sector increased. The Egyptian state was now involved in tourism development strategies to a much greater degree and the state-owned company *Misr Travel* dominated the market. In addition, from the 1970s onwards partnerships between foreign capital and state-owned associations as well as merely private investments contributed to the expansion of the tourism sector, targeting European, American and Arab tourists.[113] For these foreign tourists, the cultural sites and the nightlife of Egypt were the main attractions until the 1980s, when Fu'ad Sultan, Minister of Tourism under Mubarak, aimed to diversify the tourism sector and promoted beach resorts as an alternative branch which attracted growing numbers of leisure tourists.[114]

Recreating Egyptian tourist spaces

Despite the competition that Egyptian proponents of tourism development faced, it remained on the political agenda of Egyptian politicians, who combined their demands for tourism development with the ambition to shape a fully sovereign Egyptian nation-state. This ambition was communicated to two audiences, one domestic and one international. On an international level, tourism served as a vehicle for claims for independence and admission among the 'advanced' nations, symbolised by membership of the League of Nations. The Egyptian actors and associations often argued from a defensive position, fiercely denying tourists' assumptions about an uncivilised, primitive and backward Egypt. On a domestic level, tourism development was related to imposing structures of governance. Measures introduced in the context of tourism regulated domestic space, defined access and borders, and incorporated different parts of the nation into a unified, coherent national body. Both agendas – to take control of the image as well as the space of the nation – redefined the spatial structure of Egypt as a national tourist space. As the borders of the nation were less controversial than in neighbouring *Bilad al-Sham*, the Egyptian *efendiyya* intended to use tourism from the outset as a resource to actively foster a national integration.

Four places on the Egyptian tourist map exemplify these dynamics. Both Cairo and Luxor were tourist destinations already defined and shaped by international tourism in the nineteenth century. Egyptian attempts at regaining authority over the interpretation of these spaces thus resulted from an interplay between European attributions and Egyptian visions. In the case of Cairo, I show that for actors on both sides, ascribing temporality was a major method of defining the Egyptian tourist space. Luxor and the Valley of the Kings, in contrast, stand for the competing claims of international archaeologists and local experts of having a share in 'universal heritage'. At the same time, the international competition required a national rallying behind the exhibition of ancient heritage, which promoted Luxor in a domestic context too. The case of southern Egypt demonstrates that the dynamics between national and international ambitions united Egyptians by excluding Others. The exclusion of parts of the population from claims of 'civilisation' and 'modernity' not only encouraged the integration of the Arab Egyptian population at the expense of identified internal Others; it also seemed to permit the nation to climb up the scales of civilisation. Finally, debates about Egyptian beach resorts demonstrated that seaside tourism was fostered mainly as a domestic tourist practice. As such, it aimed either at strengthening the national economy or at integrating the 'masses' into the national project, thereby linking Egyptian tourism development to global debates of the interwar period.

Cairo: Spatialising past and present

One day in 1925, George H. Williams and his father got up early and made their way to the Qasr al-Nil bridge in Cairo, where they waited patiently for the perfect shot. The bridge was a widespread motif among tourists, not because of its monumental appearance – statues of lions decorated the pillars on both sides of the wide bridge – but because it allowed them to take photographs of Egyptians transporting agricultural produce from the countryside to Cairo.[115] In the photographs in Williams' album, donkeys tow carriages across the bridge, sometimes heavily loaded with groups of women and men in long *galabiyyas*. The photographs seemed to provide a glimpse into the daily life of ordinary Cairenes. As the use of animals for transportation stood for an archaic lifestyle close to nature, the photographs suggested that the photographer had witnessed the allegedly authentic and unspoilt daily life of pre-industrial Egypt.

Williams' friends and relatives contemplating the photographs would not see that the scenes in his album were far from representative of Cairene street scenes. In order to capture the motifs, Williams had carefully planned his day and waited patiently for the right moment. After all, the motif had become rare by the 1920s and Williams had to get up early in the morning in order to catch sight of the carriages. In his diary, he revealed that, in fact, another means of transport had come to dominate the streets of Cairo: the bicycle.

> One sees a great many bicycles about in the streets. The long dress that is the usual outer garment here does not do well for this exercise. The great attraction seems to be the bicycle bell. They usually carry two, one on each handle bar [sic] (I saw one man with 4). They ring them on the slightest provocation or apparently to pass away the time.[116]

Despite Williams' apparent fascination with the scenario, none of the photographs in his album represented the cyclists of Cairo. The gap between Williams' description and what he considered worth capturing suggests that the photographs showed a carefully produced 'authenticity',[117] while skipping a number of other possible views of daily life in Cairo.

Whereas, as his diary shows, Williams was unable to ignore the transformations of Egyptian society, photographic acknowledgement of these changes was not an option. Signifiers of a universal modernity[118] – signposted across the globe by similarly shaped grand hotels, railway stations and even urban architecture – were a problem for the amateur photographer, as they had not existed when nineteenth-century photographs visually defined what 'authentic' Egypt looked like. The search for the 'unmodern', generally equated with the 'authentic', was often described as the paradox at the heart of tourism: processes of 'modernisation' enhanced the accessibility of a country, but caused problems for tourist imaginaries, since the global modernity of the early twentieth century annihilated the difference that tourists were looking for.[119] The active search for clichés explains why George H. Williams captured the wooden carriages pulled by animals, while the memorable presence of bicycles on the streets of Cairo remained unacknowledged in his album. The photographer George H. Williams – unlike Williams as a tourist – was able to select the elements that stood for an Oriental Egypt as it had been shaped in narratives and photographs of the nineteenth century from an Egyptian reality.[120] Therefore, he waited for the occasion to capture scenes showing donkeys and carriages, unspoilt by the presence of 'modern' means of transport.

This carefully framed authenticity was one of three major strategies by which European tourists reconciled their expectations of local backwardness and the visible presence of 'modern civilisation'. While Williams created an 'authentic Egypt' by selecting adequate scenes, other visitors were grateful for prefabricated impressions. The bazaars of Cairo were among these sights. The products for sale and the atmosphere fascinated most tourists.[121] Even if visitors perfectly understood that the bazaar was "made up for British and American tourists", as the Scottish tourist Mary Steele-Maitland noted, it was a must-see for her, and she spent an enjoyable time exploring the stalls and buying souvenirs.[122] Hence, even if the staged character of the bazaar was obvious to the travellers, they enjoyed the experience of difference, confirming Urry's observation that in tourism difference mattered rather than the assumption of authenticity.[123]

As a second strategy, tourists mentally reconciled observations of change with prefabricated images of a timeless Orient by means of ascribing agency. Tourists, or mediators such as guidebook authors, interpreted the identified modernity as a result of European intervention, thereby attributing the agency in transformation processes to European actors. This strategy was built on the idea of the colonial enclave.[124] According to such reasoning, the presence of Europeans guaranteed the maintenance of certain living conditions in foreign countries. The existence of European quarters permitted travellers to access familiar (and allegedly trustworthy) people and comfortable, well-ordered places.[125]

By attributing processes of change in the host countries to European agents, tourists and guidebook authors stabilised the presumed cultural difference between Europeans and local inhabitants. Roy Elston, the author of *Cook's* tourist guidebook, attributed the transformations the tourist witnessed to the presence of European actors, claiming that the 'essential spirit' of the country and its inhabitants had not changed:

> The process of Westernisation in the Cairo bazaars has been comparatively slight [...]. One is justified in assuming that, except for hygienic improvements, the great bazaar is in structure very much the same as it has always been. Certainly its merchants are the same; for if the rest of the world change and a thousand catastrophies [sic] leave their mark on the characters of nations, still will the merchants of the Cairene bazaars sit cross-legged and impervious, finding their spiritual sustenance in contemplation of the prophet's words, and their physical sustenance in the prodigal naïveté of tourists.[126]

Elston argued in this passage that the Egyptian merchants had shifted their clients over time, now depending on the purchases of tourists; yet that they had maintained their 'traditional' practices of selling products. From the perspective of tourists, the scenes were still sufficiently exotic to justify a visit, and the temporal framing of Elston's description relegated the merchants to a less advanced stage of civilisation than their European visitors. Change was supposedly imported by Western visitors and administrators, and only passively experienced by the local population.

If agency could not be denied to Egyptians, a third strategy of tourists and their European mediators ridiculed and denigrated the results as failed attempts at copying the European model. This strategy was adopted, for example, by guidebook authors describing the large French-style quarters of Ismailiyya; in the comments of Williams on cyclists in *galabiyyas*; or in the contemptuous remark of the British traveller Edwin W. Smith on the mixing of clothing styles among Egyptian women: "Yesterday I saw a woman veiled to the eyes & wearing a big pair of coloured goggles with black reins[?]! Old & new together, if you like!"[127] The author mocked the perceived hubris of local actors. The imitation of European habits or fashions was interpreted as poor attempts, essentially incompatible with local customs and producing hopeless and ridiculous results – the veiled woman with additional sunglasses, the cyclists stumbling over their *galabiyyas*.[128] From his point of view, backwardness could not be overcome by willpower and the adoption of modern ways of life was doomed to fail if attempted by non-European protagonists. The irony discernible in such comments shows that Smith and other observers did not take local actors seriously as participants in what they perceived to be the universal, modern way of life.

Yet the Egyptians promoting and fostering the modernisation of the country rejected such interpretations and communicated their active participation in the transformations occurring in Egypt. Not only did the changing habits among the Cairene population complicate tourist aspirations of photographing the much-anticipated donkeys and camels; partly at least, the challenge also derived from new regulations established by Egyptian legislators. In 1927 the Governor of Cairo, Mahmud Sidqi, decreed that during the daytime, no carts, whether drawn by men or animals, were to be permitted on the major bridges, the Qasr al-Nil Bridge and the Pont des Anglais. Moreover, laden animals were also prohibited from passing these bridges between 8 a.m. and 8 p.m.[129] Tourists had to get up very early indeed if they intended to document the presence of animals on the streets of Cairo!

Although animals gradually disappeared from daily life in the Egyptian capital – increasingly replaced by cars, buses and bicycles on the Qasr al-Nil bridge and elsewhere (fig. 7)[130] – they remained important attractions for tourists. Instead of pulling carriages, donkeys and camels carried tourists, yet continuing to symbolise an imagined 'authentic Egypt' and illustrate the allegedly picturesque Oriental life. While municipalities across Egypt banned the carriages from ordinary street traffic, they assigned official 'parking spaces' to the animals generally situated next to the major hotels and rail stations.[131] From these stations, the owners would pick up tourists and offer rides to the major attractions. Regulations of urban space thus considered the expectations of tourists while simultaneously affecting their photographic practices. The Governorate of Cairo allocated times and spaces to different participants in street traffic. Such regulations of movement appear as one element within larger projects of regulating urban space launched by Egyptian actors after the conditional independence granted by Great Britain in 1922.[132] Egyptian municipalities did not ban the animals but confined

Figure 7: Street traffic in Cairo as seen by the *Tourist Development Association of Egypt*, 1939 (The Tourist Development Association of Egypt, Egypt: A Travel Quarterly 2/3 (1939), jd private collection)

them to delimited spaces of tourism. Over time, the symbolic essence of donkeys and camels would shift – from depicting an archaic, backward present to representing the tourism sector and its nostalgic visions of a bygone past.

Like European observers, Egyptian actors identified signs of different temporalities – past (i.e., tradition, authenticity) and present (i.e., modernity, progress) – in contemporary culture. Rather than attributing these elements to European impact (present) and local features (past) respectively, Egyptian authors attributed both temporal categories to Egyptian agency. Being aware of the tourist attraction of a pre-industrial Egypt, local authors of guidebooks and brochures drafted and spread visions of an idealised Egyptian past, framed as the authentic origins of 'modern Egypt'.[133] Yet, in contrast to European guidebook narratives, these local brochures always added proof of Egyptian modernity and progress, underlining that pre-industrial society was but one facet to local society. The authors of the 1930s guidebook *Egypt*, published by the *Egyptian State Tourist Administration* and the *Tourist Development Association of Egypt*, for example, directed the awareness of potential visitors to the coexistence of the picturesque ancient and modern Egypt:

> Which Egypt is it that you wish to visit? [...] Is it the land of sunshine, the Egypt of golden days and starry nights? [...] Is it the Egypt of ancient lore, the oldest country in the world?

> Nowhere else is there so much to delight the lover of the past. [...] Is it perhaps the magic of the East which attracts you? [...] the city of Cairo, treasure-house of Arab art and architecture, where a thousand minarets stand out in silhouette against the clear Eastern sky? Or, it may be, you delight in contrast and wish to see, in its age-old setting, the new-born Egypt of today. Here, for you to behold, is a nation advancing by leaps and bounds in the path of modern progress. Here are cities with wide modern streets, tall modern buildings, electric trams, broadcasting stations and all the adjuncts of twentieth-century civilisation.[134]

The authors thus identified wide streets, multistorey buildings, electricity cables, public transport and new forms of translocal communication as signifiers of modernity. These elements were distinguished from symbols of a romantic East, such as mosques and Arab architecture, and from Egypt's ancient history. Rather than establishing a cultural hierarchy between the three kinds of monuments, the authors attributed the elements to different times, connected by an assumed evolutionary process. The notion of linear progress was highlighted by the structure of the passage, which introduced the reader first to Egypt's eternal, unchanging natural conditions, before tracing different stages of cultural development and culminating in "twentieth-century civilisation", thus emphasising the modernity of contemporary life in Egypt.

Egyptian tour guides and guidebooks offered not only textual but also visual alternatives to nineteenth-century Orientalist photographs.[135] Such a shift in perspectives is exemplified, for example, by the photograph of A. Kalfayan reprinted in the guidebook *Cairo – How to see it*. On the text pages, the first photograph showed a view of the Pyramids. Whereas this choice is hardly surprising, the framing of the illustration was certainly unusual (fig. 8). In the background of the black-and-white photograph rose silhouettes of two pyramids, positioned at roughly half of the height of the picture. In the foreground on the left, the attentive observer perceived a water canal, while to the right, a line of young trees separated an asphalted road from two lines of rails. The rails at the centre of the photograph directed the spectator's gaze to the Pyramids in the background.[136]

This representation of the Pyramids contrasted with most photographs in albums and postcard collections of tourists, which generally positioned the Pyramids at the centre of the photographs, as large and rather isolated monoliths surrounded by desert sands.[137] In contrast to these views, Kalfayan and the guidebook author Alexander Khoori, a tour guide and former employee of the Egyptian Department of Antiquities, presented Egypt as an inhabited and cultivated, well organised space – despite the absence of persons on the photograph: the trees were planted in neat rows, the rails ran in straight lines. In case the observer overlooked the recently reorganised landscape, Khoori added a telling sub-title

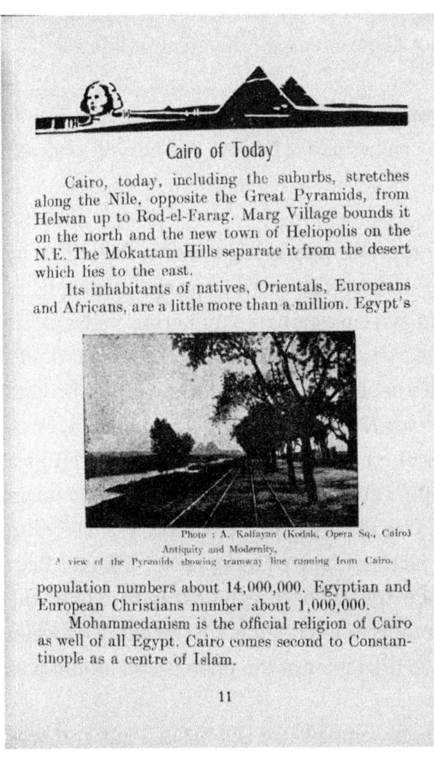

Cairo of Today

Cairo, today, including the suburbs, stretches along the Nile, opposite the Great Pyramids, from Helwan up to Rod-el-Farag. Marg Village bounds it on the north and the new town of Heliopolis on the N.E. The Mokattam Hills separate it from the desert which lies to the east.

Its inhabitants of natives, Orientals, Europeans and Africans, are a little more than a million. Egypt's

Photo : A. Kalfayan (Kodak, Opera Sq., Cairo)
Antiquity and Modernity.
A view of the Pyramids showing tramway line running from Cairo.

population numbers about 14,000,000. Egyptian and European Christians number about 1,000,000.

Mohammedanism is the official religion of Cairo as well of all Egypt. Cairo comes second to Constantinople as a centre of Islam.

11

Figure 8: Guidebook page showing the Pyramids, from Khoori's *Cairo – How to see it* (Photographer: A. Kalfayan, Cairo. Reproduced by kind permission of the British Library)

to the photograph: *Antiquity and Modernity: A view of the Pyramids showing tramway line running from Cairo*. In this way, Khoori modified the general implications of the antithetical antiquity/modernity. In Kalfayan's photograph, the features did not indicate two segregated spaces (for example the ancient monuments and the archaic bazaars on the one hand and the neat, clean and modern spaces of hotels and European quarters on the other); rather, they were attributed to the *same* space. Simultaneously, Khoori's choice revealed that the infrastructural modernity of the tramway was the condition for the tourist's access to the ancient heritage. In doing so, Khoori and Kalfayan denied assumptions of Oriental backwardness. Like the authors of the aforementioned guidebook, they demonstrated the co-spatiality of past and present, claiming a connection between the Egyptian past and contemporary Egyptian society.

While European observers of transformation processes in Egypt attributed them to the effects of imperial connections and rejected local processes of change as inauthentic, ridiculous attempts at appropriation, Egyptian politicians, intellectuals and publishers contested these interpretations of Cairo's urban space. Although actors in tourism took the expectations of travellers as their starting point, providing them with views ranging from allegedly exotic animals to the monumental Pyramids, these visions were self-confidently embedded in narratives of modernity. This modernity was claimed to be grounded in local tradition and development, rather than based on the appropriation of a foreign culture. The coexistence of 'ancient' and 'modern' in the same space, among members of the same society, was unthinkable for the essentialising views of European spectators, yet self-confidently claimed by their Egyptian interlocutors.

Luxor and the Valley of the Kings: National grandeur by universal standards

> We walked through the town: as in all eastern places the shops are one large open door no
> windows. Camels – donkeys – carriages & motors pass at intervals and Luxor is evidently
> a large place.[138]

The hustle and bustle of a flourishing tourist town at high season described by
Mary Steele-Maitland in 1932 stood in curious contrast to most photographs of
Luxor. The photograph album of Major Lawrence gathered numerous views of the
town and its temples, yet none of these views illustrated Mary Steele-Maitland's
comment. His photographs showed the neat and straight path leading through
a lush garden to the entrance of the *Luxor Hotel*, the ruins in their natural environment – palm trees, hills bordering the horizon, and the Nile – as well as a
white mosque emerging from amidst the ruins. The last photograph of Luxor in
the album portrayed a man standing next to his camel in the shadow of a tree, with
fields in the background. Although the pictures showed more facets of Luxor than
the common views of ancient monuments, none of the photographs evoked the
idea of a "large place". Apart from some tourists and their guides strolling between
the temples and the man with his camel, the Luxor of the photograph album was
an empty place.[139]

Yet, by the interwar period, the famous temples had established Luxor firmly
on the tourist map. While in the 1870s Luxor had indeed been a small village knitted
in and around the ruins of the still largely buried temples, the settlement grew rapidly after the opening of the first hotel in 1877, which allowed *Cook's* Nile tourists
to disembark and spend some days in Luxor before proceeding further up the Nile.
A mere ten years after, Luxor had turned into a village of about 3,600 inhabitants
with access to public services such as a post office "with a postmaster who speaks
English fluently", shops and other amenities.[140] By the mid-1920s, the authorities of
Luxor registered almost 19,000 inhabitants, who received 30,000 to 40,000 visitors
per year.[141] The hotels offered the same luxurious services and entertainment as the
prestigious grand hotels in Cairo, and, as in the capital, the tourism sector remained
in the hands of European entrepreneurs.[142] *Cook's* agents welcomed the disembark-
ing tourists, and the largest and most famous hotels, the *Winter Palace* and the
Luxor Hotel, were managed by A. R. Badrutt, a Swiss manager from St. Moritz.[143]

Major Lawrence and his wife thus enjoyed the amenities of a vivid tourist
town, modern hotels and the sociability of an international community; however,
his photographs showed Luxor as a village frozen in time. Narratives about a
backward Egypt and Egyptian claims of agency clashed even more fiercely in the
Valley of the Kings than in the case of Cairo. In the age of empires, calls of archaeologists and historians, collectors and tourists for access to the 'universal' heritage

of Pharaonic Egypt were easily turned into justifications for foreign intervention and control.

The idea of the civilising mission was another rhetorical pattern justifying imperial interventions. The following paragraph on the development of Luxor from *Cook's Handbook for Travellers* of 1929 shows how the British tour operator adopted the narrative in order to highlight the benefits of tourism for the inhabitants.

> Not more than thirty-five years ago Luxor was a cluster of poorly built mud houses of a kind with which the traveller becomes familiar in his journey up the Nile. They stood close to the edge of the river bank and even sprawled among the courts and on the roof of the temple of Luxor. The village was ill-kept and ill-scavanged[sic], its roads were alleys of insufferable dirt, and its natives were unprosperous. The advent of the Nile steamer in 1886, with its cargoes of tourists eager to observe the wonders of Thebes, was a godsend to this community, which gradually developed in numbers and prosperity. At the same time, Mr John Cook, whose enterprise had brought about a new era in the history of Nile travel, undertook the transformation of the village into something less offensive to the European taste. He caused steps to be built up the bank, improved the river front, and induced the local authorities to clean the streets and alleys. He rebuilt the old Luxor Hotel and gave impetus to the creation of a modern and progressive resort, earning the gratitude of the natives, not only because of the prosperity which had followed in his wake, but also by the founding of a hospital which proved to be of incalculable benefit.[144]

The passage above described a civilisational gap between the inhabitants of Luxor and the European visitors. The building techniques of their houses and the alleged insalubrious living conditions identified the local inhabitants as primitive, and the fact that they had constructed their homes amidst the ruins of the temples served as proof of their ignorance. Whereas visitors from all over the world admired the awe-inspiring grandeur of Thebes, the population seemed to overlook or even destroy the treasures surrounding them.[145] Given their purportedly primitive living conditions, the author Roy Elston equated the "advent" of tourists to a "godsend" revelation for the local population. Even though the modernisation taking place first and foremost targeted European tourists, it supposedly benefited the local population as a side effect. In this narrative, John Cook appeared both as the prophet of modernity and as a good governor: it was Cook who exhorted the allegedly phlegmatic local authorities to work for the communal good, and he provided the population with prosperity and sanity, the hospital symbolising the beneficial aspect of progress. *Cook's* economic activity, imperial rule and tourism were understood in this passage as a confluence of events mutually benefitting each other and the population. Tourism and prosperity went hand in hand, according to Roy Elston, and his description likened the tourist to the missionary, as both allegedly contributed to the enlightenment of the world's remote places. The negative effects of the "advent" of archaeology and tourism for the inhabitants of Luxor, notably forced labour and the expulsion of the inhabitants living around the ruins, passed without mention.[146]

Not only modern standards of living and hygiene, but also scientific exploration and expertise remained arguments in favour of the European presence in the imagination of travellers. The presence of European archaeologists – whether in person or in the heroic narratives of discovery – at the point of encounter with pharaonic antiquities enhanced the relevance attributed to the tourist's visit. While Mary Steele-Maitland had severe doubts about the competence of her guide in Luxor, Georges Michail (not even the fact that he was officially recommended by the *Baedeker* could ease her doubts),[147] she felt in safe hands when she was guided around the Egyptian Museum by Reginald Engelbach, the Chief Inspector of Antiquities in Upper Egypt.[148] George H. Williams, who had heard a lecture by Howard Carter in London, visited the National Museum in January 1925 in order to see "the new things recently found by Lord Carnavaron & Mr. H. Carter in Tut-ankh-Amen's tomb".[149] While visitors like him did appreciate the refined monuments and pieces of art from ancient Egypt, scientific exploration, explanation and expertise allegedly distinguished European from Egyptian approaches.

Egyptians did, however, express their dissent about the European presence, in addition to European claims of superiority, seeking ways to contest the European dominance in archaeology, tourism and the narratives circulated in this context. The discovery of the tomb of Tutankhamun in 1922 gave an additional impetus to these aims. Donald Reid argued that the discovery aroused interest in archaeology among Egyptian upper and middle classes and contributed to a renewed consciousness about the pharaonic past. Having occurred in a time of heated protests against the British presence in the country, the pharaonic past was incorporated as an essential element into the emerging Egyptian national identity.[150] The ancient past and the internationally uncontested grandeur of the historical and archaeological remains provided Egyptians with a seemingly objective proof of the cultural and civilisational potential of the country, systematically questioned by British domination. While the archaeological findings served as arguments, tourism provided actors with a possibility of actively reaching out to Europeans and Americans in order to extinguish European claims of civilisational superiority.

Among the members of the Egyptian *efendiyya* who promoted the exploration and the exhibition of the pharaonic heritage, two main interests were at stake. The first concern was to profit from the pharaonic heritage in order to position Egypt among the civilised nations, thus addressing an international audience. Second, local politicians hoped to profit from this atmosphere in order to position their constituencies at a domestic, Egyptian level. In the context of the early

1920s, the first aim was barely contested among the deputies. For the Egyptian *efendiyya*, pride in the pharaonic heritage served as proof of Egyptian grandeur and civilisation, thereby justifying demands for full independence. Hence, deputies tended to approve of expenditures regarding exhibitions of Egyptian heritage. The discovery of the tomb of Tutankhamun, for example, required expansion of the Egyptian Museum, but none of the deputies contested the expenses. The deputies agreed that these findings had to be displayed, expecting the exhibition not only to attract international tourists, but also to establish Egypt as a centre in the "world of scientists".[151]

Tourism mattered because it created a space for communication and thus the condition for a shift in narratives. To international visitors, actors in tourism offered new narratives aimed at rewriting, or at least modifying, Western narratives of discovery. In Alexander Khoori's guidebook, *Luxor – How to see it*, the second bilingual edition of which was published in 1925 (in English and French), the author provided a remarkable account of the discovery of Tutankhamun's tomb. He not only questioned European claims to superiority, but also the cherished notion of the solitary explorer by pointing out local collaborators, staff and informants of the European archaeologists. While common knowledge among tourists held two British gentlemen, Lord Carnavaron and Howard Carter, responsible for the discovery of the tomb, Khoori introduced another protagonist: Husayn Ahmad al-Sayyid, overseer of the workers. According to Khoori, Lord Carnavaron was in England and "his English agent" (i.e. Carter) was absent from the site when al-Sayyid advised the workers to continue digging in another direction, without prior permission of his employers. There, the tomb was found.[152] Thus, Khoori awarded the credit for the seminal discovery to the Egyptian overseer, to whom he dedicated a portrait in his guidebook. The caption of the photograph, in both English and French, introduced al-Sayyid as a scion of a family of archaeologists that had been involved in the important archaeological discoveries at Deir al-Bahari. While this information could not be verified, the implication of this genealogical information is obvious.[153] By characterising al-Sayyid as a member of a family of archaeologists, Khoori attributed professionalism to the Egyptian explorer, rejecting suppositions that the discovery might have resulted from pure chance.[154]

Indeed, the Tutankhamun discovery had stirred trouble between the Egyptian Government and the British explorers Lord Carnavaron and Carter over rights in the excavation and the exhibition of the treasure. The major newspapers in Egypt had reported on the case as, particularly for anti-British observers, Egyptian independence was at stake. Influenced by the Revolution of 1919, they denounced foreign dominance in excavations of the Egyptian 'national heritage'. Journalists not only disputed the transfer of antiquities to European museums; they also criticised the attitude of Carter and his team, who were accused of lacking respect for the Egyptian authorities and monopolising access to the tomb for private interests.[155]

The conflict was described by an enraged Alexander Khoori, who dedicated several pages of his guidebook to an extensive summary of the events. He harshly attacked Carter's conduct and criticised the exclusive coverage of the expedition granted to *The Times* of London, thereby withholding access and information from Egyptian scientists, the Egyptian public and the international community.[156] He did not leave room for doubt that the tomb was part of the national Egyptian and universal cultural heritage, which had been usurped by two British would-be explorers. The Egyptian authorities, in contrast, were presented as the rightful administrators of the heritage, fulfilling their duties before the scientific and interested lay community (represented by tourists).[157] Addressing European tourists, the guidebook denied British claims of superiority and aimed at the acknowledgement of the Egyptian share in their heritage.

In addition, the heated British-Egyptian conflict and the nationalist atmosphere allowed local proponents of tourism to promote the interests of their constituencies on a domestic level. The representatives of touristic constituencies in particular argued that the development of these sites was of national and anti-imperial interest. The international prominence of towns such as Luxor thereby reshaped the map of Egypt. They attracted international visitors, tourists, and groups of Egyptian pupils[158] – as well as the attention and funds of the Egyptian Government. Demands for the embellishment of Luxor could be framed as nationalist action and justified the financial support of the state.[159] In the case of Luxor, the deputy of the constituency, Tawfiq Andraous Bishara, did not fail to mention the significance of tourism, which had turned the tiny village into a town of international interest, beneficial to the reputation of Egypt.[160]

In August 1926, Bishara enthusiastically fought for a credit of 20,000 £.E. intended for measures aimed at the embellishment of Luxor. When doubts were voiced about the necessity of the expenditure, Bishara defended the proposal, arguing that Luxor brought both material and political benefits for Egypt. Economically, the investment would benefit the railway company, hotels and other related branches all over Egypt. In particular, however, he considered that lighting in public places, a water system and broad boulevards for the town were of absolute necessity in order to maintain a good reputation among visitors from "the entire universe".[161]

Hence, Bishara argued that beyond monetary benefits, Luxor was of invaluable use for the prestige of the nation. Tourists would perceive the town as a symbol for the intellectual and moral state of the country. Dr Naguib Iskandar, another deputy, agreed, reminding the Chamber that the famous monuments of Luxor were a source of glory and pride for all Egyptians, and therefore had to be presented adequately. After all, most tourists *"de marque"* from Europe and America wintering in Egypt headed to Luxor before even stopping in Alexandria or in Cairo. Iskandar therefore advocated measures of urban renewal for Luxor, which would come to represent the state of civilisation in Egypt:

> The tourist leaving the station and seeing all these miserable dwellings and the curvy alleys had to feel himself in the unexplored parts of Africa, whereas he believed to visit the symbol of glory and pride of Egypt. As soon as he arrived at the banks of the Nile, seeing the hotels in European style, he probably told himself: "This is the quarter of the Europeans, it shows their civilisation and progress. On the other shore is the quarter of the natives, which corresponds to their backward state compared to other peoples." This is a shame, isn't it?[162]

Iskandar seized on rhetorical patterns attributing elements of modernity and progress to European actors, meanwhile associating local inhabitants with Orientalist residues. The remedy was obvious to him: the Egyptian Government had to falsify assumptions about the unsanitary and backward living conditions in Egypt by pursuing the modernisation of the country, beginning with the places most exposed to the eyes of tourists.[163] Indeed, by summer, works on lighting and water systems in Luxor were being executed.[164] The reasoning of Bishara and Iskandar demonstrated that, even in a domestic context, the major argument for tourism development was not economic, but political. In the context of restricted sovereignty, considerations regarding the prestige of Egypt among foreign countries mattered to a large extent to the Egyptian modernist *efendiyya*. Only a few years after the anti-British Revolution of 1919, references to the national heritage testified to the grandeur of the Egyptian civilisation. From another angle, ambitions to gain full independence justified a programme of urban modernisation in the tourist centres of Egypt. The international attention turned them – mentally and physically – into centres of the modern nation.

Redirecting the gaze: Aswan and beyond

Members of the Egyptian *efendiyya* conceived of themselves as modern subjects and refused to accept assumptions about Egyptian inferiority. Yet the interventions of Khoori, Iskandar and others did not question the validity of the basic assumptions underpinning such narratives. The Egyptian *efendiyya* shared the notion that nations underwent an evolutionary process from a 'primitive' state toward 'civilisation'.[165] The inferior Others they identified were notably lower classes, as well as dark-skinned Egyptians and Sudanese beyond the borders of Egypt.[166] In

A FISHERMAN

UN PECHEUR

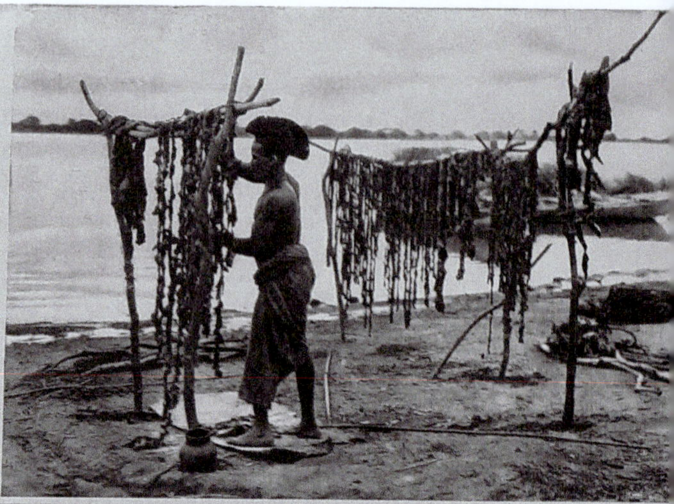

A VILLAGE GIRL

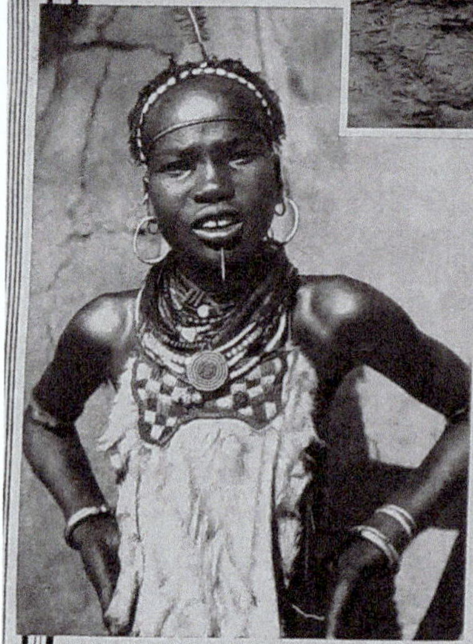

SUDAN

Beyond Aswan lies the Sudan, gateway to Central Africa, land of strange and interesting tribes, of thatched huts and big game. And yet at Khartoum, the capital, where the Blue Nile and the White Nile join, the visitor can make his headquarters at one of the hotels with the same comfort and assurance as at home. There he may stay and enjoy the amenities of social life in a warm winter climate, or venture farther afield with tent and gun into the mountains and marshes of a vast land.

A SHOP UNE BOUTIQUE

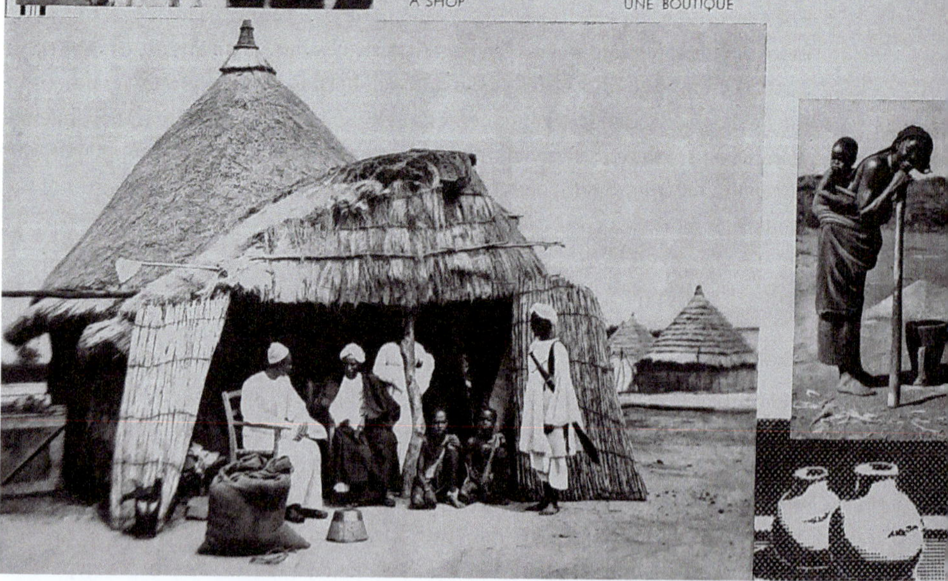

Figure 9: The Sudan represented in the magazine of the *Tourist Development Association of Egypt* (The Tourist Development Association of Egypt, Egypt: A Travel Quarterly 1/1 (1937), jd private collection)

the context of tourism, the *efendiyya* presented the dark-skinned population in particular as Others, thereby satisfying the exotic desires of visitors and deterring their gaze from Arab Egyptian inhabitants. The establishment of a civilisational hierarchy visualised in tourist brochures allowed its creators to position themselves among the allegedly advanced nations.

The illustrated magazine *Egypt: A Travel Quarterly* published by the Egyptian State Tourist Department in collaboration with the *Tourist Development Association of Egypt* exemplifies this differentiation. The issue of autumn 1937 contained a large black-and-white photograph showing a man at work at the riverside (fig. 9). He was shown in profile from the left, so his black hair in 'Mohican' hairstyle was clearly visible. Bare-chested and bare-footed, the man was putting ropes on a simple construction of wooden poles. The caption contextualising the photograph, "A fisherman/Un pêcheur", allowed the spectator to identify the ropes as fishing nets the man was hanging on a rack to dry. From the perspective of a European spectator familiar with colonial photographs, such simple tools and the barely covered body (with only a loincloth) identified the person as a representative of a 'primitive' culture. The double page on which the Egyptian State Department published the photograph was dedicated to the Sudan, formally a British-Egyptian condominium yet largely controlled by Great Britain. The Egyptian Government, which laid claim to the territory, underpinned its colonial ambitions through the illustrations on this magazine page.

The photograph stood in marked contrast to the rest of the magazine. 10 of the 14 double-pages of the magazine gathered photographs of 'modern life' in Egypt, such as gardens, apartment buildings, hotels, the airport and sporting clubs. On the double-page about the Sudan, however, 12 photographs showed dark-skinned persons in simple clothing, with conspicuous piercings and elaborately styled hair. The photographers showed them at manual labour, or else added codified elements to the image such as straw huts or earthen pottery. The brief introductory text presented the Sudan as "the gateway to Central Africa", specifying that it was a "land of strange and interesting tribes, of thatched huts and big game" – repeating stereotypical European imperialist imaginations of Africa. In this way, the Sudan was clearly distinguished from the Egypt the reader had seen on the preceding pages. It was presented as a part of Africa, supposedly tribal and primitive.

A closer look at the portrait photographs in the magazine demonstrates the difference in the representation of the Egyptian *efendiyya* and the Sudanese population. On the first pages of the magazine, prior to the introduction, large, official portraits introduced potential tourists to major Egyptian dignitaries. King Faruq of Egypt, the President of the Council of Ministers, the Minister of Finances and the Minister of Commerce and Industry were portrayed in dignified poses, wearing black suits and tarbushes. On the pages illustrating leisure activities, the spectator encountered persons in European-style clothing at beaches and sports clubs, as well as tourist groups at ancient sites. The section "Our distinguished guests" presented a selection of members of the global upper classes, from famous musicians to politicians and noblemen. In contrast, the dark-skinned Sudanese or Nubians were portrayed in the rural contexts described above, despite the fact that a large number of Nubian Egyptians lived and worked in the expanding urban centres of Egypt at the time.[167] Like Williams' photographs, the editors intended to represent them in contexts considered authentic or traditional, turning them into curiosities, "strange and interesting" even in their ordinary daily life, because they supposedly represented a bygone past.[168]

In accordance with such visual stereotypes, the Alexandrian lawyer, journalist and intellectual Gaston Zananiri published an article on *Nubia* in which he described the curiosities and attractions of the region. In the 1937 issue of the magazine *Le Réveil de l'Egypte*, Zananiri characterised Nubia in several paragraphs. Next to crocodiles, the Nubian inhabitants were considered a major attraction due to the 'primitive' conditions of life, their special outer appearance (Zananiri mentioned the tattoos decorating the women's faces) and curious customs and beliefs. From the perspective of the educated city-dweller Zananiri, Nubian Egyptians constituted a curious and exotic attraction on Egyptian territory.[169] The colonial gaze of the "colonized colonizer" Egypt not only legitimated claims for Egyptian rule, but also brought the Egyptian 'civilisers' closer to the European great powers.[170]

This is at least what the introduction to the *Réveil de l'Egypte* suggested. Its author, Guido Bertero, proudly introduced Egypt as a new member of the League of Nations, having thereby joined the illustrious circle of the developed great powers.[171] Bertero's remark suggests that the political relevance of representations of the Nubian or Sudanese Other stemmed from the specific context of nation-state formation under international observation. Since the Egyptian *efendiyya* needed the support of the imperial powers in order to be admitted as a member state of the League of Nations, they had to pay special attention to outside perceptions of Egypt. Hence, the editors of the magazine drew the civilisational line of demarcation between Egypt and its former colony in the Sudan, defining Egypt as non-African and implicitly shifting the border of political and civilisational maturity "beyond Aswan".

The editors of the magazine differentiated between the ethnographic or exotic gaze that allowed the tourist (and the photographer) to distinguish himself from the subject of the photograph, and the gaze of sociability that invited the spectator to take part. The leisure society united tourists with the dignitaries patronising the magazine, but also with photographers, editors and authors associated with the *Tourist Development Association*. The *efendiyya* edited the magazine, fostered the infrastructural accessibility of Egypt, promoted sporting clubs, organised guided tours, issued multilingual guidebooks and brochures, had their tea on the terrace of *Shepheard's* or *Groppi's* and spent the weekends on the beach. Their publications focused on lifestyle and leisure practices familiar to the foreign visitors or presented the exotic Other. Put differently, the members of the various societies for tourism development, who carefully and consciously shaped the images communicated to visitors, did not consider themselves an attraction to be gazed at. They refused to present Egypt as a backward, uncivilised, archaic country.

The different visualisations of life in Egypt and life in the Sudan were an attempt of the new Egyptian middle classes to orchestrate a shift in the tourist gaze. I argued in the section on Williams' album that the Egyptian girl and other individuals he portrayed in the Egyptian countryside controlled their self-representation by adopting a dignified posture and returning the gaze of the tourist. Compared to these individual processes of interaction and negotiation, however, the Egyptian *efendiyya* in the *Tourist Development Association*, the Egyptian Chamber and other institutions formed a collective of considerable political and cultural influence. They had a more powerful means at their disposal to counter the Orientalist gaze of the tourist: they managed not only to return, but to redirect the gaze. The authors and editors denied tourists the gaze on their own social group, while simultaneously redirecting it towards marginalised 'internal Others'.

As a result of their efforts, images of the exotic Other in Egypt changed. The 'Oriental type' that European photographers had depicted in postcards and travel photographs prior to World War I no longer dominated the visualisation of the Other in Egyptian tourism. Such representations of women in erotic poses, or of men and children in small workshops, handling curious and 'primitive means', ceased to circulate by the mid-1920s.[172] Henceforth, the attributes of the exotic Other gazed at by European travellers shifted from 'Orientals' to black 'Africans'. While the Egyptian *efendiyya* attempted to divert European gazes from the Egyptian population, the same *efendiyya* did not mind the parallel construction of clichés if they were attributed to Egypt's southern periphery.

The gradual disappearance of Orientalist depictions of Egyptian society was a conscious political choice. Initiatives to ban 'type' photographs of orientalised Egyptians complemented the redirection of the tourist gaze towards the Nubian and the Sudanese Other. In 1927, the Egyptian deputy Ahmad al-Sawi harshly criticised the "obscene and immoral publications" circulating in Egypt. He feared the "infection" of the Egyptian people with immoral thoughts, but was equally afraid of a negative image of Egypt in other countries. Perceiving tourists as a major intermediary in transmitting images, al-Sawi attacked the photographs circulating on postcards:

> Among the numerous postcards widespread in this country, there are some that contradict the manners of the Orient; others show Egypt in a primitive and barbarian state. These

postcards, which are sold to tourists, provide them with a negative impression of Egypt, without which they would have properly known her.[173]

It seemed that he referred to *type* photographs, as well as erotic images of North African women. The contemptuous implications of such images, the colonising gaze of the – predominantly European – photographers and the construction of an inferior Other by such visualisations have been largely analysed in scholarship.[174] Al-Sawi's statement demonstrates that such a sentiment was in effect shared by contemporary observers. To remedy the problem, the deputy suggested harsh controls of the customs administrations. He demanded that the importation of similar postcards be monitored and their entry onto Egyptian lands prevented. Their destruction should be arranged "when they are obscene or constitute a bad propaganda for Egypt". The Minister of the Interior replied that the necessary measures had been taken already: the Customs and Postal Administrations were prohibiting the entry and shipping of obscene postcards. In the case that they arrived as contraband goods, the police would confiscate the cards and the postcard seller would be judged in court.[175] While al-Sawi's intervention thus shows that the bans on circulating erotic-type postcards were circumvented, the fact that Egyptian legislation prohibited the distribution of obscene erotic photographs of women indicates that deputies and legislators were aware of their implications. By defining the representation of Egyptians in visual media, they intended to shape the image of the country in a global context. Alternative exotic motifs redirected the tourist gaze, yet the modification of tourists' habits of perception was – early on – based on political intervention and regulation.

In the context of nation-state formation, the *efendiyya* redirected the European tourist gaze. While shared practices of leisure underscored the belonging of the Egyptian middle class to a Mediterranean modernity, the redirection of the tourist gaze towards an allegedly inferior Other emphasised Egyptian claims for civilisation and sovereignty. The tourist's curiosity for the exotic was deflected from the Oriental types to images of alternative 'primitive' populations, meanwhile rendering the recognition of a modern Egypt acceptable to the same eyes.

Egypt by the sea: Nationalising the Mediterranean beaches

Beyond the aim of shaping international public opinion, Egyptian tourists were identified as another target group by advocates of tourism development and led deputies to explore the potential of tourism for national integration. Egyptian deputies debated whether the government should provide incentives for Egyptians to spend their vacations in Egypt. It was Ismaʿil Sidqi, a politician from Alexandria and later prime minister, infamous for his authoritarian rule, who first triggered the debate in 1927.[176] Sidqi argued that state subsidies and other benefits should create incentives for making tourists stay inside the country, rather than travelling abroad. He assumed that visits to Egyptian summer resorts and historical sites presented opportunities for the population to get to know the country, and thereby contributed to developing a national consciousness.[177] The foundation of new associations to launch tourism development or advertisement campaigns, as in France, he concluded, might serve as inspiration for the Egyptian deputies.[178]

Among the places that would profit from such a strengthening of national resorts was Alexandria – the municipality in which Isma'il Sidqi had formerly held a seat. Similar to Tawfiq Bishara in Luxor, Sidqi apparently attempted to use the stage of the national parliament for the benefit of his constituency. Already as a member of the municipal council, Sidqi had played a major role in fostering the embellishment of Alexandria and in promoting the city as a summer resort. At the time, his projects – and notably the extraordinary cost of the newly constructed corniche – had raised suspicions about corruption.[179] As argued by the historian Robert Ilbert, the new national order helped Sidqi to pursue his political (and economic) ambitions on a new scale.[180] Sidqi continued to ask favours for his constituency, yet he also reshaped his arguments. Rather than putting forward the interest of Alexandria, he emphasised the national interest in strengthening seaside tourism to the Mediterranean coast.

Certainly, the Egyptian beach resort had British, French, Belgian and other predecessors. Yet, in an Egyptian context, the beach, in contrast to the antiquities, could be claimed as a national tourist space par excellence, as it had neither been claimed nor defined by European explorers, Orientalists, photographers, travellers and tourists. Egyptian actors, deputies and interest groups thus had the chance to shape the Egyptian beach, and they defined it as a space of modernity. The Egyptian beach would become a site for the leisure of the masses; a well-organised beach which welcomed women too, as well as a place associated with physical exercise and the exhibition of youthful bodies (fig. 10). Not only in Egypt, the beach was, as Hazbun put it, "a blank slate" at which new meanings could be projected.[181]

The promotion of the Egyptian beaches followed a different logic than the cultural tourism that had established Egypt's fame as a tourist destination. In cultural tourism, Egyptian tour guides, authors and associations mainly reacted to an established European tourism and attempted to appropriate, shift and modify the narratives projected on the country. The development of leisure spaces at beaches, by contrast, could be undertaken 'from scratch'. The absence of Orientalist expectations allowed its proponents to exhibit an Egyptian participation in trends of international modernity, without even referring to assumptions about Oriental backwardness. That the Egyptian *efendiyya* in parliaments, municipal administrations and media shaped these resorts along the lines of the European

Figure 10: On the beach (The Tourist Development Association of Egypt, Egypt: A Travel Quarterly 2/3 (1939), jd private collection)

bathing resorts expressed their conviction of participating in a universally shared modernity. Debates revolved around facilitating nationwide access and recreation for all social classes (including workers), while prescribing certain forms of conduct, clothing and hygiene standards.[182] Seen from such a perspective, the deputies actively shaped the tourist space as a laboratory for a future, modern Egypt.

The emergence of bathing in Egypt was reputedly related to the father of Egypt's modernity. At Muhammad ʿAli's palace at Ra's al-Tin, on the future corniche of Alexandria, the first bathing establishment had already been founded in 1821.[183] Around 1908, a more systematic development of Alexandria's beaches at Ramleh was launched. The seaside would be turned into a place of leisure with casinos and a seafront, as well as pleasure ports, sports facilities, and cabins at the bathing beaches.[184] By the time Mary Steele-Maitland visited Egypt in 1931, the beaches of Alexandria were reputed to be élite resorts, while Port Said emerged as a resort for summer guests of more modest means.[185]

The institutionalisation and the regulation of seaside tourism successively spread from the Mediterranean coast to the Red Sea. In Alexandria, the municipal commission publicly declared rules for the 12 beaches in autumn 1927. While initially the authorities modified the ordinances each time a new beach was established, they soon amended the ordinance, stating that it would be applicable to any newly opened beach in the district, which suggests an ongoing increase in the number of beaches.[186]

In spatial terms, the decrees and ordinances thus show how the leisure practices related to beaches spread along the coasts of Egypt. Given that the governors and municipalities established similar regulations throughout the country, yet at different moments, the decrees indicated how local developments urged district administrations to act. Moreover, the assignment of certain practices to specially defined zones on the beach suggested a high frequentation. This, in turn, indicated that the beach as a space of leisure and recreation was increasingly generalised among its visitors.

The public beach, as opposed to private clubs, emerged in the late 1920s. At this time, district administrations controlled the beaches and established regulations for the correct behaviour. The rules generally concerned three major fields: hygiene, security and morals, suggesting that they also had an educational impetus. The sanitary measures were in line with regulations from Ottoman times and often reacted to outbreaks of diseases.[187] The decrees stipulated particularly how to treat refuse, especially around the years 1930–1934, when actual outbreaks of typhus and other venereal diseases occurred in Egypt.[188]

Another set of rules concerned the security of the bathers. Professional fishermen were prohibited from entering the swimming sections of the beaches, and orders prescribed that rubbish, in particular bottles and broken glass, had to be placed in bins. In addition, the prohibition of bringing horses, donkeys and other large animals to the beach, of playing "football or any other violent game [...]", as well as that of driving cars in the walking area, intended to guarantee the safety of the visitors.[189]

The greatest concern of the decrees, however, which was applied gradually across the country, was the correct and decent conduct of the bathers. In particular, the necessity of wearing "complete" and buttoned bathing costumes was repeated frequently – a rule directed against male swimmers in trunks. Moreover, it was forbidden to sit in an "indecent manner" on the banks at the beach.[190] Yet the repetition of similar ordinances, notably on decent attire, suggests that the authorities faced numerous complaints. The Governor of the Canal region, which included the beaches of Port Said, released an additional decree clarifying that bathers had not only to cover their chests on the beaches, but also to cover themselves with more than bathing costumes when leaving the beach area.[191] As a whole, the regulations suggest a broadening of activities and governmental responsibilities. The Ottoman focus on guaranteeing hygienic standards and sanity had shifted towards regulations and control of citizens' behaviour as a domain of governmental action.[192]

While the regulation of beaches was administered mainly at district level, national politics became increasingly involved in the decision-making processes, at least in the case of large resorts. This growing attention indicated that the deputies and ministers identified a national interest in seaside tourism. It was the Minister of the Interior, Mustafa al-Nahas, who created an "administrative committee" for the organisation of the famous resort at Ra's al-Bar on the Mediterranean coast in spring 1928.[193] The committee was charged with the organisation of the services necessary to run the resort, and it administered the finances. In contrast to the *Tourist Development Association* or the committee preparing the International Congress of Tourism, the members of this local committee were recruited from public administrative bodies, including the Governor of Damietta, who served as president.[194] Although the structure was modified several times – centralising the body or re-establishing autonomy at a local level – the representatives of the international tourism business or foreign officials were not involved.[195] At the time, the beach resorts were mainly of domestic interest.

The main actors promoting beach tourism in the mid-to-late 1920s were the national deputies as well as members of government. Members of parliament discussed how beach resorts could be made accessible to a broader range of social classes. Deputies such as ʿAli Ibrahim Radwan (from al-Tallein in the Eastern Delta) aimed to increase the traffic to the beach resorts. He argued that it would not only contribute to the prosperity of the resorts, but also improve the state of health of the Egyptian population. Radwan suggested a reduction of railway tickets to the resorts by fifty per cent during the summer season, from June to September. According to him, the larger number of travellers resulting from reduced ticket prices would compensate the eventual losses for the railway company, and this would allow members of the lower middle classes, who were otherwise unable to afford a summer holiday, to participate in the tourist movement.[196]

However, his assumptions were far from uncontested. The commission in charge of infrastructures came to the conclusion that the number of travellers was not sufficient to compensate the losses to the railway company resulting from possible reductions. In addition, the commission expressed doubts about the positive effects of the measure on public health. They argued that weekend tickets had already been introduced (offering a reduction of 37.5% on certain lines), but it was mainly business travellers who profited from the initiative, rather than the lower middle-class leisure tourists it had targeted. Consequently, the members of the commission recommended offering reduced tickets only to obvious summer resorts such as Ra's al-bar, but including more cities of departure. The reduction of 37.5% was not to be increased further.[197]

Although most deputies supported the idea of fostering seaside tourism, they did so for different reasons.[198] Some deputies argued in favour of more inclusive solutions, for example by extending the reduction beyond first and second class tickets, while others rejected the proposal on the grounds that these social groups could not afford a stay at the seaside anyway.[199] Muhammad Fikri Abaza, for example, criticised the project as such, objecting that it was socially biased. He pointed out that the price reduction only benefited travellers who could afford summer holidays by the sea and, consequently, a reduction of tickets would ultimately do damage to the Egyptian people: rich Egyptians and foreigners, who would spend holidays by the sea with or without the reduction, would be privileged at the expense of the poor.[200]

Some deputies sought to foster tourism as it contributed to strengthening the national economy; others hoped that it would encourage Egyptians to discover and identify with their nation. Deputies claiming to represent the interests of the working classes insisted on lowering prices for recreational practices, and most deputies agreed that sports and leisure increased the productivity of the nation. Isma'il Sidqi, representing Alexandria and thus the Egyptian high-class resorts, argued that tourism mattered as a contribution to the national economy. Even if wealthy Egyptians profited from subsidies, he argued, these incentives encouraged the wealthy summer guests to stay in Egypt and to spend large sums in Egyptian resorts, rather than in neighbouring countries.[201] In particular, however, fostering tourism enhanced the identification of the travellers with their home country. Therefore, he argued, municipalities and other actors had to encourage summer guests to visit Egyptian resorts:

> The aim of the Railway Administration cannot be limited to a lucrative goal. The Administration has to encourage travelling. It is in the interest of the Administration that summer guests visit the Egyptian summer resorts in great numbers during the summer season, and the towns with historical monuments in winter. Other than that: We have to consider the national interest, and to make the country and its antiquity known to Egyptians is of great use for education. I can assure you, and I do it with regret, that many of our fellow countrymen know the foreign countries better than they know their own.[202]

What emerges from these debates is that tourism was considered a relevant factor in the formation of a national Egyptian society, united in their practices as well as the knowledge and appreciation of their country.[203] While the deputies behind the proposals often pursued the interests of their respective constituencies, in the case of seaside tourism, other deputies considered the question worthy of serious debate. Domestic tourism was an economic potential for some maritime localities concerned, yet it was also linked to questions vital to a modern Egyptian society.

The debates which ensued in 1927 did not lead to a decision in favour of the reduced train tickets.[204] Only in October 1931 and thus in the context of the Great Depression did the Minister of Communications, Ibrahim Fahmi Karim, finally introduce "week-end tickets" for the following summer season. The tickets were to be available for the first and second classes between 1 June and 30 September 1932. The subsidies reduced the prices of journeys on Thursdays, Fridays and Saturdays from Cairo, Tanta, Kafr el-Zayat, Mehalla-Kébir, Mansoura and Zagazig to Port Said, Alexandria and Damietta; additionally, for a trip from Ismailieh to Port Said.[205] Whether the minister hoped to compensate losses from international tourism, was not explicitly stated. At least from the perspective of individual observers, the measures proved to be a success – in the 1937 issue of the magazine *Le Réveil de L'Egypte,* the authors described weekly arrivals of young pleasure tourists on the beaches of Alexandria.[206]

Modern Egypt was an Egypt by the sea. The emergence of seaside tourism changed the landscape of tourism in Egypt, as its proponents addressed an Egyptian audience, rather than an international one. The Egyptian beach resorts, emerging on a large scale after World War I, provided Egyptian deputies with room for manoeuvre in shaping this tourist space, and thereby shaping society. As the main target group were Egyptian tourists, rather than foreigners, the dynamics differed from the administration and promotion of cultural sites. In heritage tourism, which largely attracted visitors from abroad, the deputies assumed the national prestige to be at stake. In contrast, domestic tourism to Egypt's beaches was of nationalist relevance with regard to shaping the body of the nation in a figurative and a concrete sense. It was part of the attempts of the *efendiyya* to integrate Egyptian society – through spreading concepts of a modern society, shaping the body of the nation, regulating corporal behaviour and inciting tourists to identify with their country. On a figurative level, questions of the integration of the nation and identification with Egypt were negotiated.[207] As a result, from the point of view of the Egyptian representatives, such leisure tourism justified economic incentives to a larger degree than sightseeing tourism of foreigners did. Heritage sites were considered a showcase that allowed the Egyptian *efendiyya* to put Egyptian civilisation on display for an international audience. The establishment of seaside tourism, by contrast, aimed at integrating the nation through shared practices and nationwide mobilities that were hoped to promote identification with Egypt.

Conclusion

The burning *Shepheard's Hotel* in Cairo in 1952 symbolised the end of imperial tourism in Egypt and the end of cooperation as a strategy to replace visions of Egyptian backwardness. At the time, foreign interests still influenced Egyptian politics, and international companies continued to dominate Egyptian tourism. The grand hotels remained associated with foreign political and economic dominance.

How far discourses of cultural alienation played a role here, as argued for other contexts,[208] remains to be examined. Yet the cooperation between Egyptian political institutions and European entrepreneurs, which had established exclusive structures of leisure and entertainment on Egyptian territory, was resented by a portion of the Egyptian population. The symbols of this type of foreign presence in Egypt – the *Shepheard's Hotel* and *Cook's* agency – burst into flames in 1952. Even after the 1952 outcry, however, international private companies remained involved in the Egyptian tourism sector. Large international chains managed major hotels and shipping lines, and airlines operated rather freely in the country. Although tourism during the 1950s and 1960s fluctuated due to the instability in the region, it remained an important sector of the Egyptian economy under President Gamal ʿAbd al-Nasser.[209]

The post-World War II policies of tourism were therefore somehow a continuation of the ambitions of the *efendiyya* to expand influence, gain control and profit from the tourism business during the interwar period. Although well-established European companies economically profited most from the business of tourism, local competitors had emerged, such as *Misr* airlines or *Pharos* transport company. Moreover, members of the Egyptian *efendiyya*, intellectuals, journalists, entrepreneurs and politicians perceived tourism as a vehicle for political aims.

To them, tourism was an opportunity to shape imagined and experienced spaces for both Egyptians and international tourists. Lobbying groups, members of the administration and deputies in the Chamber addressed two audiences. The message communicated to an international audience, consisting mainly of tourists from the European great powers and the US, aimed at the recognition of Egypt as a civilised nation and international recognition of its full sovereignty on equal terms. In delivering proof, the Egyptian *efendiyya* argued with the same vocabularies as their mainly European visitors: ancient monuments and the Pharaonic civilisation were presented as the historical foundation of Egyptian society, whereas urban infrastructures and leisure spaces, notably beaches, visualised the ideal of a modern Egypt participating in global trends and fashions. Such claims of belonging were mirrored by the drawing of a racially defined border between a Mediterranean Egypt and Africa. While the significance of religious belonging was downplayed by referring to the pharaonic ancestors, the racial and geographical border was drawn sharply and equated with a civilisational border. Dark-skinned Nubians in Egypt's south, as well as in the Sudan, redirected the tourist gaze to an allegedly primitive Other.

Moreover, the narratives and practices related to tourism addressed a domestic Egyptian audience. To the advocates of tourism development, the Egyptian *efendiyya*, practices of tourism bore the potential to instil a sense of belonging among Egyptian citizens, but also to establish rules for an envisioned modern society.

This ambition became particularly pronounced in the 1930s: during the 1920s, mostly deputies from relevant tourism localities lobbied in favour of domestic tourism. They aimed at attributing national relevance to their constituencies and promoting them as idealised centres of the Egyptian nation, which justified additional resources for the embellishment and improvement of infrastructures in these towns. This said, in the 1930s, deputies from constituencies renowned for their cultural heritage, such as Luxor, lost ground in the debates in comparison to proponents of the seaside resorts.

The shift from a widespread cultural nationalism in the 1920s, represented in the *efendiyya* Pharaonism, towards an increasing perception of the nation in terms of class differences in the 1930s appears to have acted as a catalyst for this development.[210] Debates about national belonging and the integration of different social groups – both allegedly cosmopolitan bourgeoisies as well as the working classes – turned tourism and recreation into concerns considered vital by many parliament members in the 1930s. As far as the upper classes were concerned, the patriotic appeal that Egyptians should tour their home country for moral reasons created an affirmative definition of Egypt in political terms. The advertisement of seaside resorts sustained claims of belonging to a united, modern Mediterranean, while simultaneously nationalising the coast: since modern resorts corresponding to the latest standards were available, crossing the borders was deemed no longer necessary. The increasing nationalism of the 1930s and a more pronounced focus on the national economy after the economic crises made such assumptions more widely acceptable. In addition, leisure practices were a means of integrating Egyptians of all backgrounds and classes into the nation. From the ambition to educate Egyptians at historical sites, attention shifted towards the seaside in the 1930s. Leisure, recreation, national mobility and the shaping of the modern bodies associated with health and productivity were key practices intended to form a united body for the modern, industrious nation.[211]

Yet among European visitors, it appears that the narratives and practices of modern Egypt were not so easily adopted. Visions of Egyptian backwardness largely prevailed in European brochures, guidebooks and photograph albums. Tourists turned their gaze from elements of modernity and actively searched for elements in line with established discourses; they attributed processes of change to European actors or ridiculed Egyptian modernity. Only the fundamental rupture that represented the coup of the Free Officers to international observers seemed to allow for an at least partial redefinition of images of Egypt. On a domestic level, in contrast, the expansion of leisure practices and thus the touristic integration of the country was a slow if steady process. The expansion of beaches and the urban crowds flocking to Alexandria at the weekends indicated a growing participation in leisure outings and recreation, and photographs from the 1950s suggested that physical activities and the exposure of bodies associated with beaches continued to spread, at least for the moment.[212] Alongside the cultural tourist space mainly frequented by Europeans and the Egyptian *efendiyya*, a second Egyptian tourist space of leisure emerged in the 1930s.

Notes

[1] THE TIMES DIGITAL ARCHIVE, *Cairo Buildings Fired, 28/01/1952*. THE TIMES DIGITAL ARCHIVE, *Extremists' Role in Riots, 29/01/1952*. On the Riots see Reynolds, *A City Consumed*, pp. 183–189. On the history of the *Shepheard's Hotel*: Humphreys, *On the Nile*, pp. 54–55. Humphreys, *Grand Hotels of Egypt*, pp. 75–98.

[2] The idea of the 1930s as a 'golden age' of travel is still very present in numerous publications, which often celebrate the leisure culture and the aesthetics of the grand hotels: Humphreys, *On the Nile*. Humphreys, *Grand Hotels of Egypt*. Gregory, *The Golden Age of Travel*. Watkin, *Grand Hotel*. Meade, Fitchett and Lawrence, *Grand Oriental Hotels*.

[3] Hunter, 'Tourism and Empire', pp. 42–43. The post-war growth in tourism was a larger trend across the region: NLI, *Tourist Traffic Decreasing, 02/01/1934*.

[4] *Galabiyya*: a wide, ankle-length garment with long sleeves, worn by Egyptians in the countryside of the Nile Valley.

[5] Mitchell, *Colonising Egypt*, pp. 21–31.

[6] Mitchell, *Colonising Egypt*, p. 29.

[7] Mitchell, *Colonising Egypt*, pp. 166–168, 174–179. Thereby Mitchell modifies Said's analysis emphasising distinction: Said, *Orientalism*, pp. 42–46, 253–254.

[8] Cf. Di-Capua, 'Common Skies Divided Horizons'. Jacob, *Working Out Egypt*. Baron, *Egypt as a Woman*. Kholoussy, *For Better, For Worse*.

[9] Ryzova, *The Age of the Efendiyya*, p. 26 et pass. The authentification of such an Egyptian modernity was based on references to a rural Egyptian culture, yet modern in the sense that scientific approaches had helped to overcome the alleged poverty and ignorance of the countryside: Ryzova, *The Age of the Efendiyya*, pp. 85–87. On such reproaches: Baron, *The Orphan Scandal*. On the ambivalence of Arab thinkers about notions of modernity: Hanssen and Weiss, 'Introduction', pp. 16–17.

[10] RCSA, Williams, *Travel Diary: Egypt, 1925*, p. 166.

[11] Hulme and Youngs, 'Introduction', p. 5.

[12] On the principles of Orientalist photography: Behdad, *Camera Orientalis*, pp. 41–66. Reproduction of early photographs from Egypt: Perez, *Focus East*. Stewart Howe and Wilson, *Excursions Along the Nile*. Bull and Lorimer, *Up the Nile*. Frith, van Haaften and White, *Egypt and the Holy Land*. Blottière, *Un voyage en Égypte*.

[13] Said, *Orientalism*. Mitchell, *Colonising Egypt*, pp. 29–33. Gregory, 'Emperors of the Gaze', pp. 204–207. Behdad, 'The Orientalist Photograph'.

[14] Gregory, 'Imaginative Geographies', p. 463. Zananiri, 'From Still to Moving Image', p. 65.

[15] Gregory, 'Emperors of the Gaze', pp. 209–212. Hazbun, 'The East as an Exhibit', pp. 17–19. Other authors argued that the shift from monumental to ethnographic photography was a result of technological progress, notably of a reduction in the required time of exposure: Schwartz, 'The Geography Lesson', pp. 30, 35.

[16] Mitchell, *Colonising Egypt*, pp. 29–33.

17   Bredekamp, *Theorie des Bildakts*. Bohnsack, 'Disziplinierte Zugänge zum undisziplinierten Bild', pp. 65–70.

18   Behdad, *Camera Orientalis*, pp. 9–15. Sheehi, *The Arab Imago*, pp. xx–xxiii.

19   Rauch, 'Du Joanne au Routard', p. 111. Behdad, 'The Orientalist Photograph', pp. 18–20. Gregory, 'Scripting Egypt', p. 135. Haldrup and Larsen, *Tourism, Performance and the Everyday*, p. 17.

20   Schwartz, 'The Geography Lesson', p. 18. Osterhammel, *Die Verwandlung der Welt*, pp. 76–80.

21   Cf. Rosenberg, 'Transnationale Strömungen', p. 986. Löfgren, *On Holiday*, p. 82. Gunning, 'The Whole World Within Reach', pp. 28–29. On the connection between 'kodakable' visions and stereotypes: Gregory, 'Scripting Egypt', p. 145.

22   Urry and Larsen, *The Tourist Gaze 3.0*, pp. 1–13. Kathleen Stewart Howe interprets „ethnographic photographs" as an expression of difference between the ways of life of the travellers and the population: Stewart Howe and Wilson, *Excursions Along the Nile*, p. 41. The search of the tourist for "authenticity" in tourism has been assumed by MacCannell: MacCannell, *The Tourist*, pp. 100–107.

23   Urry and Larsen, *The Tourist Gaze 3.0*, pp. 178–179.

24   Behdad, *Camera Orientalis*, pp. 82–83, 98.

25   Weber, 'Le Genre romanesque du récit', p. 60. Gregory, 'Scripting Egypt', p. 135. The *topos* as understood by Weber and in this work is thus related to the idea of a collective memory, cf. Fulda, 'Topik'.

26   Mitchell, *Colonising Egypt*, pp. 171–173. Weber, 'Le Genre romanesque du récit', pp. 64, 69–70. Korte, *Der englische Reisebericht*, pp. 14–17.

27   Cardon and Jackson, 'Everyday Empires'. Urry and Larsen, *The Tourist Gaze 3.0*, pp. 155–156.

28   Irwin, *Camel*, pp. 118–134, 173–178, 195. Bough, *Donkey*, pp. 7–20. Cf. Huber, *Channelling Mobilities*, pp. 164–166. For interpictorial references cf. Stewart Howe and Wilson, *Excursions Along the Nile*, pp. 42–43. On the daily life as sight: Gregory, 'Scripting Egypt', pp. 137–139.

29   Urry and Larsen, *The Tourist Gaze 3.0*, pp. 178–179. Cf. Behdad, *Camera Orientalis*, pp. 26–28.

30   RCSA, Williams, *Travel Diary: Egypt, 1925*, p. 191.

31   Behdad, *Camera Orientalis*, p. 46.

32   I adopt Behdad's categories, because the distinction between exotic and erotic photography seems relevant in tourist photographs, and because Gregory's binary distinction does not allow including landscape photography and panoramic views popular in the interwar photographs.

33   RCSA, Williams, *Travel Diary: Egypt, 1925*, pp. 19, 25–26, 37, 39.

34   E.g. BNF RICHELIEU, *Album de photographies, 1931*. Urry distinguishes between the "collective tourist gaze" and the "family gaze": Urry and Larsen, *The Tourist Gaze 3.0*, pp. 19–20.

35   Compare for example: Haag, *Vintage Alexandria*, or the photographic collection of Moustafa El Moghazi, published on Facebook in his digital album "Vintage Egypt": https://www.facebook.com/elmoghazi.me/photos_albums, last accessed 18/06/2022 (content is only visible for users who are logged in on Facebook).

36   RCSA, Williams, *Travel Diary: Egypt, 1925*, pp. 18, 22, 38, 56, 64–66.

37   On the emergence of the „tourist gaze" and typical objects of the gaze: Urry and Larsen, *The Tourist Gaze 3.0*, pp. 15–16.

38   Urry and Larsen, *The Tourist Gaze 3.0*, pp. 167–171.

39   Cf. BNF TOLBIAC, The Tourist Development Association of Egypt, *Egypt and the Sudan*, p. 35.

40   Baron, *Egypt as a Woman*, p. 95.

41   On textual anchorage: Behdad, *Camera Orientalis*, pp. 29–31.

42   Micklewright, *A Victorian Traveler*, pp. 181–184. Micklewright, 'Orientalism and Photography', pp. 104–106.

43   Pinney, 'What's Photography Got to Do with It?', p. 49. Woodward, 'Between Orientalist Clichés and Images of Modernization', pp. 371–373. Behdad and Gartlan, 'Introduction', pp. 3–4.

44   Urry and Larsen, *The Tourist Gaze 3.0*, pp. 186–194.

45   Çelik and Eldem, *Camera Ottomana*. Roberts, 'The Limits of Circumscription', pp. 69–70. Nassar, 'On Photographs', p. 4. Sheehi, *The Arab Imago*, xx-xxiii. A brief overview on photography in the Middle East and particularly Egypt, with a focus on local uses of photography: Baron, *Egypt as a Woman*, pp. 83–88.

46   DeRoo, 'Colonial Collecting', pp. 165–169. Sheehi, *The Arab Imago*.

47   Staging such scenes was common among professional photographers, cf. Gavin, *The Image of the East*. Bopp, 'Orientalismus im Bild'. Debbas, *Des photographes à Beyrouth*. Erdoğdu, 'Picturing Alterity'.

48   Gregory on photography as a "predatory act": Gregory, 'Emperors of the Gaze', pp. 212–217. Pinney postulated this participation: Pinney, 'What's Photography Got to Do with It?', p. 46. Ali Behdad pointed out that the „archive" of Orientalist photography did not result from one-sided visions, but emerged as the consequence of inter-human relationships: Behdad, *Camera Orientalis*, p. 14.

49   Urry calls it the 'mutual gaze'. Urry and Larsen, *The Tourist Gaze 3.0*, pp. 204–205.

50   Behdad, *Camera Orientalis*, p. 61.

51   RCSA, Williams, *Travel Diary: Egypt, 1925*, pp. 114–115. At first sight, Mahmud seems to correspond to a typical "cultural broker" described for colonial contexts (cf. Osterhammel, 'Kulturelle Grenzen in der Expansion Europas', p. 227), yet, since in organised tourism the relationship between tourist (group) and guide differs in significant regards (a temporarily limited encounter, the dynamics of tourist groups, the guide as competitor of other sources of information, the phenomenon of staging authenticity) from situations of early colonial contact, I prefer not to use the term.

52   RCSA, Williams, *Travel Diary: Egypt, 1925*, pp. 114–115.

53   Derek Gregory interpreted the Egyptian 'commercialisation' of the photograph as a capitulation to large-scale tourism: Gregory, 'Emperors of the Gaze', pp. 222–223.

54   RCSA, Williams, *Travel Diary: Egypt, 1925*, pp. 56–57, 123, 129–130, 137. Edwin W. Smith also reported on the positive posture of the local population with regard to having pictures taken: SOAS, Smith, *Travel Diaries, Diary 1929, I*, 05/11/1929. He made the experience that bribery sometimes allowed him to take a picture of monuments where it was usually forbidden, as in a mosque: SOAS, Smith, *Travel Diaries, Diary 1929, I*, 22/11/1929.

55   RCSA, Williams, *Travel Diary: Egypt, 1925*, p. 94.

56   RCSA, Williams, *Travel Diary: Egypt, 1925*, p. 123.

57   NLI, *International Tourist Congress, 12/12/1932*.

58   Eg Norval, *The Tourist Industry*.

59   AUB, *Rescrit Royal No. 39, 05/07/1932*. AUB, *Rescrit Royal No. 45, 09/08/1932*. The president of the committee changed when a new Minister of Communications came into office, namely Ibrahim Fahmi Karim Pasha: AUB, *Rescrit Royal No. 10, 09/01/1933*.

60   FRENCH LINES, Compagnie des Messageries Maritimes, *Agence d'Alexandrie, Rapport général de Service 1920*, pp. 4–7.

61   Alexandria, for example, was hit by economic crises in 1921, 1926, 1931: Ilbert, *Alexandrie*, II, p. 599. Hunter, 'Tourism and Empire', pp. 42, 50.

62   FRENCH LINES, Compagnie des Messageries Maritimes, *Agence d'Alexandrie, Rapport général de Service 1921*, p. 4. Ilbert, *Alexandrie*, II, p. 599.

63   FRENCH LINES, Compagnie des Messageries Maritimes, *Agence d'Alexandrie, Rapport général de Service 1933*, p. III.

64   FRENCH LINES, Compagnie des Messageries Maritimes, *Agence d'Alexandrie, Rapport général de Service 1928*. FRENCH LINES, Compagnie des Messageries Maritimes, *Agence d'Alexandrie, Rapport général de Service 1929*. FRENCH LINES, Compagnie des Messageries Maritimes, *Agence d'Alexandrie,*

Rapport général de Service 1930. FRENCH LINES, Compagnie des Messageries Maritimes, *Agence d'Alexandrie, Rapport général de Service 1931*. FRENCH LINES, Compagnie des Messageries Maritimes, *Agence d'Alexandrie, Rapport général de Service 1932*. FRENCH LINES, Compagnie des Messageries Maritimes, *Agence d'Alexandrie, Rapport général de Service 1933*, p. IV. FRENCH LINES, Compagnie des Messageries Maritimes, *Agence d'Alexandrie, Rapport général de Service 1934*, p. V. FRENCH LINES, Compagnie des Messageries Maritimes, *Agence d'Alexandrie, Rapport général de Service 1936*, p. IV. FRENCH LINES, Compagnie des Messageries Maritimes, *Agence d'Alexandrie, Rapport général de Service 1937*, p. IV. FRENCH LINES, Compagnie des Messageries Maritimes, *Agence d'Alexandrie, Rapport général de Service 1938*, p. V.

[65]   AUB, *Moudirieh de Kéneh, Arrêté, 03/04/1932*.

[66]   Hunter, 'Tourism and Empire', p. 31. Zuelow, *A History of Modern Tourism*, pp. 99–102. Walton, 'British Tourism', pp. 110, 122–123.

[67]   E.g. Sir George Newnes, 1898, quoted in Hunter, 'The Thomas Cook Archive', p. 163. Gray, 'Economic Reform, Privatization and Tourism', p. 92. Cf. also Hazbun, *Beaches, Ruins, Resorts*, pp. xx–xxii.

[68]   Hunter, 'Tourism and Empire', pp. 37–38.

[69]   Anderson, 'The Development of British Tourism', pp. 260–262.

[70]   Hunter, 'Tourism and Empire', pp. 33–34, 39–44, 48–50. Hazbun, 'The East as an Exhibit', pp. 24–26. On the role of Cook in the British expedition to Sudan: Humphreys, *On the Nile*, pp. 67–82.

[71]   Hunter, 'Tourism and Empire', pp. 34–37, 48–49.

[72]   Hunter, 'Tourism and Empire', pp. 50–51.

[73]   E.g. NLS, Steele-Maitland, *Diary of Visit to Egypt*, 30/12/1931, 02/01/1932.

[74]   Schölch, *Ägypten den Ägyptern!* Lapidus, *A History of Islamic Societies*, p. 565. Gelvin, *The Modern Middle East*, pp. 205–207.

[75]   Goldschmidt Jr, *Modern Egypt*, pp. 71–72.

[76]   Khoori, *Luxor: How to See It*, pp. 144–145. Cf. the files on employees in the French Lines Archives, e.g. FRENCH LINES, Compagnie des Messageries Maritimes, *Agence d'Alexandrie, Rapport général de Service 1931*.

[77]   AUB, *Parlement. Cinquième séance ordinaire, 17/11/1927*, pp. 2–4.

[78]   E.g. AUB, *Discours du Trône du Roi, 12/11/1924*. AUB, *Discours du Trône du Roi, 10/06/1926*. AUB, *Discours du Trône du Roi, 17/11/1927*. Enas Fares Yehia has recently demonstrated that King Fu'ad actively supported tourism development in Egypt, yet the activities King Fu'ad also suggest that he fostered tourism as a diplomatic stage to present Egypt rather than as a major economic potential: Yehia, 'Promotion of tourism in Egypt'. I am grateful to the anonymous reviewer who pointed out this article to me.

[79]   AUB, *Décision portant reconstitution du Conseil Economique, 15/02/1925*.

[80]   AUB, *Arrêté Ministériel No. 2, 04/01/1928*, pp. 10–11.

[81]   Cf. AUB, *Sénat, 22e Séance publique, 23/04/1930*.

[82]   AUB, *Séances de la Chambre des Députés, 18/11/1926-28/03/1927*.

[83]   Dawletschin-Linder, 'Neue Wege in der Wirtschaftspolitik', pp. 71–78. On *Banque Misr*: Di-Capua, 'Common Skies Divided Horizons', p. 923.

[84]   Dawletschin-Linder, 'Neue Wege in der Wirtschaftspolitik', pp. 80–82. Still at the end of the 1930s, cotton and cotton products provided around 80% of the Egyptian exportation value: Dawletschin-Linder, 'Neue Wege in der Wirtschaftspolitik', p. 86.

[85]   AUB, *Sénat, 4e Séance publique, 21/01/1930*.

[86]   AUB, *Chambre des Députés, 67e séance publique du 30/05/1927*, pp. 940–943.

[87]   Similar debates were triggered by the question of granting subsidies to the local opera house. AUB, *Chambre des Députés, 52e séance publique, 21/06/1924*, pp. 545–546.

88   Badrawi, 'Financial Cerberus?', pp. 98–99.

89   AUB, *Chambre des Députés, 67e séance publique du 30/05/1927*.

90   AUB, *Chambre des Députés, 67e séance publique du 30/05/1927*, pp. 940–943.

91   Mairs and Muratov, *Archaeologists, Tourists, Interpreters*, pp. 13–20.

92   AUB, *Chambre des Députés, 52e séance publique, 03/05/1927*.

93   AUB, *Moudirieh de Kéneh, Arrêté, 16/05/1928*. AUB, *Moudirieh de Kéneh, Arrêté, 12/06/1928*.

94   AUB, *Arrêté, 17/08/1926*. AUB, *Moudirieh de Guirgueh, Arrêté, 26/02/1927*. AUB, *Gouvernorat du Caire, Arrêté, 12/03/1931*. AUB, *Gouvernorat d'Alexandrie, Arrêté, 19/08/1931*. AUB, *Moudirieh de Guizeh, Arrêté, 17/08/1931*. AUB, *Moudirieh d'Assouan, Arrêté, 03/04/1932*.

95   AUB, *Gouvernorat du Caire, Arrêté, 07/04/1928*. AUB, *Gouvernorat du Caire, Arrêté, 07/04/1928*.

96   A comparison with neighbouring Lebanon suggests the importance of the latter argument. There, around the same time measures such as publicly announced price lists in hotels and at offices of transport companies were introduced in order to reassure foreign tourists.

97   AUB, *Ministère de l'Intérieur, Arrêté, 10/07/1929*. AUB, *Gouvernorat du Caire, Arrêté, 07/09/1932*. AUB, *Gouvernorat du Canal, Arrêté, 13/07/1932*. AUB, *Moudirieh de Fayoum, Arrêté, 12/07/1932*. AUB, *Moudirieh d'Assiout, Arrêté, 17/07/1933*. AUB, *Moudirieh d'Assouan, Arrêté, 25/07/1933*. AUB, *Moudirieh de Guizeh, Arrêté, 21/08/1933*. AUB, *Moudirieh de Minieh, Arrêté, 13/09/1933*. AUB, *Moudirieh de Kéneh, Arrêté, 04/06/1934*. AUB, *Gouvernorat d'Alexandrie, Arrêté, 18/01/1936*.

98   AUB, *Moudirieh de Guizeh, Arrêté, 30/08/1936*.

99   FRENCH LINES, Compagnie des Messageries Maritimes, *Agence d'Alexandrie, Rapport général de Service 1933*, Trafic. Cf. also Yehia, 'Promotion of tourism in Egypt', pp. 10–14. On British and American decline: CADN, *Le Gouvernement et le projet, 13/11/1936*. Italy was an important reference against which Egypt positioned itself also in the context of tourism, cf. Muhammad Hafiz Ramadan, AUB, *Chambre des Députés, 19e séance publique, 12/01/1927*, p. 200.

100   BNF GALLICA, Sidky, *L'Egypte économique*.

101   AUB, *Décret portant création d'un Ministère du Commerce et de l'industrie, 20/12/1934*. The new Minister of Trade and Industry was named in June 1935: AUB, *Décret portant nomination d'un Ministre pour le Commerce et l'industrie, 18/06/1935*.

102   AUB, *Décret portant constitution d'une Société Anonyme, 13/09/1928*.

103   JD PRIVATE COLLECTION, Tourist Development Association of Egypt, *Le Nil d'Alexandrie à Assouan*. In 1926, the group changed its statutes. Cf. NLI, *Statutes: Tourist Development Association of Egypt*.

104   NLI, *International Tourist Congress, 12/12/1932*. The TDA did not address Egyptian tourists, but targeted an exclusively European audience: AUB, *Chambre des Députés, 19e séance publique, 12/01/1927*, p. 200.

105   JD PRIVATE COLLECTION, Taylor and Tourist Development Association of Egypt, *Egypt and the Sudan*, p. 11.

106   World Biographical Information System, 's.v. Perera, Guy U'. JD PRIVATE COLLECTION, Taylor and Tourist Development Association of Egypt, *Egypt and the Sudan*, p. 11.

107   JD PRIVATE COLLECTION, Perera, Otto and Tourist Development Association of Egypt, *Un voyage en Egypte et au Soudan*. The photographs were reprinted in JD PRIVATE COLLECTION, Taylor and Tourist Development Association of Egypt, *Egypt and the Sudan*. On Lehnert and Landrock cf. Degroise, 'Photographes d'Asie'.

108   CADN, *Note du Ministère des Affaires étrangères, 07/09/1935*. CADN, *L'attaché commercial, 09/1935*.

109   CADN, *Le Directeur général de l'Office de tourisme, 30/12/1935*.

110   CADN, *Une heureuse initiative*. CADN, *Ministre de France en Egypte, 26/11/1936*. CADN, *Le Gouvernement et le projet, 13/11/1936*.

111 The networks leading to the marketing campaign offered by *Cook* to the Egyptian Government, it seems, were developed long before this cooperation in 1936. Already in June 1934, Sir Edward Grigg, former Governor of Kenya and at the time administrator of *Thomas Cook & Son* intended to visit Egypt at the end of the month, where he hoped to present a large marketing campaign for travels to Egypt to the Government: CADN, Compagnie Internationale des Wagons-Lits, *Cabinet du Directeur Général, 02/06/1934*.

112 Gray, 'Economic Reform, Privatization and Tourism', p. 93.

113 Gray, 'Economic Reform, Privatization and Tourism', p. 94.

114 Gray, 'Economic Reform, Privatization and Tourism', p. 97.

115 RCSA, Williams, *Travel Diary: Egypt, 1925*, pp. 18–19, 39–40.

116 RCSA, Williams, *Travel Diary: Egypt, 1925*, p. 38.

117 Cf. MacCannell, 'Staged Authenticity', p. 602.

118 On the claim to participate in a modernity of a shared global standard: Osterhammel, *Die Verwandlung der Welt*, pp. 1093–1095.

119 Urry and Larsen, *The Tourist Gaze 3.0*, p. 13.

120 The nostalgic impulse in tourism has been analysed by Groebner, *Retroland*.

121 E.g. SOAS, Smith, *Travel Diaries, Diary 1929, II*, 19/11/1929.

122 NLS, Steele-Maitland, *Diary of Visit to Egypt*, 21/12/1931; 05/01/1932; 13/01/1932.

123 Urry and Larsen, *The Tourist Gaze 3.0*, p. 13.

124 As described by Maurizio Peleggi in his study on colonial hotels: Peleggi, 'The Social and Material Life of Colonial Hotels'.

125 E.g. Elston, *Handbook for Egypt and the Sûdân*, pp. 36–37.

126 Elston, *Handbook for Egypt and the Sûdân*, p. 48.

127 SOAS, Smith, *Travel Diaries, Diary 1929, II*, 21/11/1929.

128 Similar observations with regard to cultural and technological incompatibility: SOAS, Smith, *Travel Diaries, Diary 1929, II*, 18/11/1929.

129 AUB, *Gouvernorat du Caire, Arrêté, 05/09/1927*.

130 JD PRIVATE COLLECTION, Egyptian State Tourist Department and Tourist Development Association of Egypt, *Photograph "In Modern Cairo"*.

131 E.g. Assuan: AUB, *Moudirieh d'Assouan, Arrêté, 01/09/1927*.

132 Volait, *Architectes et architectures*, pp. 199–200.

133 Cf. also Ryzova, *The Age of the Efendiyya*, pp. 85–87.

134 JD PRIVATE COLLECTION, Egyptian State Tourist Administration and Tourist Development Association of Egypt, *Egypt: Tourist Guide, Cairo*, pp. 6–7.

135 Edwin W. Smith reported on conversations with Egyptians and noted the "progress" made in the country. One of his interlocutors on a train, ʿAziz Mikhail Gabriel, served as officer in the Ministry of Finance. SOAS, Smith, *Travel Diaries, Diary 1929, II*, 10/11/1929.

136 BL, Khoori, *Cairo: How to See It*, p. 11. Picture credit: A. Kalfayan, Kodak, Cairo.

137 An exception are views of the pyramids taken from the terrace of the *Mena House Hotel*, yet they framed the pyramids in a touristic rather than Egyptian context.

138 NLS, Steele-Maitland, *Diary of Visit to Egypt*, 09/01/1932.

139 NA, Lawrence, *Travel Photographs: Voyage to Egypt*.

140 Mary Steele-Maitland, for example, sent postcards home from here: NLS, Steele-Maitland, *Diary of Visit to Egypt*, 09/01/1932. Humphreys, *On the Nile*, p. 62.

141 Khoori, *Luxor: How to See It*, p. 39. AUB, *Chambre des Députés, 29e séance publique, 10/08/1926*, p. 375.

142 Khoori, *Luxor: How to See It*, pp. 119–129.

[143] NLS, Steele-Maitland, *Diary of Visit to Egypt*, 06/01/1932. Khoori, *Luxor: How to See It*, pp. 93–95, 123, 133.

[144] Elston, *Handbook for Egypt and the Sûdân, 1929*, p. 241.

[145] A common assumption among archaeologists at the time, causing numerous forced expulsions of locals. Reid, 'Nationalizing the Pharaonic Past', p. 139.

[146] For Egypt cf. Reid, 'Nationalizing the Pharaonic Past', p. 139.

[147] NLS, Steele-Maitland, *Diary of Visit to Egypt*, 07/01/1932.

[148] NLS, Steele-Maitland, *Diary of Visit to Egypt*, 03/01/1932. Reid, *Contesting Antiquity in Egypt*, p. 59.

[149] RCSA, Williams, *Travel Diary: Egypt, 1925*, pp. 19–20.

[150] Reid, 'Nationalizing the Pharaonic Past', pp. 129, 137, 140–141. Nuancing Gershoni/Jankowski's claim that the 1930s were widely dominated by a renewed Islamic/Arabist discourse, Reid pointed out the ongoing institutionalisation of ancient history and archaeology, partly as a reaction to the imperialist threat. Reid, 'Indigenous Egyptology'.

[151] AUB, *Chambre des Députés, Projet de loi, 07/06/1924*, pp. 401–402.

[152] Khoori, *Luxor: How to See It*, pp. 75–81.

[153] In these discoveries, the 'Abd al-Rasul family from Qurna was prominently involved. Family relations between them and al-Sayyid or any other involvement of al-Sayyid can neither be verified nor excluded. Reid, 'Nationalizing the Pharaonic Past', p. 139.

[154] Khoori, *Luxor: How to See It*, p. 77.

[155] An account of the case on the basis of press reports has been reconstructed by Parkinson, 'Tutankhamen on Trial'. On the British-Egyptian rivalry cf. also Reid, 'Nationalizing the Pharaonic Past', p. 129.

[156] Khoori, *Luxor: How to See It*, pp. 93–95. On the opening: Wood, 'The Use of the Pharaonic Past', p. 183.

[157] Parkinson, 'Tutankhamen on Trial', pp. 2, 5. The French involvement in the case, advising on the Egyptian Government, was not mentioned: Reid, 'Nationalizing the Pharaonic Past', pp. 133–135.

[158] Reid pointed to the fact that in 1929, the Antiquities Service shifted from the Ministry of Public Works to the Ministry of Education: Reid, 'Nationalizing the Pharaonic Past', p. 138.

[159] Badrawi, 'Financial Cerberus?', pp. 96–97.

[160] AUB, *Chambre des Députés, Première séance publique, 10/06/1926*, p. 6.

[161] AUB, *Chambre des Députés, 25e séance publique, 04/08/1926*, pp. 315–319.

[162] AUB, *Chambre des Députés, 25e séance publique, 04/08/1926*, p. 317.

[163] AUB, *Chambre des Députés, 25e séance publique, 04/08/1926*, pp. 315–319.

[164] AUB, *Chambre des Députés, 29e séance publique, 10/08/1926*, p. 376.

[165] Cf. also Makdisi, 'Ottoman Orientalism', pp. 787–792. Deringil, '"They Live in a State of Nomadism"', pp. 317–318.

[166] Eve Trout Powell coined the term of Egypt as a "colonized colonizer": Powell, *A Different Shade of Colonialism*, p. 6.

[167] BNF Tolbiac, Zananiri, *Nubie*, p. 116.

[168] JD PRIVATE COLLECTION, Egyptian State Tourist Department and Tourist Development Association of Egypt, *Egypt: A Travel Quarterly, Autumn 1937*.

[169] BNF Tolbiac, Zananiri, *Nubie*.

[170] Powell, *A Different Shade of Colonialism*, p. 6.

[171] BNF Tolbiac, Bertero, *Introduction*, p. 25. Pedersen, *The Guardians*, p. 369.

[172] Alloula, *The Colonial Harem*.

[173] AUB, *Chambre des Députés, 40e séance publique, 28/03/1927*.

[174] Alloula, *The Colonial Harem*. DeRoo, 'Colonial Collecting', pp. 162–165. Moors, 'Presenting People', pp. 94–98.

[175] AUB, *Chambre des Députés, 40e séance publique, 28/03/1927*.

176 Botman, 'The Liberal Age, 1923–1952', pp. 293–294.

177 Löfgren described similar debates in the Swedish context: Löfgren, 'Know Your Country'.

178 AUB, *Chambre des Députés, 19e séance publique, 12/01/1927*, p. 200. On the corniche and the role of Ahmad Sadiq, General Director of the Tourist Office: BNF Tolbiac, Bernard, *Plages, 1937*, p. 123.

179 Badrawi, 'Financial Cerberus?', pp. 108–110.

180 Ilbert, *Alexandrie*, II, pp. 597–603.

181 Hazbun, 'Modernity on the Beach', pp. 207–208.

182 On tourism as a marker of modernity cf. Urry and Larsen, *The Tourist Gaze 3.0*, p. 6.

183 ʿAbd al-Hafiz, 'Les hammams dʾAlexandrie', p. 358.

184 Ilbert, *Alexandrie*, I, pp. 324–325.

185 NLS, Steele-Maitland, *Diary of Visit to Egypt*, 30/12/1931. On the beach resorts see for example: BNF Tolbiac, The Tourist Development Association of Egypt, *Egypt and the Sudan*, pp. 23–24. JD PRIVATE COLLECTION, Egyptian State Tourist Department and Tourist Development Association of Egypt, *Egypt: A Travel Quarterly, Autumn 1937*. BNF Tolbiac, Bernard, *Plages, 1937*.

186 AUB, *Commission Municipale d'Alexandrie, Arrêté, 13/09/1927*. Cf. also AUB, *Municipalité d'Alexandrie, Arrêté, 01/08/1929*.

187 Ilbert, *Alexandrie*, I, pp. 313–316.

188 AUB, *Gouvernorat de Damiette, Arrêté, 22/06/1930*. Additional measures were taken in 1933: AUB, *Gouvernorat de Damiette, Arrêté, 03/08/1933*. AUB, *Gouvernorat du Canal, Arrêté, 04/02/1932*.

189 AUB, *Gouvernorat de Suez, Arrêté, 11/06/1934*. AUB, *Gouvernorat de Suez, Arrêté, 08/10/1934*.

190 AUB, *Gouvernorat du Canal, Arrêté, 01/07/1930*. AUB, *Gouvernorat du Canal, Arrêté, 20/09/1932*. AUB, *Gouvernorat de Suez, Arrêté, 11/06/1934*.

191 AUB, *Gouvernorat du Canal, Arrêté, 25/11/1935*.

192 On educational approaches to workers' tourism in a French context cf. Furlough, 'Making Mass Vacations', p. 253.

193 The new body joined the already existing "committee of hygiene".

194 AUB, *Ministère de l'intérieur, Arrêté, 07/05/1928*.

195 AUB, *Ministère de l'intérieur, Arrêté, 13/01/1931*. AUB, *Ministère de l'intérieur, Arrêté, 12/03/1934*. AUB, *Ministère de l'intérieur, Arrêté, 10/03/1935*.

196 AUB, *Chambre des Députés, 27e séance publique, 08/08/1926*. AUB, *Chambre des Députés, Première séance publique, 10/06/1926*, p. 5. AUB, *Chambre des Députés, 19e séance publique, 12/01/1927*, p. 202.

197 AUB, *Chambre des Députés, 19e séance publique, 12/01/1927*.

198 This mirrored debates in Europe and the US: Zuelow, *A History of Modern Tourism*, p. 135.

199 AUB, *Chambre des Députés, 18e séance publique, 11/01/1927*, pp. 195–196.

200 AUB, *Chambre des Députés, 19e séance publique, 12/01/1927*, p. 202.

201 AUB, *Chambre des Députés, 19e séance publique, 12/01/1927*, pp. 201–202. Sidqi criticised the lavish spending of Egyptians abroad in other contexts as well: Badrawi, *Isma'il Sidqi*, p. 48. On sports clubs and the upper class: Di-Capua, 'Common Skies Divided Horizons', pp. 924–925.

202 AUB, *Chambre des Députés, 19e séance publique, 12/01/1927*, p. 200.

203 Drawing on Turner, Anderson has highlighted the experience of shared journeys to shared centres (as in pilgrimage or educational/professional trajectories) as a meaning-creating experience at a collective level: Anderson, *Imagined Communities*, pp. 53–54, 121.

204 AUB, *Chambre des Députés, 19e séance publique, 12/01/1927*, p. 203.

205 AUB, *Ministère des Communications, Arrêté ministériel, 22/10/1931*.

206 BNF Tolbiac, Bernard, *Plages, 1937*.

207 On movement and action as integral elements of nationalism cf. Baron, *Egypt as a Woman*, pp. 95–96.

208 Baron, 'The Port Said Orphan Scandal', pp. 135–138.

[209] Gray, 'Economic Reform, Privatization and Tourism', p. 93.

[210] Beinin, *Workers and Peasants*, pp. 88–90.

[211] Jacob, *Working Out Egypt*, pp. 65–91. Jacob points out, though, that qualifications have to be made regarding a nationalist reading of shaping bodies, and that notably the desires of the individual have to be taken into account as well: Jacob, *Working Out Egypt*, p. 155.

[212] Bassil et al., *Mapping Sitting*, 3.427–3.590.

References

Sources

Bibliothèque nationale de France, Paris

Album de photographies, Un Voyage en Orient (Egypte, Syrie, Liban) en 1931, BNF Richelieu/VZ 1082.

BERNARD, JEANNE, *Plages*, in: Le Réveil de l'Egypte, ed. Société d'édition pour le tourisme, Alexandrie (1937), pp. 121–123, BNF Tolbiac/2013-301184.

BERTERO, GUIDO, *Introduction*, in: Le Réveil de l'Egypte, ed. Société d'édition pour le tourisme, Alexandrie (1937), pp. 25–26, BNF Tolbiac/2013-301184.

SIDKY, ABDUL HAMID, *L'Egypte économique d'aujourd'hui*, Thèse pour le doctorat en droit (sciences économiques), Université de Poitiers, 1931, BNF Gallica/NUMM-6202296.

THE TOURIST DEVELOPMENT ASSOCIATION OF EGYPT, *Egypt and the Sudan*, 1929, Cairo, BNF Tolbiac/2014-92173.

ZANANIRI, GASTON, *Nubie*, Le Réveil de l'Egypte, ed. Société d'édition pour le tourisme, Alexandrie (1937), pp. 115–118, BNF Tolbiac/2013-301184.

British Library, London

KHOORI, ALEXANDER R., *Cairo: How to See It*, Alexandria: The Anglo-Egyptian Supply Association, 8th edn., 1928, BL.

Centre des Archives diplomatiques de Nantes

COMPAGNIE INTERNATIONALE DES WAGONS-LITS, *Cabinet du Directeur Général à M. Gaillard, 02/06/1934*, Ambassade de France au Caire, CADN/353 PO 2/230.

Le Directeur général de l'Office du tourisme de l'Etat Egyptien, L. Hakim, à Pierre de Witasse, Ministre de France au Caire, 30/12/1935, Ambassade de France au Caire, CADN/353 PO/2/230.

Le Gouvernement et le projet de la Société Cook, Le Journal d'Egypte, 13/11/1936. Ambassade de France au Caire, CADN/353 PO 2/230.

Le Programme d'action du nouveau bureau du tourisme, L'Attaché commercial près la Légation de France en Egypte, 2ème Note: Le programme d'action du nouveau bureau du tourisme. Les suggestions du Ministère du Commerce, Sept. 1935, Ambassade de France au Caire, CADN/353 PO/2/230.

Ministre de France en Egypte, au Ministre des Affaires étrangères Yvon Delbos, 26/11/1936, Ambassade de France au Caire, CADN/353 PO 2/230.

Note du Ministère des Affaires étrangères à la Légation de France, 07/09/1935, Ambassade de France au Caire, CADN/353 PO 2/230.

Une heureuse initiative: L'encouragement du tourisme en Egypte, Coupure de presse, Ambassade de France au Caire, CADN/353 PO 2/230.

Association French Lines, Le Havre

COMPAGNIE DES MESSAGERIES MARITIMES, *Rapport général de Service 1920, Chapitre Trafic, Section Passagers*, Rapports de l'Agence d'Alexandrie, French Lines/1997 002 4392.

COMPAGNIE DES MESSAGERIES MARITIMES, *Rapport général de Service 1921, Chapitre Trafic, Section Passagers*, Rapports de l'Agence d'Alexandrie, French Lines/1997 002 4392.

COMPAGNIE DES MESSAGERIES MARITIMES, *Rapport général de Service 1928, Chapitre Trafic, Section Passages*, Rapports de l'Agence d'Alexandrie, French Lines/1997 002 4392.

COMPAGNIE DES MESSAGERIES MARITIMES, *Rapport général de Service 1929, Chapitre Trafic, Section Passages*, Rapports de l'Agence d'Alexandrie, French Lines/1997 002 4392.

COMPAGNIE DES MESSAGERIES MARITIMES, *Rapport général de Service 1930, Chapitre Trafic, Section Passages*, Rapports de l'Agence d'Alexandrie, French Lines/1997 002 4392.

COMPAGNIE DES MESSAGERIES MARITIMES, *Rapport général de Service 1931, Chapitre Personnel*, Rapports de l'Agence d'Alexandrie, French Lines/1997 002 4392.

COMPAGNIE DES MESSAGERIES MARITIMES, *Rapport général de Service 1931, Chapitre Trafic, Section Passages*, Rapports de l'Agence d'Alexandrie, French Lines/1997 002 4392.

COMPAGNIE DES MESSAGERIES MARITIMES, *Rapport général de Service 1932, Chapitre Trafic, Section Passages*, Rapports de l'Agence d'Alexandrie, French Lines/1997 002 4392.

COMPAGNIE DES MESSAGERIES MARITIMES, *Rapport général de Service 1933, Chapitre Trafic, Section Passages*, Rapports de l'Agence d'Alexandrie, French Lines/1997 002 4392.

COMPAGNIE DES MESSAGERIES MARITIMES, *Rapport général de Service 1934, Chapitre Trafic, Section Passages*, Rapports de l'Agence d'Alexandrie, French Lines/1997 002 4393.

COMPAGNIE DES MESSAGERIES MARITIMES, *Rapport général de Service 1936, Chapitre Trafic, Section Passages*, Rapports de l'Agence d'Alexandrie, French Lines/1997 002 4393.

COMPAGNIE DES MESSAGERIES MARITIMES, *Rapport général de Service 1937, Chapitre Trafic, Section Passages*, Rapports de l'Agence d'Alexandrie, French Lines/1997 002 4393.

COMPAGNIE DES MESSAGERIES MARITIMES, *Rapport général de Service 1938, Chapitre Trafic, Section Passages*, Rapports de l'Agence d'Alexandrie, French Lines/1997 002 4393.

Jafet Library, American University of Beirut

Arrêté Ministériel No. 2 de 1928 relatif à l'institution d'un Conseil Consultatif auprès du Département du Commerce et de l'Industrie, 04/01/1928, Journal officiel du gouvernement égyptien, 12/01/1928, AUB.

Arrêté portant modification du tarif des drogmans et guides publics au Bandar d'Assouan, 17/08/1926, Journal officiel du gouvernement égyptien No. 91, 23/09/1926, p. 3, AUB.

Chambre des Députés, Compte rendu de la 18e séance publique, 11/01/1927, Journal officiel du gouvernement égyptien No. 16, 24/02/1927, Supplément, AUB.

Chambre des Députés, Compte rendu de la 19e séance publique, 12/01/1927, Journal officiel du gouvernement égyptien No. 17, 28/02/1927, Supplément, pp. 199–203, AUB.

Chambre des Députés, Compte rendu de la 27e séance publique, 08/08/1926, Journal officiel du gouvernement égyptien No. 5, 17/01/1927, Supplément, pp. 340 341, AUB.

Chambre des Députés, Compte rendu de la 29e séance publique, 10/08/1926, Journal officiel du gouvernement égyptien No. 15, 21/02/1927, Supplément, AUB.

Chambre des Députés, Compte rendu de la 40e séance publique, 28/03/1927, Journal officiel du gouvernement eyptien No. 41, 09/05/1927, Supplément, pp. 503–504, AUB.

Chambre des Députés, Compte rendu de la 52e séance publique, 03/05/1927, Journal officiel du gouvernement égyptien No. 54, 23/06/1927, Supplément, pp. 673–674, AUB.

Chambre des Députés, Compte rendu de la 52e séance publique, 21/06/1924, Journal officiel du gouvernement égyptien, No. 16, 09/02/1925, AUB.

Chambre des Députés, Compte rendu de la 67e séance publique, 30/05/1927, Journal officiel du gouvernement égyptien, No. 97, 17/11/1927, Supplément, AUB.

Chambre des Députés, Projet de loi portant fixation du Budget général 1924–1925, présenté dans la chambre le 07/06/1924, Journal officiel du gouvernement égyptien, 1924, Supplément, AUB.

Chambre des Députés, Troisième législature, Compte-rendu de la 1ère séance publique, 10/06/1926, Journal officiel du gouvernement égyptien No. 62, 26/06/1926, Supplément, AUB.

Chambre des Députés, Troisième législature, Compte-rendu de la 25e séance publique, 04/08/1926, Journal officiel du gouvernement égyptien No. 1, 03/01/1927, Supplément, AUB.

Commission Municipale d'Alexandrie, Arrêté prescrivant l'observation de certaines mesures sur les plages d'Alexandrie, 13/09/1927, Journal officiel du gouvernement égyptien No. 85, 10/10/1927, pp. 1–2, AUB.

Comptes rendus des séances de la Chambre des Députés, 18/11/1926 – 28/03/1927, Journal officiel du gouvernement égyptien, No. 55, 27/06/1927, Supplément, AUB.

Décision portant reconstitution du Conseil Economique, 15/02/1925, Journal officiel du gouvernement égyptien No. 21, 26/02/1925, pp. 1–2, AUB.

Décret portant constitution d'une Société Anonyme sous la dénomination de "Société de Transports, Expéditions et Assurances 'Pharos'", 13/09/1928, Journal officiel du gouvernement égyptien, No. 92, 22/10/1928, Supplément, pp. 1–2, AUB.

Décret portant création d'un Ministère du Commerce et de l'Industrie, 20/12/1934, Journal officiel du gouvernement égyptien No. 112, 24/12/1934, p. 4, AUB.

Décret portant nomination d'un Ministre pour le Commerce et l'Industrie, 18/06/1935, Journal officiel du gouvernement égyptien No. 54, 18/06/1935, p. 1, AUB.

Discours du Trône du Roi Fouad, Compte rendu de la séance inaugurale du parlement, 10/06/1926, Journal officiel du gouvernement égyptien, No. 10, 27/01/1930, AUB.

Discours du Trône du Roi Fouad, Compte rendu de la séance inaugurale du parlement, 17/11/1927, Journal officiel du gouvernement égyptien, No. 104, 08/12/1927, Supplément, pp. 2–4, AUB.

Discours du Trône du Roi Fouad, Séance inaugurale du parlement, 12/11/1924, Journal officiel du gouvernement égyptien No. 68, 09/07/1925, Supplément, AUB.

Gouvernorat d'Alexandrie, Arrêté fixant le nombre des autorisations à délivrer aux drogmans ou guides publics à Alexandrie, 18/01/1936, Journal officiel du gouvernement égyptien No. 13, 03/02/1936, p. 6, AUB.

Gouvernorat d'Alexandrie, Arrêté fixant le tarif des drogmans publics dans la circonscription du Gouvernorat d'Alexandrie, 19/08/1931, Journal officiel du gouvernement égyptien No. 90, 10/09/1931, p. 6, AUB.

Gouvernorat de Damiette, Arrêté prescrivant certaines mesures d'hygiène pour la sauvegarde des villégiateurs à Ras el Bar, 03/08/1933, Journal officiel du gouvernement égyptien No. 74, 17/08/1933, pp. 7–8, AUB.

Gouvernorat de Damiette, Arrêté prescrivant certaines mesures sanitaires pour la protection des villégiateurs à Ras el Bar, 22/06/1930, Journal officiel du gouvernement égyptien No. 64, 03/07/1930, p. 5, AUB.

Gouvernorat de Suez, Arrêté portant interdiction de la circulation des voitures publiques et automobiles dans la zone balnéaire à Port Tewfik, 08/10/1934, Journal officiel du gouvernement égyptien No. 90, 18/10/1934, p. 5, AUB.

Gouvernorat de Suez, Arrêté prescrivant l'observation de certaines mesures durant la saison balnéaire à Port-Tewfik, 11/06/1934, Journal officiel du gouvernement égyptien No. 54, 25/06/1934, pp. 1–2, AUB.

Gouvernorat du Caire, Arrêté fixant le tarif des drogmans publics dans la circonscription du Gouvernorat du Caire, 12/03/1931, Journal officiel du gouvernement égyptien No. 29, 23/03/1931, p. 6, AUB.

Gouvernorat du Caire, Arrêté interdisant le passage des charrettes, des voitures à bras, des chameaux, etc., sur le Pont de Kasr el Nil et le Pont des Anglais dans la ville du Caire, 05/09/1927, Journal officiel du gouvernement égyptien No. 81, 26/09/1927, p. 1, AUB.

Gouvernorat du Caire, Arrêté modifiant le règlement sur les professions autorisées aux Pyramides, 07/04/1928, Journal officiel du gouvernement égyptien No. 36, 23/04/1928, p. 9, AUB.

Gouvernorat du Caire, Arrêté portant limitation du nombre des autorisations de drogmans et guides publics dans la circonscription du Gouvernorat du Caire, 07/09/1932, Journal officiel du gouvernement égyptien No. 85, 26/09/1932, p. 6, AUB.

Gouvernorat du Caire, Arrêté portant modification du règlement sur les âniers à la ville du Caire, 07/04/1928, Journal officiel du gouvernement égyptien No. 36, 23/04/1928, p. 9, AUB.

Gouvernorat du Canal, Arrêté fixant le nombre d'autorisations et le tarif des drogmans et guides publics dans la ville de Port-Said, 13/07/1932, Journal officiel du gouvernement égyptien No. 69, 31/07/1933, p. 2, AUB.

Gouvernorat du Canal, Arrêté portant certaines interdictions à observer durant la saison balnéaire au bord de la mer Méditerranée à Port-Saïd, 20/09/1932, Journal officiel du gouvernement égyptien No. 87, 02/10/1932, p. 3, AUB.

Gouvernorat du Canal, Arrêté prescrivant certaines mesures dans les zones des bains, 25/11/1935, Journal officiel du gouvernement égyptien No. 109, 05/12/1935, p. 3, AUB.

Gouvernorat du Canal, Arrêté relatif à certaines interdictions durant la saison balnéaire au bord de la mer Méditerranée à Port Said, 01/071930, Journal officiel du gouvernement égyptien No. 68, 14/07/1930, p. 4, AUB.

Gouvernorat du Canal, Arrêté relatif à la propreté des rues à Port-Saïd, 04/02/1932, Journal officiel du gouvernement égyptien No. 16, 25/02/1932, p. 8, AUB.

Ministère de l'Intérieur, Arrêté instituant un Comité Administratif pour la Station Estivale de Ras el-Bar, 07/05/1928, Journal officiel du gouvernement égyptien No. 42, 14/05/1928, p. 1, AUB.

Ministère de l'Intérieur, Arrêté ministériel relatif à l'administration de la Station Estivale de Ras el-Bar, 12/03/1934, Journal officiel du gouvernement égyptien No. 23, 17/03/1934, p. 4, AUB.

Ministère de l'Intérieur, Arrêté modifiant la composition et les attributions du Comité Administratif de la Station Estivale de Ras el-Bar, 13/01/1931, Journal officiel du gouvernement égyptien No. 6, 19/01/1931, pp. 5–6, AUB.

Ministère de l'Intérieur, Arrêté portant institution d'un Comité Administratif pour la Station Estivale de Ras el-Bar, 10/03/1935, Journal officiel du gouvernement égyptien No. 24, 21/03/1935, p. 2, AUB.

Ministère de l'Intérieur, Arrêté réglementant la profession de drogman ou guide public, 10/07/1929, Journal officiel du gouvernement égyptien No. 64, 22/07/1929, pp. 3–4, AUB.

Ministère des Communications, Arrêté ministériel No. 12 de 1931 fixant les prix de voyage entre certaines gares des Chemins de fer de l'Etat par les billets dits billets de "Week-End", 22/10/1931, Journal officiel du gouvernement égyptien No. 104, 26/10/1931, p. 5, AUB.

Moudirieh d'Assiout, Arrêté fixant le nombre d'autorisations et le tarif des drogmans et guides publics dans la Moudirieh d'Assiout, 17/07/1933, Journal officiel du gouvernement égyptien No. 69, 31/07/1933, p. 3, AUB.

Moudirieh d'Assouan, Arrêté désignant les lieux de stationnement des ânes au Bandar d'Assouan, 01/09/1927, Journal officiel du gouvernement égyptien No. 78, 15/09/1927, p. 13, AUB.

Moudirieh d'Assouan, Arrêté fixant le nombre d'autorisations des drogmans et guides publics dans la Moudirieh d'Assouan, 25/07/1933, Journal officiel du gouvernement égyptien No. 73, 14/08/1933, p. 3, AUB.

Moudirieh d'Assouan, Arrêté fixant le tarif des drogmans publics dans la circonscription de la Moudirieh d'Assouan, 03/04/1932, Journal officiel du gouvernement égyptien, No. 33, 21/04/1932, p. 5, AUB.

Moudirieh de Fayoum, Arrêté fixant le nombre d'autorisations et le tarif des drogmans et guides publics dans la Moudirieh de Fayoum, 12/07/1932, Journal officiel du gouvernement égyptien No. 69, 31/07/1933, p. 2–3, AUB.

Moudirieh de Guirgueh, Arrêté du 26/02/1927, Journal officiel du gouvernement égyptien No. 21, 14/03/1927, p. 5, AUB.

Moudirieh de Guizeh, Arrêté fixant le nombre d'autorisations des drogmans et guides publics dans la Moudirieh de Guizeh, 21/08/1933, Journal officiel du gouvernement égyptien No. 78, 31/08/1933, p. 4, AUB.

Moudirieh de Guizeh, Arrêté fixant le tarif des drogmans publics dans la circonscription de la Moudirieh de Guizeh, 17/08/1931, Journal officiel du gouvernement égyptien No. 90, 10/09/1931, p. 6, AUB.

Moudirieh de Guizeh, Arrêté portant règlement sur les chameliers à Guizeh, 30/08/1936, Journal officiel du gouvernement égyptien No. 103, 21/09/1936, p. 4, AUB.

Moudirieh de Kéneh, Arrêté fixant le nombre d'autorisation et tarif des drogmans et guides publics dans la Moudirieh de Kéneh, 04/06/1934, Journal officiel du gouvernement égyptien No. 57, 05/07/1934, p. 7, AUB.

Moudirieh de Kéneh, Arrêté fixant le tarif des âniers à Louxor, Korna et Karnak, 03/04/1932, Journal officiel du gouvernement égyptien No. 32, 14/04/1932, pp. 2–3, AUB.

Moudirieh de Kéneh, Arrêté portant modification au tarif des drogmans et guides publics à Kaft, 12/06/1928, Journal officiel du gouvernement égyptien No. 53, 25/06/1928, pp. 7–8, AUB.

Moudirieh de Kéneh, Arrêté portant modification au tarif des drogmans et guides publics au Bandar de Louxor, 16/05/1928, Journal officiel du gouvernement égyptien No. 46, 28/05/1928, pp. 3–4, AUB.

Moudirieh de Minieh, Arrêté limitant le nombre de permis et fixant le tarif des drogmans et guides publics dans la Moudirieh de Minieh, 13/09/1933, Journal officiel du gouvernement égyptien No. 85, 25/09/1933, p. 4, AUB.

Municipalité d'Alexandrie, Arrêté prescrivant l'observation de certaines mesures sur les plages d'Alexandrie, 01/08/1929, Journal officiel du gouvernement égyptien No. 72, 19/08/1929, pp. 4–5, AUB.

Parlement. 5e séance ordinaire. Compte rendu de la 5e séance inaugurale, Discours du Trône, 17/11/1927, Journal officiel du gouvernement égyptien, No. 104, 08/12/1927, Supplément, AUB.

Rescrit Royal No. 10, portant nomination du Président du comité local pour l'organisation du Congrès International de Tourisme qui se réunira au Caire en 1933, 09/01/1933, Journal officiel du gouvernement égyptien No. 7, 16/01/1933, pp. 1–2, AUB.

Rescrit Royal No. 39, portant formation d'un comité local pour l'organisation du Congrès International de Tourisme qui se réunira au Caire en 1933, 05/07/1932, Journal officiel du gouvernement égyptien No. 59, 11/07/1932, pp. 1–2, AUB.

Rescrit Royal No. 45, portant nomination de membres au comité local pour l'organisation du Congrès International de Tourisme qui se réunira au Caire en 1933, 09/08/1932, Journal officiel du gouvernement égyptien No. 71, 15/08/1932, p. 1, AUB.

Sénat, Compte rendu de la 22e Séance publique, 23/04/1930, Journal officiel du gouvernement égyptien, No. 45, 15/5/1930, Supplément, AUB.

Sénat, Compte rendu de la 4e Séance publique, 21/01/1930, Journal officiel du gouvernement égyptien, No. 19, 24/02/1930, Supplément, AUB.

jd private collection

EGYPTIAN STATE TOURIST ADMINISTRATION and TOURIST DEVELOPMENT ASSOCIATION OF EGYPT, *Egypt: Tourist Guide*, Cairo, s.d. *[1938?]*, jd private collection.

EGYPTIAN STATE TOURIST DEPARTMENT and TOURIST DEVELOPMENT ASSOCIATION OF EGYPT, *Egypt: A Travel Quarterly, Autumn 1937*, Cairo: Al-Hilal Publishing House, 1937, jd private collection.

EGYPTIAN STATE TOURIST DEPARTMENT and TOURIST DEVELOPMENT ASSOCIATION OF EGYPT, *Egypt: A Travel Quarterly*, Spring 1939, jd private collection.

PERERA, GUY U., OTTO, A., and TOURIST DEVELOPMENT ASSOCIATION OF EGYPT, *Un voyage en Egypte et au Soudan*, Le Caire, s.d. [1930s], jd private collection.

TAYLOR, PHILIP, and TOURIST DEVELOPMENT ASSOCIATION OF EGYPT, *Egypt and the Sudan*, 1934 edition, jd private collection.

TOURIST DEVELOPMENT ASSOCIATION OF EGYPT, *Le Nil d'Alexandrie à Assouan: Guide illustré*, s.d., jd private collection.

National Archives, Kew

LAWRENCE, *Travel photographs: Voyage to Egypt*, 1937, NA/CO 1069/797.

National Library of Israel, Jerusalem

International Tourist Congress, The Palestine Post, 12/12/1932, p. 2. Historical Jewish Press, NLI.

Statutes: Tourist Development Association of Egypt (Association pour favoriser le tourisme en Égypte), founded on May 18th 1912, under the name of Egypt Promotion Association (Association egyptienne de propagande), Cairo: J. Parladi, 1926, NLI/2014 C 6464.

Tourist Traffic Decreasing: Adverse Effects of Government Measures Feared, The Palestine Post, 02/01/1934, p. 1. Historical Jewish Press, NLI.

National Library of Scotland, Edinburgh

STEELE-MAITLAND, MARY, *Diary of Visit to Egypt, Palestine and Turkey, 24/12/1931 – 29/01/1932*, NLS/ACC 5553.

Royal Commonwealth Society Archives, Cambridge University Library

WILLIAMS, GEORGE H., *Travel Diary: Egypt, 1925*, RCSA/RCMS 53/3/1.

School of Oriental and African Studies, London

SMITH, EDWIN W., *Travel Diaries. Diary. 1929. Vol. I.*, SOAS/MMS/17/02/05/02/09/02/02.

SMITH, EDWIN W., *Travel Diaries. Diary. 1929. Vol. II.*, SOAS/MMS/17/02/05/02/09/02/02.

The Times Digital Archive

Cairo Buildings Fired: Widespread Damage, The Times, 28/01/1952, p. 4, The Times Digital Archive.

Extremists' Role in Riots: British Loss Estimated at £5M., The Times, 29/01/1952, p. 4, The Times Digital Archive.

Other published sources

ELSTON, ROY, *The Traveller's Handbook for Egypt and the Sûdân: With Maps and Plans* (London: Simpkin Marshall, 1929).

KHOORI, ALEXANDER R., *Luxor: How to See It* (2nd edn., Alexandria: The Anglo-Egyptian Supply Association, 1925).

NORVAL, ARTHUR JOSEPH, *The Tourist Industry: A National and International Survey*, Dissertation (Rotterdam, 1936).

THOMAS COOK & SON, ELSTON, ROY, LUKE, HARRY CHARLES et al., *The Traveller's Handbook for Palestine and Syria* (2nd edn., London: Simpkin Marshall, 1929).

Secondary Literature

ʿABD AL-HAFIZ, MOHAMMED, 'Les hammams d'Alexandrie à l'époque de Muhammad ʿAli', in Marie-Françoise Boussac, Thibaud Fournet, and Bérangère Redon (eds.), *Le bain collectif en Égypte. Balaneia, Thermae, Hammāmāt* (Le Caire: Institut Français d'Archéologie Orientale, 2009), pp. 353–359.

ALLOULA, MALEK, *The Colonial Harem* (Minneapolis: University of Minnesota Press, 1986).

ANDERSON, BENEDICT R., *Imagined Communities: Reflections on the Origin and Spread of Nationalism* (2nd edn., London, New York: Verso, 2016).

ANDERSON, MARTIN, 'The Development of British Tourism in Egypt, 1815 to 1850', *Journal of Tourism History*, 4/3 (2012), pp. 259–279.

BADRAWI, MALAK, 'Financial Cerberus? The Egyptian Parliament, 1924–52', in Arthur Goldschmidt, Amy Johnson, and Barak Salmoni (eds.), *Re-Envisioning Egypt, 1919–1952* (Cairo: The American University in Cairo Press, 2005), pp. 94–122.

BADRAWI, MALAK, *Ismaʿil Sidqi, 1875–1950: Pragmatism and Vision in Twentieth Century Egypt* (Hoboken: Taylor and Francis, 2014).

BARON, BETH, *Egypt as a Woman: Nationalism, Gender, and Politics* (Berkeley: University of California Press, 2005).

BARON, BETH, 'The Port Said Orphan Scandal of 1933: Colonialism, Islamism, and the Egyptian Welfare State', in Heather J. Sharkey (ed.), *Cultural Conversions. Unexpected Consequences of Christian Missionary Encounters in the Middle East, Africa, and South Asia* (Syracuse: Syracuse University Press, 2013), pp. 121–138.

BARON, BETH, *The Orphan Scandal: Christian Missionaries and the Rise of the Muslim Brotherhood* (Stanford: Stanford University Press, 2014).

BARRELL, JOHN, 'Death on the Nile: Fantasy and the Literature of Tourism, 1840–1860', in Catherine Hall (ed.), *Cultures of Empire: A Reader. Colonizers in Britain and the Empire in the Nineteenth and Twentieth Centuries* (Manchester: Manchester University Press, 2000), pp. 187–206.

BASSIL, KARL, MAASRI, ZEINA, RAAD, WALID et al., *Mapping Sitting: On Portraiture and Photography* (2nd edn., Beirut: Mind the gap/ Arab Image Foundation, 2005).

BEHDAD, ALI, 'Orientalist Tourism', *Peuples méditerranéens*, 50 (1990), pp. 59–73.

BEHDAD, ALI, 'The Orientalist Photograph', in Ali Behdad and Luke Gartlan (eds.), *Photography's Orientalism. New Essays on Colonial Representation* (Los Angeles: Getty Publications, 2013), pp. 11–32.

BEHDAD, ALI, *Camera Orientalis: Reflections on Photography of the Middle East* (Chicago, London: University of Chicago Press, 2016).

BEHDAD, ALI, and GARTLAN, LUKE, 'Introduction', in Ali Behdad and Luke Gartlan (eds.), *Photography's Orientalism. New Essays on Colonial Representation* (Los Angeles: Getty Publications, 2013), pp. 1–10.

BEININ, JOEL, *Workers and Peasants in the Modern Middle East* (Cambridge, New York: Cambridge University Press, 2001).

BLOTTIÈRE, ALAIN, *Un voyage en Égypte: Au temps des derniers rois* (Paris: Flammarion, 2003).

Bohnsack, Ralf, 'Disziplinierte Zugänge zum undisziplinierten Bild: Die Dokumentarische Methode', in Thomas Abel and Martin Roman Deppner (eds.), *Undisziplinierte Bilder. Fotografie als dialogische Struktur* (Bielefeld: Transcript, 2013), pp. 61–103.

Bopp, Petra, 'Orientalismus im Bild: 1903: Rudolf Lehnerts erste Photoexkursion nach Tunesien und die Tradition reisender Orientmaler', in Alexander Honold and Klaus R. Scherpe (eds.), *Mit Deutschland um die Welt. Eine Kulturgeschichte des Fremden in der Kolonialzeit* (Stuttgart: J.B. Metzler, 2004), pp. 288–299.

Botman, Selma, 'The Liberal Age, 1923–1952', in M. W. Daly (ed.), *The Cambridge History of Egypt. Volume 2: Modern Egypt, from 1517 to the End of the Twentieth Century* (Cambridge, New York: Cambridge University Press, 1998), pp. 285–308.

Bough, Jill, *Donkey* (London: Reaktion Books, 2011).

Bredekamp, Horst, *Theorie des Bildakts* (Berlin: Suhrkamp, 2010).

Bull, Deborah, and Lorimer, Donald, *Up the Nile: A Photographic Excursion, Egypt 1839–1898* (New York: Clarkson N. Potter, 1979).

Cardon, Nathan, and Jackson, Simon, 'Everyday Empires: Descriptive or Analytical Category?', Past & Present Blog, 2017 <http://pastandpresent.org.uk/everyday-empires-descriptive-analytical-category/>, updated 17 Jul 2017, accessed 27 Feb 2019.

Çelik, Zeynep, and Eldem, Edhem (eds.), *Camera Ottomana: Photography and Modernity in the Ottoman Empire, 1840–1914* (Istanbul: Koç University Press, 2015).

Dawletschin-Linder, Camilla, 'Neue Wege in der Wirtschaftspolitik Ägyptens: Das Streben nach wirtschaftlicher Unabhängigkeit in der Zwischenkriegszeit', in Alexander Schölch and Helmut Mejcher (eds.), *Die ägyptische Gesellschaft im 20. Jahrhundert* (Hamburg, 1992), pp. 71–93.

Debbas, Fouad, *Des photographes à Beyrouth, 1840–1918* (Paris: Marval, 2001).

Degroise, Marie-Hélène, 'Photographes d'Asie (1840–1944)' <http://photographesenoutremerasie.blogspot.de/>, accessed 10 Sep 2021.

Deringil, Selim, '"They Live in a State of Nomadism and Savagery": The Late Ottoman Empire and the Post-Colonial Debate', *Comparative Studies in Society and History*, 45/2 (2003), pp. 311–342.

DeRoo, Rebecca J., 'Colonial Collecting: French Women and Algerian Cartes Postales', in Eleanor M. Hight and Gary D. Sampson (eds.), *Colonialist Photography. Imag(in)ing Race and Place* (London: Routledge, 2002), pp. 159–171.

Di-Capua, Yoav, 'Common Skies Divided Horizons: Aviation, Class and Modernity in Early Twentieth Century Egypt', *Journal of Social History*, 41/4 (2008), pp. 917–942.

Erdoğdu, Ayshe, 'Picturing Alterity: Representational Strategies in Victorian Type Photographs of Ottoman Men', in Eleanor M. Hight and Gary D. Sampson (eds.), *Colonialist Photography. Imag(in)ing Race and Place* (London: Routledge, 2002), pp. 107–125.

Frith, Francis, van Haaften, Julia, and White, Jon Manchip, *Egypt and the Holy Land in Historic Photographs: 77 Views by Francis Frith* (New York: Dover Publications, 1980).

Fulda, Daniel, 'Topik', in Stefan Jordan (ed.), *Lexikon Geschichtswissenschaft. Hundert Grundbegriffe* (Stuttgart: Reclam, 2007), pp. 285–287.

Furlough, Ellen, 'Making Mass Vacations: Tourism and Consumer Culture in France, 1930s to 1970s', *Comparative Studies in Society and History*, 40/2 (1998), pp. 247–286.

Gavin, Carney E. S., *The Image of the East: Nineteenth-Century Near Eastern Photographs by Bonfils: From the Collections of the Harvard Semitic Museum* (Chicago: University of Chicago Press, 1982).

Gelvin, James L., *The Modern Middle East: A History* (4th edn., New York, Oxford: Oxford University Press, 2016).

Goldschmidt Jr, Arthur, *Modern Egypt: The Formation of a Nation-State* (2nd edn., Boulder: Westview Press, 2004).

GRAY, MATTHEW, 'Economic Reform, Privatization and Tourism in Egypt', *Middle Eastern Studies*, 34/2 (1998), pp. 91–112.

GREGORY, ALEXIS, *The Golden Age of Travel, 1880–1939* (London: Cassell, 1998).

GREGORY, DEREK, 'Imaginative Geographies', *Progress in Human Geography*, 19/4 (1995), pp. 447–485.

GREGORY, DEREK, 'Scripting Egypt: Orientalism and the Cultures of Travel', in James S. Duncan and Derek Gregory (eds.), *Writes of Passage. Reading Travel Writing* (London, New York: Routledge, 1999), pp. 114–150.

GREGORY, DEREK, 'Emperors of the Gaze: Photographic Practices and Productions of Space in Egypt, 1839–1914', in Joan M. Schwartz and James R. Ryan (eds.), *Picturing Place. Photography and the Geographical Imagination* (London: I.B. Tauris, 2009), pp. 195–225.

GROEBNER, VALENTIN, *Retroland: Geschichtstourismus und die Sehnsucht nach dem Authentischen* (Frankfurt a. M.: S. Fischer, 2018).

GUNNING, TOM, '"The Whole World Within Reach": Travel Images Without Borders', in Jeffrey Ruoff (ed.), *Virtual Voyages. Cinema and Travel* (Durham: Duke University Press, 2006), pp. 25–41.

HAAG, MICHAEL, *Vintage Alexandria: Photographs of the City, 1860–1960* (Cairo, New York: American University in Cairo Press, 2008).

HALDRUP, MICHAEL, and LARSEN, JONAS, *Tourism, Performance and the Everyday: Consuming the Orient* (London, New York: Routledge, 2010).

HANSSEN, JENS, and WEISS, MAX, 'Introduction: Language, Mind, Freedom and Time: The Modern Arab Intellectual Tradition in Four Words', in Jens Hanssen and Max Weiss (eds.), *Arabic Thought beyond the Liberal Age. Towards an Intellectual History of the Nahda* (Cambridge: Cambridge University Press, 2017), pp. 1–37.

HAZBUN, WALEED, 'The East as an Exhibit: Thomas Cook & Son and the Origins of the International Tourism Industry in Egypt', in Philip Scranton and Janet F. Davidson (eds.), *The Business of Tourism. Place, Faith, and History* (Philadelphia: University of Pennsylvania Press, 2007), pp. 3–33.

HAZBUN, WALEED, *Beaches, Ruins, Resorts: The Politics of Tourism in the Arab World* (Minneapolis: University of Minnesota Press, 2008).

HUBER, VALESKA, *Channelling Mobilities: Migration and Globalisation in the Suez Canal Region and Beyond, 1869–1914* (Cambridge, New York: Cambridge University Press, 2013).

HULME, PETER, and YOUNGS, TIM, 'Introduction', in Peter Hulme and Tim Youngs (eds.), *The Cambridge Companion to Travel Writing* (Cambridge, New York: Cambridge University Press, 2002), pp. 1–13.

HUMPHREYS, ANDREW, *Grand Hotels of Egypt in the Golden Age of Travel* (Cairo: American University in Cairo Press, 2011).

HUMPHREYS, ANDREW, *On the Nile: In the Golden Age of Travel* (Cairo: The American University in Cairo Press, 2015).

HUNTER, F. ROBERT, 'The Thomas Cook Archive for the Study of Tourism in North Africa and the Middle East', *Middle East Studies Association Bulletin*, 36/2 (2002), pp. 157–164.

HUNTER, F. ROBERT, 'Tourism and Empire: The Thomas Cook & Son Enterprise on the Nile, 1868–1914', *Middle Eastern Studies*, 40/5 (2004), pp. 28–54.

ILBERT, ROBERT, *Alexandrie, 1830–1930: Histoire d'une communauté citadine*, 2 vols. (Cairo: Institut Français d'Archéologie Orientale, 1996).

IRWIN, ROBERT, *Camel* (London: Reaktion Books, 2010).

JACOB, WILSON CHACKO, *Working Out Egypt: Effendi Masculinity and Subject Formation in Colonial Modernity, 1870–1940* (Durham: Duke University Press, 2011).

KHOLOUSSY, HANAN, *For Better, For Worse: The Marriage Crisis that Made Modern Egypt* (Stanford: Stanford University Press, 2010).

KORTE, BARBARA, *Der englische Reisebericht: Von der Pilgerfahrt bis zur Postmoderne* (Darmstadt: WBG, 1996).

LAPIDUS, IRA M., *A History of Islamic Societies* (3rd edn., Cambridge, New York: Cambridge University Press, 2014).

LÖFGREN, ORVAR, *On Holiday: A History of Vacationing* (Berkeley: University of California Press, 1999).

LÖFGREN, ORVAR, 'Know Your Country: A Comparative Perspective on Tourism and Nation Building in Sweden', in Shelley Baranowski and Ellen Furlough (eds.), *Being Elsewhere. Tourism, Consumer Culture, and Identity in Modern Europe and North America* (Ann Arbor: University of Michigan Press, 2001), pp. 137–154.

MACCANNELL, DEAN, 'Staged Authenticity: Arrangements of Social Space in Tourist Settings', *American Journal of Sociology*, 79 (1973), pp. 589–603.

MACCANNELL, DEAN, *The Tourist: A New Theory of the Leisure Class* (Berkeley: University of California Press, 1999).

MACKENZIE, JOHN M., 'Empires of Travel: British Guide Books and Cultural Imperialism in the 19th and 20th Centuries', in John K. Walton (ed.), *Histories of Tourism. Representation, Identity, and Conflict* (Clevedon, Buffalo: Channel View Publications, 2005), pp. 19–38.

MAIRS, RACHEL, and MURATOV, MAYA, *Archaeologists, Tourists, Interpreters: Exploring Egypt and the Near East in the Late Nineteenth and Early Twentieth Centuries* (London: Bloomsbury, 2015).

MAKDISI, USSAMA, 'Ottoman Orientalism', *The American Historical Review*, 107/3 (2002), pp. 768–796.

MEADE, MARTIN, FITCHETT, JOSEPH, and LAWRENCE, ANTHONY, *Grand Oriental Hotels* (New York: Vendome, 1987).

MICKLEWRIGHT, NANCY, *A Victorian Traveler in the Middle East: The Photography and Travel Writing of Annie Lady Brassey* (Aldershot: Ashgate, 2003).

MICKLEWRIGHT, NANCY, 'Orientalism and Photography', in Zeynep İnankur, Reina Lewis, and Mary Roberts (eds.), *The Poetics and Politics of Place. Ottoman Istanbul and British Orientalism* (Istanbul, Seattle: Pera Müzesİ; University of Washington Press, 2011), pp. 98 110.

MITCHELL, TIMOTHY, *Colonising Egypt* (Berkeley: University of California Press, 1988).

MOORS, ANNELIES, 'Presenting People: The Politics of Picture Postcards of Palestine/Israel', in David Prochaska and Jordana Mendelson (eds.), *Postcards. Ephemeral Histories of Modernity* (University Park: Pennsylvania State University Press, 2010), pp. 93–105.

NASSAR, ISSAM, 'On Photographs and Returning the Gaze', *Jerusalem Quarterly*, 31 (2007), pp. 3–5.

OSTERHAMMEL, JÜRGEN, *Die Verwandlung der Welt: Eine Geschichte des 19. Jahrhunderts* (München: C.H. Beck, 2009).

OSTERHAMMEL, JÜRGEN, 'Kulturelle Grenzen in der Expansion Europas', in Jürgen Osterhammel (ed.), *Geschichtswissenschaft jenseits des Nationalstaats. Studien zu Beziehungsgeschichte und Zivilisationsvergleich* (2nd ed., Göttingen: Vandenhoeck & Ruprecht, 2011), pp. 203–239.

PARKINSON, BRIAN R., 'Tutankhamen on Trial: Egyptian Nationalism and the Court Case for the Pharaoh's Artifacts', *Journal of the American Research Center in Egypt*, 44 (2008), pp. 1–8.

PEDERSEN, SUSAN, *The Guardians: The League of Nations and the Crisis of Empire* (Oxford, New York: Oxford University Press, 2015).

PELEGGI, MAURIZIO, 'The Social and Material Life of Colonial Hotels: Comfort Zones as Contact Zones in British Colombo and Singapore, ca. 1870–1930', *Journal of Social History*, 46/1 (2012), pp. 124–153.

PEREZ, NISSAN, *Focus East: Early Photography in the Near East (1839–1885)* (New York: Abrams, 1988).

PINNEY, CHRISTOPHER, 'What's Photography Got to Do with It?', in Ali Behdad and Luke Gartlan (eds.), *Photography's Orientalism. New Essays on Colonial Representation* (Los Angeles: Getty Publications, 2013), pp. 33–52.

POWELL, EVE TROUTT, *A Different Shade of Colonialism: Egypt, Great Britain, and the Mastery of the Sudan* (Berkeley: University of California Press, 2003).

RAUCH, ANDRÉ, 'Du Joanne au Routard: Le style des guides touristiques', in Gilles Chabaud, Évelyne Cohen, Natacha Coquery et al. (eds.), *Les guides imprimés du XVIe au XXe siècle. Villes, paysages, voyages* (Paris: Belin, 2000), pp. 95–112.

REID, DONALD M., 'Indigenous Egyptology: The Decolonization of a Profession?', *Journal of the American Oriental Society*, 105/2 (1985), pp. 233–246.

REID, DONALD M., 'Nationalizing the Pharaonic Past: Egyptology, Imperialism, and Egyptian Nationalism, 1922–1952', in James P. Jankowski and I. Gershoni (eds.), *Rethinking Nationalism in the Arab Middle East* (New York: Columbia University Press, 1997), pp. 127–149.

REID, DONALD M., *Contesting Antiquity in Egypt: Archaeologies, Museums, and the Struggle for Identities from World War I to Nasser* (Cairo, New York: The American University in Cairo Press, 2015).

REYNOLDS, NANCY Y., *A City Consumed: Urban Commerce, the Cairo Fire, and the Politics of Decolonization in Egypt* (Stanford: Stanford University Press, 2012).

ROBERTS, MARY, 'The Limits of Circumscription', in Ali Behdad and Luke Gartlan (eds.), *Photography's Orientalism. New Essays on Colonial Representation* (Los Angeles: Getty Publications, 2013), pp. 53–74.

ROSENBERG, EMILY S., 'Transnationale Strömungen in einer Welt, die zusammenrückt', in Emily S. Rosenberg (ed.), *Weltmärkte und Weltkriege. 1870–1945*, Akira Iriye and Jürgen Osterhammel (Geschichte der Welt, München: C.H. Beck, 2012), pp. 815–998.

RYZOVA, LUCIE, *The Age of the Efendiyya: Passages to Modernity in National-Colonial Egypt* (Oxford, New York: Oxford University Press, 2014).

SAID, EDWARD WILLIAM, *Orientalism* (London: Penguin Books, 2003 [1978]).

SCHÖLCH, ALEXANDER, *Ägypten den Ägyptern!: Die politische und gesellschaftliche Krise der Jahre 1878–1882 in Ägypten* (Zürich: Atlantis, 1972).

SCHWARTZ, JOAN M., 'The Geography Lesson: Photographs and the Construction of Imaginative Geographies', *Journal of Historical Geography*, 22/1 (1996), pp. 16–43.

SHEEHI, STEPHEN, *The Arab Imago: A Social History of Portrait Photography, 1860–1910* (Princeton: Princeton University Press, 2016).

STEWART HOWE, KATHLEEN, and WILSON, MICHAEL G., *Excursions Along the Nile: The Photographic Discovery of Ancient Egypt* (Santa Barbara: Santa Barbara Museum of Art, 1993).

URRY, JOHN, and LARSEN, JONAS, *The Tourist Gaze 3.0* (3rd edn., Los Angeles, London: Sage, 2011).

VOLAIT, MERCEDES, *Architectes et architectures de l'Egypte moderne (1830–1950): Genèse et essor d'une expertise locale* (Paris: Maisonneuve et Larose, 2005).

WALTON, JOHN K., 'British Tourism Between Industrialization and Globalization: An Overview', in Hartmut Berghoff, Barbara Korte, and Ralf Schneider (eds.), *The Making of Modern Tourism. The Cultural History of the British Experience, 1600–2000* (New York: Palgrave, 2002), pp. 109–131.

WATKIN, DAVID, *Grand Hotel: The Golden Age of Palace Hotels. An Architectural and Social History* (London: Dent, 1984).

WEBER, ANNE-GAËLLE, 'Le Genre romanesque du récit de voyage scientifique au xixe siècle', *Sociétés & Représentations*, 21/1 (2006), pp. 59–77.

WOOD, MICHAEL, 'The Use of the Pharaonic Past in Modern Egyptian Nationalism', *Journal of the American Research Center in Egypt*, 35 (1998), pp. 179–196.

WOODWARD, MICHELLE L., 'Between Orientalist Clichés and Images of Modernization', *History of Photography*, 27/4 (2003), pp. 363–374.

World Biographical Information System, 's.v. Perera, Guy U.' <https://wbis.degruyter.com/biographic-document/X170-296-5; https://wbis.degruyter.com/biographic-document/R63546>, accessed 26 Apr 2020.

Yehia, Enas Fares, 'Promotion of Tourism in Egypt during the Reign of King Fuad I', *Journal of Tourism History*, 13/3 (2021), pp. 1–15.

Zananiri, Sary, 'From Still to Moving Image: The Shifting Representation of Jerusalem and Palestinians in the Western Biblical Imaginary', *Jerusalem Quarterly*, 67 (2016), pp. 64–81.

Zuelow, Eric G. E., *A History of Modern Tourism* (London, New York: Palgrave Macmillan, 2016).

CHAPTER 3

New tourists, new attractions, 'New Palestine'

Implementing a new tourist space in Mandate Palestine

Abstract

The competition between Yishuvi and Palestinian actors shaped tourism in the British Mandate of Palestine. Among European Christian tourists, the idea of visiting the 'Holy Land' faded into the background when, from the late 1920s, conflict narratives defined their perceptions of the Mandate. Both Arab Palestinian and Yishuvi actors aimed to profit from this shift to generate support for their state-building projects. Yet the mandate treaty put them in highly unequal positions, which translated into different strategies. The Palestinian middle classes, largely excluded from political participation, were not able to fashion a national space. Instead, the *Supreme Muslim Council* established Jerusalem as a symbol of Arab Palestine, whereas Zionist tourism development strategies merged with their territorial ambitions, thereby defining the future nation-state.

Keywords: tourism; Palestinian history – 1917-1948; colonialism; Zionism; British imperialism; politics and culture

The PALESTINE TOUR was a very wonderful and successful holiday. I feel I got value, and more than value, for every penny I spent. The whole trip was delightful. We had nearly 500 miles of motoring in Palestine and we had the services of a thoroughly competent guide and a chauffeur second to none. Your representative gave of his best in service, attention and consideration for each one of the party. I could wish for nothing better than to repeat the holiday. – Miss M.B. (Bamber Bridge)[1]

In 1934, the British *Workers' Travel Association* (WTA) began offering tours to Palestine. For the potential travellers of rather modest means targeted by the WTA, the tour to Palestine was an exceptional event. Other tours announced in the magazine of the association, *The Travel Log*, advertised excursions around Great Britain or to European destinations such as Belgium or the Black Forest.[2] Aware of the limited financial means of its members, the WTA stated that the "Palestine tour [...] is the sort of holiday that one plans to have once in a lifetime." Despite the financial limitations of the intended audience, the tour proved so popular that it was quickly sold out and the WTA organised additional tours to the ones initially scheduled in 1934 – now reminding their members to book the upcoming tours as early as possible.[3] Moreover, the association extended the programme and offered new tours to "Palestine & Egypt" as well as a "study tour" that comprised excursions to Damascus and Baalbek.[4] Compared to *Cook's* tours, the prices offered by the WTA were much lower and a 25-day voyage to Palestine cost 39 guineas (c. £41) in 1934.[5]

The review of Miss M. B. in the WTA magazine indicated how travels to Palestine changed both in quantity and quality after World War I. The growth of "'cruise' and cheap tour traffic", as described by the British commercial agent in Haifa, allowed more travellers to visit the region.[6] A growing number of travel agencies and tour operators targeted a wider range of visitors, including solo female travellers like Miss M. B., an unmarried woman identifying herself as member of the working-class, thereby differing from *Cook's* bourgeois customers.[7]

The practices during and around these voyages, however, resembled the bourgeois model. Although the WTA stressed that workers on holidays preferred encounters with the local population to the admiration of monuments, the photograph competition *Holidays with a camera* conducted by the WTA suggests that these tourists were 'kodaking' the same impressions as bourgeois tourists did. Sightseeing remained an essential part of the tour.[8] In addition, as in *Cook's* travel groups, sociability among travellers played an important role. Reunion meetings of travel groups were popular among members of bourgeois associations such as the *Automobile Club* or the *Ladies' Alpine Club*, and they inspired the first gathering organised by the WTA for the participants of the Palestine tour. It took place in London in 1934, including speeches and discussions about the country.[9]

Yet the expansion of organised tourism around the Eastern Mediterranean marks a qualitative shift, insofar as it superseded religious pilgrimage as the dominant form of travel to Palestine. Miss B. classified her visit to Palestine as a "holiday" rather than a pilgrimage, and she did not mention any spiritual or religious intentions behind her journey, commenting instead on the comfort of the tour. Certainly, the distinction between pilgrims and travellers can hardly be drawn precisely. Individual beliefs notwithstanding, the places European travellers considered noteworthy in Palestine generally derived meaning from their significance in the Bible. Tourists mentioned places if they sounded familiar from stories of the Old Testament; they visited the Church of the Holy Sepulchre and the Garden of Gethsemane, the Tomb of the Virgin and the Sea of Galilee, as well as the Church of the Nativity in Bethlehem and Nazareth. The sites attributed to the life of Jesus marked major attractions and defined the tourist routes, even if the expectations of religiously motivated travellers to Palestine were often disappointed when they were confronted with local religious customs as well as architecture they experienced as unfamiliar and strange.

Even for tourists who did not perceive themselves as Christian believers, the reference to the Old and New Testament was a major point of reference. In fact, in the mandate administration report for the year 1939, which listed tourists and pilgrims separately, only 169 visitors out of 12,300 identified as pilgrims.[10] The emergence of organised tourism for reasons other than pilgrimage was the precondition for political uses of tourism. Whereas pilgrims assumed a bond between themselves and their destination, tourists do not necessarily assume that the monuments and places they visit are culturally meaningful to themselves. Tourism thus allowed actors to establish communication with travellers who did not necessarily identify with the convictions of local communities.

This chapter argues that actors in Palestine who were involved in organising tourism attempted to reformulate the imaginings tied to the 'Holy Land'. Yet in Palestine, as compared to Egypt, Arab Palestinian actors on the one hand and Zionist actors on the other, did not only disagree on which sites and places tourists should visit, but they used tourism in different ways in their attempts to define the future state. Zionist organisations in particular made efforts to reframe Palestine as a Jewish country of European modernity and presented the country as a productive unit. Arab Palestinian actors conversely highlighted cultural continuities connecting Palestinians with the land, focusing thus on Palestinian history and directing visitors' attention to the administrative and cultural centres of Palestine. As both sides aimed to disseminate their respective interpretations among visitors, tourists served as a resource in the political dispute over the land of Palestine.

In the initial section, analyses of – mainly British – travel diaries indicate the complexity of tourism as a political asset. Although visitors were aware of contemporary disputes about Palestine and searched for new narratives to interpret it, they did not simply adopt the explanations Zionist or Arab Palestinian guides presented. The diaries reveal that intentions to use tourism as an asset in political disputes did not easily translate into practice and that tourists distinguished between tour guides' expertise on Palestinian history and their remarks and explanations of the contemporary situation, which tourists often judged as partial and not trustworthy.

The second section introduces the major actors and their interests in promoting tourism. The British mandate officials perceived tourism development as an argument in favour of the successful administration of the Mandate. Yet, torn between competing interests, British policies were often inconsistent and their support for the sector remained limited. By contrast, Zionist and Arab Palestinian actors grasped the opportunities tourism provided much more decidedly. They aimed to win over an international audience, arguing in the Zionist case in favour of a 'New Palestine', and in the Palestinian case highlighting the legitimacy of the Arab Palestinian presence. To place their message more effectively, Zionist advocates of tourism development intended to establish a separate, exclusively Jewish, tourism sector that would allow them to accompany tourists from their arrival in Palestine until their departure.

Whether and how Arab Palestinian and Zionist actors used tourism as an instrument in defining a spatio-political order will be discussed in the four case studies of the final section. The *Supreme Muslim Council*, as the major Palestinian institution fostering tourism, focused on the city of Jerusalem. They hoped that the sanctuaries of Palestine's cultural and religious capital would attract a global audience and testify to the existence of an Arab culture and civilisation deeply rooted in the land of Palestine. Moreover, they intended to draw attention to Palestinian demands by defining and controlling access to the holy places. Zionist organisations and individual Yishuvi entrepreneurs pursued a more territorial approach in advertising tourism to Palestine and they adapted their strategies to the shifting international context. The fact that Zionist entrepreneurs aimed to obtain the concession for the exploitation of the hot springs of Tiberias already in the 1920s, shows the importance Zionist actors attributed to the tourist industry. By mobilising networks and capital as well as by presenting narratives likely to impress the Council of the League of Nations, they gained the support of the British administration for their vision of an international tourist resort on the Sea of Galilee. A more radical understanding of tourism can be observed in the case of the Jezreel Valley. Exhibited as a showpiece of Zionist modernity, tourism development policies not only aimed at redefining narratives of the mandate territory, but also at the mobilisation of settlers by means of tourism. The example demonstrates how tourism contributed to the making of a territory. Finally, the case of seaside tourism at the Dead Sea reveals the growing spatial segregation that occurred in the sector of tourism. The *Kallia* resort at the Dead Sea had initially been conceived of as a symbol of fruitful Arab-Yishuvi cooperation, but the increasing role of domestic tourism led the managers to abandon this vision. As in Egypt, domestic tourism allowed for the inclusion of more popular segments of the Yishuvi society, while at the same time such excursions enhanced the productivity of Palestine's geographical peripheries and tied them to the emerging nation-state. In Palestine, however, this integration implied the adoption of Zionist visions, and the initial vision of a joint leisure society was abandoned.

With respect to terminology, I use the following terms to describe the major groups advertising tourism to Mandate Palestine: 'Yishuv', or 'Yishuvi' (adj.), refers to the Jewish immigrant community in the Mandate.[11] 'Zionist' describes a political ambition of a part of the Yishuv: those actors promoted the immigration of Jewish individuals in order to establish a Jewish state in the Mandate.[12] 'Palestinian', or 'Arab Palestinian' is applied to the Arabic-speaking population of Palestine, regardless of their religious beliefs. The many non-Arab members of religious communities living in Palestine, and notably in Jerusalem, often offered accommodation facilities to visitors;[13] yet because they did not pursue active, politically motivated strategies of tourism development, they are not considered as separate groups in this chapter.

Relegating the 'Holy Land' to the past: New mediators for the present

The 1920s was a decade of transition regarding the imaginings European travellers associated with Palestine. Two major perceptions shaped the views of European Christian visitors. First, the idea of Palestine as the 'Holy Land' continued to define the approach of most tourists. Its main attractions were biblical sites familiar to travellers with a Christian upbringing. Second, British visitors in particular discussed the conflicts in Palestine that had been triggered by the Balfour Declaration and the establishment of the mandate. Particularly towards the end of the 1920s, when European newspapers covered the violent incidents of the so-called 'Wailing Wall Riots', a growing number of visitors commented on the conflict, interpreting their visit in a context of contemporary politics.[14] In this section I will trace the transition from the 'Holy Land' (and 'empty land') narrative towards a narrative of contemporary conflict. This reformulation of narratives not only implied a new framing of Palestine as a contested space, but also enabled new groups of actors to provide tourists with their interpretations of the current situation.

The idea of the 'Holy Land' in the minds of European and American visitors since the nineteenth century has been studied thoroughly. In research on travel and travellers to Palestine, most analyses and contributions focused on Western perceptions of Palestine and the creation of a Palestinian Other. These works have examined Palestine as a place of pilgrimage and of a religiously defined tourism, analysing how ideas of visiting the 'Holy Land' shaped views of Palestinian lived realities. A major topic these authors identified was the assumed dichotomy between a 'traditional' Palestinian society and an imported modernity that characterised Palestine from the point of view of observers throughout the late nineteenth and twentieth centuries. This dichotomy was not specific to Palestine, but it had two particular implications for tourists to the alleged 'Holy Land'.[15] First, as in the case of Egypt, the attribution of progress and modernity to European agency implied a European responsibility to raise local society up to Western standards. In the Palestinian context, this responsibility was transferred to Jewish settlers conceived of as "middlemen of empire", who would develop and modernise the country to the benefit of the Arab population.[16] This special position the British mandatory attributed to the Yishuv provided it with the opportunities to create, as Joel Migdal has argued, a "strong state" in a colonial context.[17] For many travellers and tourists, this argument generally legitimised the Jewish settler movement.[18] Second, this same discovery of an 'asynchronous country' led travellers to transform the travel experience to an inner one.

The 'Holy Land' as an inner experience

It has been argued that the transfiguration of the travel experience to an inner one was a strategy for coping with disillusions among visitors to Palestine. The existence of a contemporary Palestinian society in profound transformation dashed previous hopes of visiting a country where the inhabitants had preserved biblical ways of life. The historian Elliott Horowitz argued that such travellers interpreted the confrontation with modernity as a test of their faith. If their inner piety was strong enough, they reasoned, it would compensate for the ordinary appearance of the sights they had come to see.[19] The insistence on the mental strength of the pious visitor corresponded to Simon Coleman's observation that the 'emptiness' of visual representations of the 'Holy Land' allowed the spectator to mentally populate the landscape with his imagination, inserting biblical figures, but also himself, in the landscape.[20]

Both monumental sights and landscapes were read in biblical terms. Scientific approaches inspired by geography, archaeology, linguistics and history made the landscapes readable and justified the attribution of significance to the sites. In contrast to journeys across neighbouring countries, tourists to Palestine dedicated a significant amount of time to the contemplation of landscapes. In their diaries, they described them in great detail, making sense of the views by referring to biblical stories and events. One particularly diligent observer was Edwin W. Smith, a former Protestant missionary in South Africa, who travelled to the 'Holy Land' in late 1929. With religious commitment and scientific zeal, he sketched maps of the places he visited, attempting to identify precisely the biblical predecessors of contemporary towns and villages. References to scholarly disputes demonstrate that he meticulously verified the information he noted down. The quasi-scientific and very detailed mapping of the 'Holy Land' was a typically Protestant approach to the country, a phenomenon Kathleen Stewart-Howe termed "geopiety". The physical presence of travellers at biblical sites allegedly allowed them to access the biblical revelation and to spiritually experience the site.[21]

Even for less fervent Christians than the former missionary Edwin W. Smith, the main attractions Palestine offered were the places associated with the Old or New Testament. Helena Harrison, the sister of the urban planner Austen St. Barbe Harrison, travelled to Jerusalem to visit her brother. Although she was not a particularly pious Christian, she assumed that this lack of background knowledge would render it more difficult for her to understand the country.[22] Thus, both pious and less pious tourists continued to reproduce tropes and techniques of approaching Palestine in terms of the Bible. From their points of view, Palestine remained a landscape formation impregnated with the presence of the divine, imagined as an 'empty land', untouched by the course of time and therefore promising access to the original sites of the biblical revelation.

Yet various encounters in the tourism sphere prevented the conclusion of such journeys in mere European soliloquies. The previous chapter showed that, despite dominant perceptions of an Egyptian Other, tourists were confronted with the self-conceptions of the local population. Such encounters challenged visions of a passive and primitive population, and the outcome of the negotiation process was not predefined.

Setting the agenda

Economic historian Nikolas Glover argued that in contemporary heritage tourism, the connection between certain sites or places and the respective narratives resulted from interactions between locals, tour guides and the tourists themselves. Expectations, ideas and prefabricated images of tourists merged with local interpretations of sites. He highlighted that the outcome of a process depended to a large degree on the specific context and the concrete actors involved – on their personalities as well as on the social and cultural background of the tourists and their guides.[23]

While Glover's work was largely based on oral interviews with actors in tourism of the late twentieth century, reconstructing the collaborative production of narratives in the past is challenging. This section explores the collaboration between travellers and local guides based on travel diaries and travelogues. A critical reading of these sources allows us to assess the influence of predefined knowledge, tour guides and other actors in shaping the tourists' experiences. Two properties render travel diaries particularly interesting in this context. First, the personal character of such accounts reveals assumptions and expectations of travellers. Second, since tourists occasionally summarised conversations with their guides in the diaries, they are among the few documents in which tour guides come to life.

To assess whether tour guides in Palestine had an agenda when showing tourists around, the diaries have to be read against the grain. Conceiving of themselves as mediators of local culture, tour guides endeavoured to fulfil the wishes of their clients, while at the same time pursuing an agenda in selecting the sites and places that they showed to tourists.[24] Some guides paid particular attention to introducing travellers to local Palestinian culture and Eastern Christianity, the beliefs and traditions of which they defended against British Protestant assumptions of Oriental superstition and 'decay'. In the case of George H. Williams and his guide, contradictory ideas of what the visitor to Palestine should see provoked tensions. George H. Williams repeatedly expressed insecurity about details of biblical stories. His interest in the ancient history of the country, its pre-Christian monuments and contemporary customs, outweighed his interest in religious sites and biblical legends. We encountered Williams in Egypt as a person fascinated by the allegedly authentic

Oriental culture, and a similar interest in local history, culture and politics guided Williams during his visit to Palestine. He was among the few tourists who visited Nablus – of minor importance from a biblical perspective, yet a flourishing town in Ottoman times. However, despite Williams' preferences, his trajectory came to resemble the itinerary of pious tourists.

It seems that his tour guide, Solomon, defined the itinerary and selected mainly biblical sites as major attractions. Solomon, a Christian ("I forgot which sort, there are many churches out here"), accompanied Williams for one Egyptian pound a day during his entire journey across Palestine, Syria and Lebanon.[25] In Williams' opinion, Solomon's assistance had been worth spending the pound per day. Williams diligently took notes of his guide's explanations concerning churches, saints and legends (even if some myths seemed hardly credible to him; he certainly liked to point out inconsistencies). Yet, although he trusted him in many regards, Williams lost his patience at one point. Frustrated by a visit to yet another church at ʿAyn Karim, he told Solomon that he did not want to see any more churches and that, if there was nothing else to see, he had better leave Palestine. Apparently, the tourist and his guide had different assumptions as to the character of the journey. While Williams hoped to see antiquities and grasp glimpses of everyday local life, Solomon had assumed that his client would be interested in visiting the sites, places and buildings associated with Christianity. The misunderstanding demonstrates the collaborative character of the itinerary. Certainly, the tourist had expectations about what to see during his trip, but he depended a great deal on the expertise of his tour guide, who brought in local knowledge and experiences concerning the major tourist attractions.

Not only by defining itineraries, but also by their explanations, guides shaped the points of view of their clients. If tourists considered them to be knowledgeable, Palestinian tour guides could indeed make a difference, as the example of the Church of the Holy Sepulchre showed. Whereas none of the tourists disputed the relevance of the Church of the Holy Sepulchre, their appreciation of the site depended significantly on the efforts of their respective guide. Generally, the initial reaction to the monument was negative. Edwin W. Smith's impression was common among European visitors:

> Gaudy images of virgin. Ghastly crucifixes. [...] The whole thing a horror. I could wish the earthquake might remove everything & restore things to what they were. The bare rock – a tomb – would move me: this tawdry magnificence leaves me cold & checks[?] all imagination.[26]

The desire for simplicity and humbleness was shared by many British tourists, as was the rejection of ornaments they perceived as expressions of 'Eastern' or

Orthodox opulence.[27] Confronted with Smith's negative view of the Church, his guide Siraganian, a Jerusalemite of Armenian origin and the depositary of the Bible Society, retorted: "Yes, it [the bare rock, jd] might appeal to you, but others are not like you & have been trained otherwise. They could not grasp the significance of it all if confronted with a bare rock – must have images & perfumes & candles, etc. etc."[28] Not accepting Smith's claim to an absolute standard of aesthetics in arts and architecture, Siraganian reminded him of the numerous congregations sharing the church. From his perspective, Jerusalem was the centre of a world religion, and Christianity not limited to its European (or Protestant) varieties. To Smith's claims of universal-yet-hierarchical standards, Siraganian reacted by proclaiming the equality of different forms of worship. Although Siraganian did not convince Smith entirely, the fact that he noted down Siraganian's reply indicates that there was room for reflection, consideration and dialogue between the tour guide and the traveller.[29] Moreover, the intervention shows that the guide perceived himself as a mediator vis-à-vis his foreign guest – a mediator who intended to enhance the visitor's understanding of local culture.

Not only Siraganian, but also other Arab Palestinian tour guides introduced sights to their clients in a way that induced respect and sometimes even managed to change the perspectives of tourists. Demetrius, who guided Helena Harrison around the Church of the Holy Sepulchre, was responsible for the very positive experience she reported. Despite certain aesthetic reservations, she enjoyed the atmosphere and visited it several times during her stay in Palestine in 1925.[30] While the guides had different backgrounds, encountered their clients in different contexts, and pursued different approaches in sharing their views of Palestine, their narratives generally emphasised the multireligious coexistence in Palestine. They attempted to make the Protestant British travellers encounter the sanctuaries with respect, openness and understanding.

Tourists did not unconditionally trust the explanations of their guides, however. Whereas most tourists appreciated the knowledge of the guides as far as monuments and local customs were concerned, they tended to have doubts about their expertise regarding the current political situation. This is significant as the interest of tourists in the emerging conflict increasingly overshadowed the biblical approach to Palestine. After the Wailing Wall Riots in 1929, all visitors of my sample commented upon the conflict in their travelogues. George H. Williams, who had a pronounced interest in politics, reported on tensions between Arab Palestinians and Jewish immigrants already in the mid-1920s.[31]

Yet even if he apparently discussed politics with his tour guide, he did not award this information the same value as his knowledge of local monuments and history.[32] When Solomon narrated his personal story to George H. Williams, the latter was annoyed that the guide bothered him with such details. The only element Williams retained was the alleged interreligious strife reigning in Palestine. According to his summary, Solomon had been a government official in Nablus, but had lost his job because of a quarrel with "the Moslems, who are very fanatical there". Since "changing sides" was not an option, "he hates Jews like poison, so he is now a guide."[33] While we do not know exactly what Solomon told his client, the latter's complaint that the guide pestered him all morning indicates that he discussed the contemporary situation in greater detail than is documented in Williams' summary.

Rather than debating with their tour guides, tourists sought other experts to inform them about contemporary Palestine. Trustworthy interlocutors, from the point of view of travellers, were characterised by an academic background or a governmental position, belonging to an allegedly 'neutral' group such as Protestant Christians or British nationals. Williams, for example, dedicated more space to dialogues with interlocutors holding administrative positions.[34] Edwin W. Smith, by contrast, though travelling during a time of political tensions, did not discuss politics with Siraganian.[35] Instead, he turned to interlocutors such as the Protestant Reverend Mansur from Nazareth, whose views on the Palestinian national movement Smith quoted in detail.[36] Thus, if the emerging conflict provided the occasion for new interpretations of Palestine, Palestinian guides did not necessarily profit from the occasion, given that travellers attributed much higher credentials to 'experts' or European expatriates.

Arab guides and the staff of transport services or hotels were therefore in close contact with European tourists, but the credibility attributed to these mediators varied among travellers and depended to a great extent on the context. Consequently, the assumption that "under the Ottomans [the Arabs had] acquired a virtual monopoly over the country's [...] tourist industry"[37] does not tell us much about the ability of Arab guides to turn access to tourists into actual political influence. Being in close contact with tourists did not necessarily qualify them from the latters' perspective as reliable interlocutors. In general, tourists preferred higher-standing representatives of local society and, if available, fellow European interpreters of the contemporary political situation in Palestine, such as expatriate residents or employees in service to the mandate government. To these actors, observers attributed the highest credibility and often recorded their points of view in their diaries. While they observed the situation on the ground, the travellers' most important interpreters remained other Europeans.

The conflicts in Palestine resulting from the British presence and the Jewish immigration contributed to a growing interest of visitors in the contemporary political situation, gradually superseding the approach to Palestine as an inner experience. While such a shift opened an opportunity for local guides to assume the role of cultural mediators, aiming to create understanding in regional culture, we have seen the limits of such a collaborative production of narratives in an imperial context. Given that numerous European experts on Palestine were available on the ground, tourists attributed limited credibility to Arab Palestinian guides in comparison. This concerned the contemporary political situation particularly, in which such guides were considered partial. The Arab Palestinian tour guides thus were at best able to coproduce narratives of the past, but not of the Palestinian present.

The 'conquest of tourism': Reaching out to tourists

The shift from religious pilgrims to sightseeing tourists after World War I allowed both Zionist and Arab Palestinian actors to compete for the attention of international tourists. In contrast to Palestinian tour guides individually addressing visitors, Zionist representatives aimed at establishing a monopoly on information. The Jewish Agency, the quasi-Government of the Yishuv, founded associations and institutions and used a broad range of media to present their visions of a modern, industrial and Jewish 'New Palestine'[38]. The officials of the British Mandate Government, presenting themselves as neutral administrators, did not present a narrative regarding the future of Palestine, but they put forward tourism development as an argument in favour of the British mandate administration. In Palestine only the Yishuv, and in particular Zionist organisations, pursued a comprehensive policy of tourism development, which went beyond shaping narratives, rather aiming to shape itineraries and create a national space. As a parallel to the Zionist strategies of the 1930s, "the conquest of labor" and "the conquest of the soil", we may term this comprehensive strategy "the conquest of tourism", another step on the way to a sovereign Jewish state.[39]

Tourist movements to Mandate Palestine

The overall development of tourism to Palestine during the 1920s and 1930s was a success story. Travel and tourism emerged as an important economic branch and, despite several setbacks due to economic crises and the emerging Palestinian-Zionist conflict, the number of leisure tourists reached unprecedented heights in the late 1920s and then again in the mid-1930s. However, the absolute numbers of tourists given in research seem to be highly exaggerated. The geographers Kobi Cohen-Hattab and Noam Shoval, for example, who based their estimates on local newspaper reports, assumed that more than 50,000 travellers visited Palestine in 1925. Michael Berkowitz, referring to Jewish newspapers from Europe and the US, stated that 70,000 visitors from Western countries visited Palestine in 1924.[40] The British administration reports, by contrast, mentioned only 11,000 tourists in 1925. Jewish newspapers had a vested interest in publishing higher figures, as high numbers of visitors directed international attention to the Zionist project, while the British administration preferred keeping numbers of immigrants low in order to avoid criticism from Arab actors. The significant discrepancy in numbers, however, might be due to the categories the British immigration authorities applied. The category "travellers" comprised all incoming visitors who did not intend to stay longer than three months in Palestine, and these numbers were probably quoted by the newspapers. In contrast, the term "tourists" in the administration reports referred to visitors on sightseeing tours. This number seems to have been an estimate, roughly based on the number of cruise ships and "tourist trains", as well as organised tours and pilgrimages. In the following overview, I will refer to the number of sightseeing tourists. The general trends I outline, however, concerned both categories.

Tourism resumed quickly after the end of the war. Already in 1923, the amount of 7,000 registered tourists surpassed the number of visitors during the record year of the pre-war days (6,800 tourists during the season 1913/14).[41] In addition, the example of the WTA indicated that tourism during the summer months became increasingly popular, adding a second tourist season to the peak season in winter.[42] Ever-larger numbers arrived in organised parties on cruise ships and special trains – although it does not emerge from the documents whether the special trains carried tourists whose destination was Palestine or rather transit travellers to Lebanon.[43] While the economic depression in autumn 1925 led to a minor setback,[44] the major downturn occurred in 1929–1931 as a consequence of the Wailing Wall Riots in August 1929 and the international economic crisis. A large number of tourists from the United States and Europe cancelled their bookings due to the financial crisis.[45] The number plunged from 16,500 tourists in 1928 to a mere 9,000 in 1931.[46]

Shortly before the Depression, the expansion of tourism had led to a boom in two segments of the hotel business. Hoteliers and investors involved in the Egyptian hotel business perceived a potential for the expansion of luxury tourism to neighbouring Palestine, while Jewish immigrants to Palestine established small hotels and pensions all over the country, catering for travellers of more modest means. The main Arab Palestinian venture during the time of the hotel boom was the prestigious *Palace Hotel*, which will be discussed in the section on Jerusalem.

As in Egypt, the growth in tourism in the 1920s led to the construction of new, grandiose hotels, which opened around 1930 in the midst of the recession. In Palestine, the most prominent example of this trend was the *King David Hotel* in Jerusalem. It was founded by the Mosseri family, a Jewish family originating from Cairo and Alexandria, who owned a bank with shares in the *Egyptian Hotels Ltd*. In 1921, the family established the *Palestine Hotels Company* with the intention to construct the *King David Hotel*, Palestine's first prestigious grand hotel in Jerusalem. The shares were sold to Egyptian businessmen as well as wealthy Jewish companies and individuals from around the world.[47]

Designed in the late 1920s, the architecture of the *King David Hotel* expressed the ambition to compete with the global grand hotels of the turn of the century. It was designed by the Swiss architect Emil Vogt in collaboration with Benjamin Chaikin, who had studied architecture in London and emigrated to Palestine in 1920. Gustave Hufschmid, a member of the Swiss Werkbund, was responsible for the interior design, for which he combined historicising pseudo-Assyrian, -Hittite and -Phoenician elements with ornaments inspired by Arab-Islamic styles. Since the opulent, historicising interior designs corresponded to the interiors of other luxurious grand hotels, rather than to the Werkbund's visions of clarity and purity, Daniella Ohad Smith concluded that the *King David* addressed a mostly European upper-class clientele, demonstrating the Jewish ability to erect and run a state in accordance with Western standards.[48]

Despite its untimely opening during the economic crisis, the *King David Hotel* was sufficiently resourced to overcome the difficulties, and on 19 January 1931, the hotel was inaugurated with great pomp.[49] As in the Egyptian grand hotels, a Swiss management enjoyed a reputation for quality and discretion, while local employees and waiters dressed in *galabiyyas* and red fezzes seemingly added a local touch.

Alongside the luxurious grand hotel catering to the needs of upper-class tourists and the local British expatriate society, a number of mid-sized and smaller hotels were established in Jerusalem, often situated outside the walls of the Old City. Cohen-Hattab and Shoval, who extensively studied these businesses, argued that religious segregation occurred in this segment. The guests of Jewish establishments came from different religious and social backgrounds, ranging from teachers, physicians and lawyers to students and military personnel, merchants and farmers. Arab hotels were generally smaller than comparable Jewish establishments and mostly accommodated Muslim merchants from Jerusalem and its environs, or neighbouring regions and countries. The Greek Orthodox Church offered two guest houses of a higher standard to its middle-class guests.[50] For other Christian visitors, various religious congregations provided places to stay. Hospices such as the Austrian Hospice in Jerusalem and the one of the Lazarist fathers at the Sea of Galilee not only provided them with lodgings, but also facilitated the exchange of knowledge and advice. The lowest and cheapest category of accommodation was boarding houses, mostly located in the Jewish garden suburbs, which rented out several rooms and typically accommodated guests from the local Jewish population for long stays, whether for vacation or business purposes.[51] This business greatly expanded as a result of the increasing number of Jewish immigrants, for whom the smaller pensions were often the first places to stay.[52]

The expansion of the Jewish accommodation sector can be traced in the census of industries, the first of which was conducted in the spring of 1928 at the initiative of Elie Eliachar, an officer at the Department of Customs, Excise and Trade. In total, the number of hotels had more than quadrupled in comparison with the pre-war period. Most of the establishments were small family businesses; more than 80% of the hotels were run without any or by a maximum of three wage-earners.[53] The report showed the spatial expansion of the accommodation business. Since the end of the war, the number of hotels had significantly increased across Palestine. Small pensions covered Palestine exhaustively and offered accommodation possibilities beyond the urban centres, suggesting that they not only catered for visitors on packaged tours, but also immigrants, to whom they offered a temporary place to stay. Thereby, they reflected the Zionist ambition of creating a Yishuvi presence across the territory of Mandate Palestine. Regarding Arab pensions and hotels, the report does not provide reliable information, because Palestinian entrepreneurs boycotted the survey – they feared the data would serve a more systematic levying of taxes.[54]

From 1932 onwards, numbers of tourists to Palestine rose steadily until 1935. Cohen-Hattab and Shoval suggested that the Maccabiah sports games and the Levant Fair, both of which took place in 1932, contributed to the relaunch of tourism after the economic depression.[55] More generally, the increase in tourists went hand in hand with a general growth in the Yishuvi sector of the Palestinian economy, a prosperity mainly due to the constant influx of capital related to the Jewish immigration. At the same time, better shipping connections, the opening of new grand hotels, and a well-organised and well-advertised tourist industry added to the attraction of Mandate Palestine as a tourist destination, with numbers of tourists reaching a peak in 1935.[56] Moreover, advertisements in newspapers and magazines indicate that domestic tourism increased during the 1930s as well. As in Egypt, these leisure tourists were attracted by the beaches, although some local operators also offered tours to the plantations of the Jezreel Valley (cf. p. 169).

The outbreak of the Great Palestinian Revolt in 1936 interrupted this renewed expansion of tourism. Compared to 1935, the number of incoming travellers decreased by about 50 per cent. In 1938/39, when violence flared up vigorously, the tourist sector experienced the sharpest decline in the number of visitors during the mandate period.[57] In contrast, the ensuing World War marked a period of great prosperity for the tourism sector in Palestine: in 1945, the authorities recorded approximately 150,000 travellers – mostly allied soldiers sent to Palestine for recovery, who were classified as "temporary visitors".[58]

Tourism development as a dilemma: The British administration in Palestine

Although the mandate period marked a significant phase in tourism development in Palestine, the British role in this development was ambivalent. While the British mandate administration initially provided organised tourist parties with easy access to visas, and the improvement of infrastructural connections facilitated tourism as a side-effect, concerns about the business grew among the British authorities over time.[59] Their position on tourism development revolved around three major issues: first, tourism development was considered a contribution to the fulfilment of the mandate obligations; second, officials were reluctant to spend money on the development of the British overseas possessions; third, the political aims of strengthening the tourism sector and controlling immigration to Palestine were increasingly difficult to reconcile.

The case of tourism development thus mirrored the general flaws of the Palestine Mandate. In July 1922, the Council of the League of Nations had confirmed the treaty on the British mandate for Palestine, which was brought into effect in September 1923. Based on the assumption that the Arab provinces of the former Ottoman Empire were "inhabited by peoples not yet able to stand by themselves under the strenuous conditions of the modern world", the British mandatory should establish a national home for Jews in Palestine, while assisting the Palestinian population in administering the territory "until such time as they are able to stand alone".[60] The treaty stipulated that the British mandatory had to safeguard the political, administrative and economic interests of the Jewish immigrant community, while guaranteeing the civil and religious rights of the Arab Palestinian inhabitants.[61] The treaty on the Palestine Mandate thus created highly unequal conditions, favouring the realisation of Yishuvi interests over political aims of the Arab Palestinian population.[62] In addition to the tensions between Jewish immigrants and the Arab Palestinian society resulting from this so-called dual obligation in the mandate treaty, Susan Pedersen demonstrated that the mechanisms of control and oversight established in the mandate system also put the British mandatory in a difficult position. She argued that the British intended to observe the aims of the League of Nations, while at the same time hoping to serve their own interests, which produced inconsistent, often incoherent policies in the Mandate.[63]

The annual reports submitted to the League of Nations suggest that, from the perspective of the British mandate officials, tourism development provided evidence of the rightful administration of the mandate. Projects related to tourism not only strengthened the local economy, but also contributed to the civilisational development of Palestine, as exemplified by the Palestine Museum of Antiquities, a showpiece of British cultural policies in Palestine. The American oil magnate John D. Rockefeller had donated two million dollars to "build, equip, and maintain an archaeological museum for Palestine."[64] Initial plans were discussed in 1925, and the cornerstone was laid on 19 June 1930. When the museum opened its doors in 1938, the approximate monthly average of visitors numbered at 600, reaching its peak of more than 4,000 in 1943.[65]

For British officials, the project was proof of their good administration of the mandate. In their reports to the League of Nations, the administrators regularly presented the prestigious museum as a major contribution to Palestinian development. The attraction of funds from private investors and donors such as Rockefeller indicated that the British rightfully administered the territory on behalf of an international community, while the emergence of cultural and scientific institutions demonstrated the fulfilment of the educational obligations of the mandate treaty.[66]

The emphasis of the administration reports on British-American support for the project largely glossed over the Ottoman origins of the museum. Beatrice St. Laurent and Himmet Taşkömür demonstrated that the museum was the successor of the Jerusalem Government Museum (*Müze-i Hümayun*), which opened in 1901. The collection of the Ottoman museum formed the core of the British Museum of Antiquities founded in 1921, which, in turn, was merged with the Rockefeller museum after its opening in the 1930s.[67] Thus, the actual contribution of the British administration to the exhibition of Palestinian heritage has to be nuanced. In general, their cultural policies in Palestine are perhaps better understood as the creation of a framework to be filled with contributions by other interested individuals and associations.

The preference for private sponsorship was partly due to the mandate obligation to guarantee equal access to investors, and partly due to a reluctance to invest substantial means in the development of Palestine. Regarding tourism, this accounted for the cultural as well as the economic domain. In contrast to the French mandate administration in Syria, which promised tax exemptions and hotel credits for the building or renovation of accommodation infrastructures, the British introduced only limited economic incentives to foster the expansion of tourism.[68] Although the lack of grand hotels equipped according to the latest standards was identified as a major hindrance to the further development of tourism in 1927, the British administration rather "hoped" for some modern hotels to be constructed, relying on private capital investment for this endeavour.[69]

The reluctance of the British administration to commit financially to Palestinian development concerned all entrepreneurs and interest groups involved in tourism.[70] It appears however that Zionist associations managed to profit politically from the British reluctance towards expenditure. Aware of the opportunities which tourism development provided, Zionist actors were seen to offer support for cultural projects on several occasions. In this way, they were able to contribute to defining ideas of Palestine. In late 1932, British officials in Jerusalem and London decided to enhance the promotion of Palestine as a tourist destination, giving in to Zionist pressure and the interest of tour operators. The Mandate Government made it clear from the outset that the promotion of tourism must not entail substantial financial contributions on its part.[71] If the Government declared that it was "prepared to cooperate with interested persons", the latter would have to bear a major part of the expenses.[72] These interested persons were notably Zionist representatives, who were ready to invest significant funds in advertisements.

Such collaborations were beneficial to both British and Zionist institutions. However, by leaving Zionist actors to harness these opportunities, the mandate government permitted them to shape the 'official' image of Palestine on an international stage. The promotion of tourism to Palestine at the *Anglo-Palestine Exhibition* in London, for example, was funded and conducted entirely by the *Jewish National Fund*.[73]

The British strategy to keep expenditures low thus opened up spaces for other actors to shape explanations and narratives. The strategy of cooperation implied that most activities in organising and promoting tourism were delegated to private, or at least non-governmental, actors. As in other cases, it was Zionist actors in particular who grasped such opportunities to propagate their vision of a modern, Jewish, industrial Palestine and establish, as Migdal put it, the "statelike" authority that would allow them to take over when the British left Palestine.[74]

Although measures related to tourism development strengthened the British argument in the mandate reports, the British authorities were confronted with a major dilemma: namely, the difficulty of distinguishing between tourism and other forms of migration. While the mandate administration in principle agreed upon the economic necessity of strengthening tourism, it appeared even more important to prevent uncontrolled immigration – whether of illegal Jewish immigrants or 'suspect individuals' such as alleged Bolsheviks or female artists.[75] I will return to the movements of 'suspect individuals' in more depth in chapter 4. The remainder of this section will be dedicated to the relationship between tourism policies and the immigration of Jewish settlers, a unique and particularly contested aspect in the context of tourism to Palestine.

From the perspective of the mandate institutions, visas for travellers constituted a risk, as they provided an opportunity for illegal immigration. In 1933, it transpired that an estimated 10,000 Jews had entered Palestine on tourist visas and stayed in the country. The subsequent Arab Palestinian outrage forced the mandate administration to react and to modify their policy of granting entry permits.[76] From that point on, travel visas were granted only on the condition of a return ticket,[77] and a deposit of 60 Palestinian Pounds (£P) had to be paid by all visitors entering on travel visas except those travelling first class. The deposit was refundable only in the event that visitors left Palestine before the expiry of their visa.[78] Mobility became a privilege for wealthier tourists, and the mandate administration faced severe criticism in this regard, from Zionist representatives, but also from tour operators.[79]

Shortly after the introduction of the new principle, British Legation officers in different countries addressed complaints to the Colonial Office in London. They had received countless requests that the travellers' fees be waived, ranging from individual demands to letters from tour agents such as *Thomas Cook*, who feared that the entire business model of the cruise would suffer if "tourist class" travellers had to deposit the sum. These fears were not unjustified. After all, the price for the WTA Cruise to Palestine stood at £40 – an amount considered a "once in a lifetime" expenditure for working and lower middle classes.[80] Angrily, the passport control officer in The Hague judged that "the Immigration Authorities in Jerusalem have lost sight of the fact that, in addition to Jews, there are a number of people who visit Palestine every year for business and religious reasons, who are obviously bona-fide tourists."[81] Moreover, he feared negative consequences for British international relations: shortly before, he argued, the Dutch Parliament had debated the case of four Catholic priests – from his point of view, evidently "bona-fide tourists" – who had been denied a visit to the 'Holy Land'. He warned that such incidents would create severe tensions between Great Britain and other countries.[82] As a reaction to these interventions, the British authorities introduced exemptions issued at the request of tour operators, eligible for which were "bona-fide first and second class passengers holding round-tour tickets on vessels visiting Palestine in the course of voyages solely intended to be pleasure cruises."[83] The exemption seems to have had the intended effect: of the approximately 100,000 travellers entering Palestine in 1936, only about half made the deposits.[84]

The British understanding of their role in Palestine as a broker of diverging interests thus impacted their tourism development policies. They opened the Mandate to investors and voluntarily supported initiatives of private individuals and associations promising to foster the cultural and economic development of Palestine. Moreover, the mandate officials took international diplomatic considerations into account and were concerned about their reputation as a mandate power. More often than not, the outcome was an incoherent policy with regard to tourism and a large degree of autonomy for private actors, to the effect that the interests of Arab Palestinian actors became one aspect among many others that had to be balanced, and a subordinate one at that.

Circulating Zionist visions of 'New Palestine'

> It is quite obvious that the tourists' visit is the living link between Palestine and the great civilized world. This contact affords the best means for us to capture the sympathy of all countries for our National Homeland, and to prove them that we are capable of building our great future. [...]
>
> It must be confessed, however, that the Jewish influence on the thousands of tourists in general and on the Jewish travellers in particular is still very small. We should create a special Jewish tourist organization which should bring over tourists and in the course of time and with patience they will acquire a thorough knowledge of the country and will interest themselves in its future.[85]

These conclusions were drawn in 1926 by the Jewish journalist and immigrant to Palestine M. Robinson in an article for *The Sentinel*, a pro-Zionist newspaper from Chicago. The urgency he expressed is surprising, since by the time the article was published the *Zionist Organization for Palestine* had already established a travel agency and an information bureau, and it lobbied among the mandate administration for the consideration of Yishuvi interests in tourism. The insistence of the article reveals the importance Zionist actors attributed to the tourism sector not only for economic, but also for political reasons.

The British implementation of the mandate treaty assigned a particularly powerful position to Zionist functionaries. During the 1920s, the *Zionist Organization* assumed the functions of the so-called Jewish Agency proclaimed in the mandate treaty, which represented the interests of the Jewish community in Palestine. The strong position held by the *Zionist Organization*, seated in London, in decisions concerning Palestine was contested within the Yishuv, since only one part of the Jewish immigrants to Palestine identified with the Zionist project. In 1929 the delegates to the Zionist Congress established a restructured Jewish Agency (also termed the *Jewish Agency for Palestine*). This Jewish Agency was based in Jerusalem, included both Zionist and non-Zionist Jewish members, and was acknowledged in 1930 by the British mandate power as the representation of the Jewish population in the Mandate. Its responsibilities increased considerably, to the extent that it became a quasi-Government parallel to the British Mandate Government.[86]

As Robinson's reflections reveal, for Zionist actors, addressing tourists was a chance to disseminate information that would raise international support for the Zionist project. The messages reached out to an international audience: the "great civilized world" would testify to the legitimacy of the projected National Homeland in Palestine and contribute to the realisation of Zionist ambitions.[87] Such reflections on the potential of tourism explain the attention Zionist institutions dedicated to reaching out to tourists. Zionist activities regarding tourism were not largely unstructured and sporadic as suggested by Michael Berkowitz; instead, the institutions in charge pursued a well-thought-out strategy.[88]

In order to spread views of a fruitful immigration movement among visitors to Palestine, Zionist associations pursued a comprehensive approach based on a broad range of media. They published information leaflets, guidebooks and brochures, and spread messages by means of visual media, such as photographs or maps. These sources informed visitors about the progress of Palestine which the authors attributed to the immigration of European Jews. The core elements of the strategy, however, to which the following parts of the section will be dedicated, were Jewish tour guides as important multipliers as well as the attempt to create a network of exclusively Jewish travel services. The idea was to accompany the tourist as closely as possible during his journey.

Already in 1922, Zionist actors identified the Arab Palestinian dominance in the business of guiding tourists as a major concern. The Arab Palestinian, mostly Christian, background of the tour guides collaborating with European tour operators was an Ottoman legacy. Often, the guides and translators had been recruited in the previous decades as *dragomans* by European travel agencies who continued to work with these approved and reliable guides well into the mandate period. This position seemed to provide Palestinian guides with a monopoly of information which Zionist bodies attempted to break, as Kobi Cohen-Hattab and Noam Shoval demonstrated. From a Zionist point of view, the long-established relations with foreign tour operators threatened employment opportunities for Jewish guides. More importantly, however, Zionist associations accused the Arab Palestinian guides of disseminating disadvantageous messages among tourists and presenting a partial selection of attractions: they presented local historical and religious sites while omitting modern Jewish settlements, critically commented on Zionist immigration, and boycotted Jewish shops and hotels.[89]

In order to break this monopoly, Zionist institutions launched an offensive aiming at the qualification of Jewish tour guides and at the professionalisation of the business. From 1922 onwards, the Zionist Trade and Industry Department offered professional courses for Jewish tour guides, thereby institutionalising the training of guides. In the same year, the *Association of Jewish Tour Guides in Eretz Yisrael* was established, a professional union and interest group which promoted Zionist content in tour programmes and lobbied for the employment of Zionist tour guides.[90] In the mid-1920s, press articles complained about the quality of Arab Palestinian tour guides. Publishers and journalists slandered Arab guides and reinforced contemptuous stereotypes circulating among European visitors. Comments ranged from familiar stereotypes about the alleged lack of reliability and competence of 'Orientals' to outright racism as in the article of Robinson, who referred to Arab Palestinians as "half-savage guides".[91] It is not evident whether the articles were part of an orchestrated negative campaign, but they coincided with further lobbying of the *Zionist Organization* among tour operators and the British Mandate Government. The *Zionist Organization* established contact with the major tour operators, demanding Jewish guides for Jewish visitors, insisting on 'Jewish sights' in tour programmes and even requesting that the manager of *Cook's* Jerusalem office be replaced, for he allegedly favoured Arab Palestinian guides.[92]

In early 1927, the British authorities agreed to the regularisation of the profession, which was finally enacted in July. Like the Egyptian model, the Palestinian *Tour Guide Ordinance* of 1927 established fixed rates for guided tours and aimed to guarantee the quality and decent behaviour of guides. While the High Commission officially stated that the regulation reacted to complaints of tourists regarding "importunate persons" or "excessive fees", Cohen-Hattab and Shoval assumed that the mandate government rather hoped to defuse further tensions between Arab and Jewish guides by defining the required qualifications of guides. From then on, tour guides had to obtain an official permit from the government, which required a certificate from the Department of Antiquities.[93]

The regularisation, however, did not end the Arab-Jewish competition. Local Jewish observers continued to complain about an ongoing preference for Arab tour guides. After the 'Wailing Wall Riots' in 1929, Zionist actors headed by the reformed Jewish Agency, pursued their attempts to offer tours conducted and organised by Jews with even greater energy.[94] Arab Palestinian religious authorities, by contrast, reacted to the growing tensions by banning Jewish guides from entering monuments such as the Church of the Holy Sepulchre or the Haram al-Sharif (the Muslim sanctuaries on the Temple Mount, including Al-Aqsa Mosque and the Dome of the Rock).[95] Since European tour agents would not hire Jewish guides if they could not access these major attractions, the *Zionist Organization* reinforced their lobbying activities among national and international tour operators. Their efforts proved successful: the director of the British office of *Thomas Cook & Son* promised to include Jewish sites and settlements in the programme.[96] The fact that conflicts over the employment of tour guides were particularly harsh during the period of the riots indicated that demands for a regularisation of the business were motivated by political considerations, rather than by concerns about quality.

After the riots of 1929, Yishuvi institutions concerned with tourism development expanded their activities, placing special emphasis on the aim to offer not only Jewish tour guides, but exclusively Jewish travel services to visitors. Cohen-Hattab and Katz showed that around this time the number of tourist businesses run by Jews increased significantly. Most entrepreneurs were immigrants from Europe or the US, who had moved to Palestine around 1930. Tourism thereby reflected the growing segregation of the Yishuvi and the Arab Palestinian economic sectors after 1929.[97]

A major step towards offering exclusively Jewish tour services was the expansion of the *Palestine and Egypt Lloyd Ltd.* It had already been founded in the early 1920s under the name *Palestine Express Travel Agency* by the Jewish Agency and the *Jewish National Fund* under the aegis of the Zionist *Anglo-Palestine Bank* (later the bank *Leumi LeIsrael*).[98] Its first guidebook (in German and in Hebrew) appeared in 1921. Under its new name (since 1928), the company expanded across the region, successively opening branches in Cairo, Alexandria, Port Said, Qantara and Beirut (in 1931). Its tours, pursuing an exclusive Zionist agenda, showcased the modern Yishuvi Palestine and guided tourists to sights such as the new agricultural settlements.[99] After the dramatic decline of tourism as a consequence of the revolt in 1936, its Zionist inclinations allowed the company to adapt its business model. It transported immigrants to Palestine and organised meetings of the *Zionist Organization*.[100]

The successful lobbying among tour operators and the establishment of a separate tourism sector allowed the *Zionist Organization* to communicate coherent messages advertising a new spatial order for Palestine. At the core of these efforts was the *Zionist Information Bureau for Tourists* (ZIB), established by the Jewish Agency, *Keren Hayesod*, as well as the *Jewish National Fund* (JNF, *Keren Kayemet Le-Yisrael*) in the early 1920s.[101] The ZIB offered guided tours around "the various aspects of Jewish constructive work" free of charge, but also handed out brochures and guidebooks to individual travellers.[102]

In a report on their achievements submitted to the Council of the Jewish Agency in the summer of 1929, the authors reported that the activities of the *Zionist Information Bureau* were growing not only in volume, but also in importance. It targeted European tourists as well as Jewish visitors from Germany and the US, "among them numerous active Zionists". Despite a rather low budget, the ZIB pursued a carefully thought-out strategy by actively approaching visitors. Its members distributed placards and a guidebook among hotels and travel agencies. Moreover, representatives of the ZIB received travellers entering Palestine by rail at Qantara or by ship at Haifa, trying to make contact while the visitors were still on board.[103] Thanks to the constant lobbying, the authors stated that they had observed an increased influence among tour agencies and a wider circulation of positive views about 'New Palestine'. The constant exchange with actors from abroad, they argued, disturbed the routine work, yet the "explanations, information and advice given […] often react[ed] upon the visitor's Zionist work in his home country" and therefore should be maintained.[104] The major attraction to be advertised among visitors was the Zionist project of 'New Palestine'.

To circulate such positive impressions about the Zionist state-building project, the actors pursued an ambitious strategy of generating publicity, particularly through literary media. Various smaller guidebooks addressed Jewish tourists and potential future settlers in the early 1920s, and in 1927 the ZIB published a more general and comprehensive guidebook on Palestine. While the title of the first edition was *Das jüdische Palästina/A Guide to Jewish Palestine*, the title of its later editions, *Guide to New Palestine*, expressed the growing ambitions of its authors.[105] The ZIB, directed by Dr Fritz Löwenstein, addressed its readers in German, French and English. Alongside chapters on geography, climate, archaeology and history, Löwenstein inserted sections for a specifically Jewish audience on "Post-biblical history", "Diaspora and Palestine", as well as "The Return (Hibbath Zion)". These sections covered around 20 pages, whereas the much larger part of the guidebook introduced important Jewish or Zionist institutions and associations, and provided a survey of the Yishuvi towns and settlements of Palestine.[106] In later editions, the space dedicated to the Zionist movement, its institutions, services and the amenities of Jewish life in Palestine grew, making up approximately 50 per cent of the content by 1934.[107] In contrast to European guidebooks focusing on historical information, the guidebooks edited by the ZIB highlighted the Zionist present.

The efforts of the *Zionist Organization* in Palestine and its affiliated institutions consciously targeted international tourists, hoping to gain sympathies or even mobilise support for the Zionist project, including financial support and investments from the Zionist community as well as the mobilisation of potential settlers. At the centre of this comprehensive strategy was the *Zionist Information Bureau for Tourists in Palestine* and its director Fritz Löwenstein. Its staff members built on a vast network ranging from photograph agencies distributing photographs and postcards, to newspapers in the US.[108] Beyond these media outlets across the world, the *Zionist Organization* provided tourists with guidebooks and specifically trained guides, and established exclusively Jewish infrastructures of accommodation, guidance and transportation. In this way, the institutions made sure that tourists anywhere in the mandate would be confronted with the same message of a progressive, evolving Zionist Palestine. The transnational networks of Zionist institutions brought them into contact with potential travellers. Moreover, the decision to offer qualified guides and to refer to seemingly objective media was likely to satisfy the desire of tourists for trustworthy information, expressed repeatedly in their travelogues.

The limits of symbolism: Communication and cooperation in Arab Palestine

Even though Kobi Cohen-Hattab and Noam Shoval rightfully stated that contemporary Zionist observers envied the decades-long cooperation of Arab Palestinian tour guides with European tour agents and their apparently advantageous position, the notes in travel diaries showed that mere access to tourists did not guarantee successful dissemination of interpretations and narratives. In general, Palestinian attempts at addressing tourists lagged behind the strategies of Zionist institutions and associations. While the latter vigorously pursued the professionalisation of the tourism business and the institutionalisation of procedures, Palestinian actors reacted slowly to such advances. Almost a decade after Zionists launched their first training programmes for guides, the branch of the Christian YMCA in Jerusalem established courses offering tour guide certification. The programme of the YMCA was likely a reaction to the tour guide ordinance, which stipulated that all tour guides had to pass an official examination of the Department of Antiquities.[109] Whereas Zionist associations pressed for legislation, Arab Palestinian actors thus adapted slowly to the new rules. In tourism, their strategies of spreading interpretations about Palestine were less comprehensive than those of their Zionist counterparts.

It seems likely that the less structured Palestinian approach to tourism development was at least partly due to a lack of Palestinian political representation.

While the mandate treaty had provided for representation of the Jewish immigrant community's interests, a Palestinian body comparable to the Jewish Agency did not exist. The main institution supposed to represent the Arab Palestinian population was the so-called *Supreme Muslim Council* (SMC), presided over by the notable Amin al-Husayni from Jerusalem. It had been established by the British mandate administration and administered the civil and religious affairs of the Muslim population. Its authority was thus restricted compared to the far-ranging political responsibilities of the Jewish Agency. Moreover, the division of the inhabitants of the Mandate along religious lines deprived Palestine's Arab Christian population of institutional representation.[110]

In opposing the British mandate and the Jewish immigration movement, Amin al-Husayni pursued a strategy of symbolic communication centred on Jerusalem, which came to symbolise the legitimacy of Palestinian nationalist ambitions. Similar to the *Zionist Organization*, the SMC seized the opportunity of tourism. It has been argued that the SMC promoted Jerusalem as a symbol in order to rally an international community of Muslims behind the Palestinian cause.[111] Yet tourism allowed the SMC to reach out beyond a Muslim audience too. By the early 1920s, the SMC had already turned towards non-Muslim tourists, particularly European and American Christians, and reorganised the modalities of access, as well as the visiting hours, of the Haram al-Sharif. Access to the Haram al-Sharif was facilitated for non-Muslims and the SMC published a brief guide containing background information on the monument in English, which was translated into several other languages (cf. p. 155).[112]

Whereas both the *Supreme Muslim Council* and Zionist institutions aimed to reach out to tourists, the measures of the SMC were less segregationist and less comprehensive. Rather than pursuing a structural approach of shaping legislation and establishing institutions, the SMC opted for symbolic communication. The case of the *Palace Hotel* in Jerusalem exemplified these differences. In 1928, prior to the erection of the *King David Hotel*, the *Supreme Muslim Council* and the Palestinian Waqf administration[113] planned a grand hotel that would meet the expectations of international guests, in particular the delegates of the World Islamic Congress of 1931.[114] After the successful restoration of the Haram al-Sharif, the Waqf administration dedicated the remainder of the donations collected for this purpose to the hotel project.[115] The *Palace Hotel*, designed in late Ottoman modernist style by the architect Mehmet Nihad bey from Istanbul, was to be able to compete with grand hotels across the East and the West, communicating a message of Palestinian pride and sovereignty to its guests.[116] The historian Uri Kupferschmidt pointed out that the inscription above the entrance of the hotel quoted words of the Umayyad poet al-Mutawakkil al-Laythi: "We will build like our forefathers did and act like they did; the *Supreme Muslim Council* in Palestine constructed this building, 1348/1929".[117] The motto demonstrated that the SMC conceived of the hotel as a symbol of Arab cultural production, reminding visitors and passers-by of the continuity linking Palestinians with centuries of Arab civilisation and urban culture.

Behind the symbolism of the façade, however, the *Palace Hotel* was neither a Muslim nor a Palestinian nationalist project. It was built by the Jewish contractor Barukh Qatinqa and already in February 1929, long before its opening, the *Palace Hotel* was leased out to the Jewish hotelier and millionaire George Barsky, manager of the *Allenby Hotel* in Jerusalem, for a period of 25 years. The building costs had by far exceeded the initial estimates and left the SMC heavily indebted. The *Palace Hotel*, intended as a symbol of Arab pride, depended on Yishuvi investments and business partners for its realisation.[118] The strictly segregationist approach of the Zionist institutions was not mirrored in the activities of the *Supreme Muslim Council*.

When the new hotel of more than 200 rooms was inaugurated on 22 December 1929, the *Palestine Bulletin* proudly reported that the hotel was "the finest in the Middle East", equipped according to the latest standards – boasting, for example, 60 telephones. Its prospects were promising: according to the newspaper, more than 10,000 tourists had already booked rooms for the next season.[119] Yet fortune did not smile on the SMC: shortly after the opening of the *Palace Hotel*, it had to face the effects of the economic crisis. The situation was aggravated by the fact that shortly before, the nearby *King David Hotel* had opened its doors and competed with the *Palace Hotel* for the same wealthy clientele.[120] In September 1932, the venture of the *Supreme Muslim Council* was bankrupt.[121] Its ambitions in tourism never materialised.

While the grandiose plans of the SMC, as well as the architecture of the *Palace Hotel*, testified to the SMC's awareness of the importance of tourism in addressing an international audience, the general approach of Arab actors in tourism appears to have been more pragmatic than ideological, at least during the 1920s. Such a pragmatic approach was reflected in guidebook narratives, differentiating between a multireligious Arab society on the one hand and Jewish immigrants on the other, yet without expressing hostility in their description of the latter community. The consideration of Jewish immigrants as one among many other communities is revealed, for example, in the guidebook of Alexander Khoori. We have encountered him already as a guidebook author in Egypt. Khoori, originally from Palestine, had also published a guidebook on Jerusalem and Greater Syria.[122] Whereas Khoori, of Arab nationalist and anti-British leanings, openly rejected the division of *Bilad al-Sham* under the mandate system, he did not criticise the presence of Jewish immigrants in Palestine. On the contrary, Khoori's guidebook included advertisements for Yishuvi companies and the Mayor of Tel Aviv contributed an article on the summer resorts of the town to the guidebook.[123]

While symbols such as the façade of the *Palace Hotel* or the insistence on a united *Bilad al-Sham* in Khoori's guidebook communicated Palestinian demands for an end to foreign rule, the Yishuvi immigrants were not described as a threat in these contexts. The collaborative stance represented by the *Supreme Muslim Council*, as well as Khoori, might have resulted from a certain dependence on Jewish investments. Regardless of the motivation, however, the cooperation between Arab Palestinian actors and Jewish entrepreneurs testified to the pragmatic, rather than ideological, approach of the former in tourism development activities.

The art of lobbying: The *Tourist Development Association of Palestine*

Among the various actors involved in tourism development, the *Tourist Development Association of Palestine* (TDA) had a special position. It gathered both Arab Palestinian and Zionist actors, and the British administration considered it an interlocutor to be taken seriously. Yet, despite this seeming impartiality, the publications of the TDA revealed pro-Zionist leanings.

The creation of the TDA in 1932 was initiated notably by the tour operator *Palestine Lloyd Ltd.* (formerly *Zionist Palestine Express Travel Agency*, cf. p. 143), and High Commissioner Herbert Samuel served as president of the association.[124] It seems to have been inspired by predecessors and models inside and outside of Palestine. Similar associations emerged across various Arab Mediterranean countries in the early twentieth century, and in Palestine several smaller groups lobbying for tourism development had been founded before the establishment of the TDA:[125] as early as 1922, the *Society for the Promotion of Travel in the Holy Land* and a local *Promotion Society* in Haifa aimed to "encourage the movement of tourists to Palestine and to advertise the country's attractions".[126]

Similar to the Egyptian *Tourist Development Association*, the TDA gathered representatives of the large companies involved in tourism to Palestine. Among its seven directors were Chairman Dimitri Salameh, manager of *Thomas Cook* in Palestine; Vice-Chairman Theodor Fast, manager of several hotels in Jerusalem; the Zionist businessman Moss Levy who had "for many years dealt with tourists to Palestine"; Edwin Salama of the *American Express* agency; as well as A. Jamal, manager of the tour agency *Jamal Brothers* in Jerusalem.[127] In 1933, its second year of operation, the following new members joined the executive committee: the secretary of the Arab Chamber of Commerce, George Assad Khadder, as Secretary; Mr Sargent, a representative of the Palestine Railways; and Walter Turnowsky, an "expert on tourism".[128] In addition, the editor of the *Palestine Post*, Gershon Agronsky, followed the activities of the TDA closely. Agronsky was not involved in official functions, but he was regularly present at TDA meetings, joined internal debates and supported its activities by close coverage in his newspaper.[129]

Upon official registration of the *Tourist Development Association of Palestine* in 1932, its founders stated that its purpose was to make Palestine and Trans-Jordan known abroad in order to attract a greater number of tourists and visitors. Its major model being the Egyptian *Tourist Development Association*, the Palestinian members of the TDA closely observed the activities in the neighbouring country and took inspiration from measures and policies introduced on the Nile.[130] Yet the financial means of the Palestinian TDA were much more limited. While the Egyptian TDA had a budget of around £.E. 30,000 at its disposal, funded by the Egyptian Government and interested private companies, demands for a budget for Palestinian tourism development were largely ignored by the mandate government.[131] As a result, the Palestinian *Tourist Development Association* focused mainly on lobbying for legislative measures, such as the regulation of tour guide licences, and issued smaller publications, among them an information brochure of seven pages, which advertised the historical and religious sights of Palestine in English, French and German.[132]

Despite the presence of several Palestinian Arab representatives on the executive committee of the *Tourist Development Association*, it was only a matter of years before the narratives disseminated in brochures and guidebooks shifted toward Zionist interpretations of Palestinian realities.[133] In 1933, Agronsky's pro-Zionist newspaper *Palestine Post* had criticised the "propaganda folder" published by the TDA in English, French and German for its allegedly narrow focus on historical and religious sites.[134] However, when the TDA published its tourist guide *Ancient and modern Palestine* in 1935, Zionist authors and narratives largely dominated both textual and visual depictions of Palestine. More than one-third (36.5%) of the illustrations in the brochure represented 'modern' Zionist Palestine, such as panoramic views of Haifa and its new port, aeroplanes, the opera and the beaches of Tel Aviv, agricultural settlements and the Hebrew University. Another third of the photographs represented sites of archaeological interest or historical monuments, while 17% showed landscape views, and eight out of a total 63 photographs presented an exotic Arab Palestine.

The written articles showed a similar preference for the presentation of a Zionist 'New Palestine'. The only Palestinian author contributing an article to the brochure was Tawfiq Canaan, a physician who had become famous for his ethnographic observations about Palestinian peasant society in the early twentieth century. His nuanced observations on Palestinian village life in times of fundamental transformations were considered as contributions to a nationalist discourse by the historian Salim Tamari and served as the basis for his contribution to the guidebook. In his article on "Mohammedan Sanctuaries", Canaan depicted local religious customs and beliefs, highlighting continuities and overlaps interlinking Islam, Christianity and Judaism in Palestine. He interpreted the relationship between "ancient and modern Palestine" as a continuous process linking the Palestinian past to its present and future. Canaan thus pleaded for a respectful approach of the traveller towards local culture and rejected colonial and Zionist assumptions of local backwardness.[135] However, his text was an exception for the guidebook: the other articles were written by Yishuvi authors and mostly adopted Zionist narratives of an empty and backward Palestine which was revived under the aegis of the new settlers.[136]

The Zionist messages in the publications of the TDA became more vigorous in late 1938, at the height of the Palestinian Revolt. In spring, the British company *Publicity Films Ltd.* produced a "travelogue film of Palestine" – in colour – on behalf of the *Citrus Fruit Advertising Committee* and the Palestinian *Tourist Development Association*. Despite the participation of the TDA in the production, the topoi characterising Palestine in the film suggested that Palestinian representatives had not been consulted during the planning of the project. Before its release, Stewart Perowne reviewed the film for the British Colonial Office. Perowne, a British historian who had worked for the mandate government in Palestine and who organised the Arabic programme of the BBC in 1938/39, strongly recommended a revision of the film, because

> [i]n the preliminary portion of the film the picturesque and the progressive are contrasted in terms of Arab and Jew: the Jews are represented as active and the Arabs as decorative. From my knowledge of Palestine I believe that this would give offence to an Arab audience.[137]

In detailed annotations, Perowne justified his recommendation. He criticised, for example, the assumption of the commentator that the Jezreel Valley had been "mainly swamps" before the Jewish colonisation and condemned the representation of Arabs working with "primitive" instruments while the Jewish settlers apparently laboured according to recent technological standards.[138] In addition, he pointed out that a sequence showing Jewish colonists dancing "might give the impression that it is only the Jews who have any energy. No Arab in the picture is shown doing anything so mobile". The travelogue film addressing European tourists thus largely repeated tropes about the Zionist revival of Palestine. Technology, development and productivity were attributed exclusively to 'New Palestine', creating the impression of two segregated communities living in different temporal stages.[139]

The producers ignored Perowne's recommendations. As a consequence of the one-sidedness of the film, British officials rather took pains to carefully select the audience. When the Colonial Office prepared the screening of the finished product a few months later, the British Under Secretary of State preferred not to present the film to the Arab delegation, afraid that its members might feel offended by "any scenes showing Jewish settlers." In contrast, the Jewish delegation invited to watch the film was "delighted, and commented upon its objectiveness."[140] During the revolt, Zionist actors reportedly took control of the contents propagated by the

TDA. It seems that either the official activities of the association had ceased due to the revolt or that the Arab Palestinian representatives were no longer involved in its undertakings. Initially a joint project, the TDA had turned into a one-sided institution disseminating ever more vigorous propagandistic messages of an exclusively Jewish Palestine.

Although both Arab Palestinian and Zionist actors used tourism to address an international audience, the conditions and capacities for the dissemination of messages differed between them. Arab Palestinian actors and institutions largely focused on creating symbols evoking the ancient Palestinian culture and civilisation to visitors. However, given that they lacked access to capital investments and to political decision-making, their room for manoeuvre was limited. In the Yishuv, notably Zionist associations pursued a systematic approach to targeting the international audience. They insisted on favourable legislation, mobilised networks outside of Palestine, and repeatedly harnessed various forms of media. The *Zionist Information Bureau for Tourists* was an important institution in this regard, as it guaranteed a coherent approach and coordinated the activities of various bodies. The *Tourist Development Association of Palestine*, by contrast, had been created as an inclusive association, yet it adopted – and was allowed to maintain – a staunchly Zionist position when the conflict turned acute in the mid-1930s.

Creating 'New Palestine': Tourism development as a territorial claim

Tourism served as an opportunity to reach out to an international audience and communicate political visions about the future of Palestine; but its proponents also used tourism to shape spatial orders experienced by foreign tourists and the local population alike. This potential of tourism will be scrutinised in four case studies, as set out below.

At the centre of Arab Palestinian efforts in shaping Palestine was Jerusalem. The numerous holy sites of the city turned it into a powerful symbol, and the *Supreme Muslim Council* used policies of regulating access in order to demonstrate Palestinian defiance against the mandate. The second case study focuses on the British-Zionist cooperation in tourism development at the Sea of Galilee. Although the idea of turning Tiberias into a bathing resort went back to Ottoman times, the British mandatory granted the concession for the exploitation of the local hot springs to a Zionist concessionary. While the story behind the transfer is not entirely known, the available documents suggest that in British mandate policies of 'developing Palestine', political considerations sometimes outweighed economic expectations and favoured Zionist visions over the regional conceptions of the former Arab concessionary. The third section on the Jezreel Valley demonstrates how notably the

Zionist Information Bureau for Tourists used tourism as a means of radically redefining Palestinian space. This redefinition went beyond mere narratives. Instead, tourism was applied as a tool to attract potential investors and future settlers, and thereby transform material realities in Palestine. The final section focuses on the establishment of a seaside resort at the Dead Sea, which complements the section on the Sea of Galilee, as it reveals a shift in the conception of bathing tourism from the 1920s to the 1930s. In contrast to ideas of a rather elitist global bathing tourism during the 1920s, its proponents targeted a more popular, domestic audience from the mid-1930s. This reorientation also modified the general approach to tourists and the tourist resort. Increasingly harsh anti-Arab overtones accompanied the project, and the leisure resort ultimately became a colonial outpost.

Symbolic struggles for Jerusalem

In the spring of 1925, the former British Secretary of State for Foreign Affairs Lord Balfour visited Jerusalem. Since the declaration associated with his name was highly resented by the Palestinian population, his visit bore certain risks. While Balfour seemingly wandered around the old town freely, in fact, security forces followed him at close distance, among them four police officers disguised as tourists. What they feared most was that Balfour would wish to visit the Haram al-Sharif, which had been closed and barred by the Waqf administration for the duration of Balfour's visit. As George H. Williams put it: "Had he done so, the various people (New Jews) who had come up for the various ceremonies would have tried to force an entrance and there <u>would</u> have been trouble."[141]

Whereas religious communities opened the doors of their sanctuaries to European Christian tourists during the mandate period, they regularly denied access to persons classified as supporters of Zionism, Lord Balfour being the most prominent example. In general, the policies of access to the Haram al-Sharif administered by the *Supreme Muslim Council* and the Waqf administration were rather liberal during the interwar period. Although travel reports of the nineteenth century frequently reported stories of travellers disguised in "local dress" in order to access the famous Dome of the Rock, this trope later disappeared from narratives as foreign travellers were easily admitted to the sanctuary.

Yet, as an expression of protest, Arab Palestinian Muslim and Christian communities occasionally reverted to denials of access. The religious communities administering the Church of the Holy Sepulchre, for example, prevented Jewish tour guides from entering the church in early 1928, at the protest of Zionist commentators who feared that this would lead foreign tour agencies to give Arab guides preference over their Jewish counterparts.[142] Since the measure excluded Jewish guides rather than Jewish visitors, it seems likely that the boycott was a reaction to the tour guide ordinance, which had been declared shortly before as a result of Zionist lobbying (cf. p. 142).[143] Similarly, Jews were excluded from access to the Haram al-Sharif in 1929 as a reaction to the Wailing Wall Riots.[144]

At first sight, the episodes seem to confirm the analysis of Kobi Cohen-Hattab and Noam Shoval, who argued that Jerusalem as a tourist destination was a highly fragmented space due to the overarching attribution of religious meaning to the city. According to them, religious ideals largely shaped the expectations of travellers who visited the city, to the degree that Jerusalem was not a single, coherent and continuous 'tourist space', but that it consisted of various religiously and nationally defined tourist spaces. According to them, while pilgrims visited the particular sites defined by their communities, tourists in the twentieth century visited "national monuments and symbols that were sharply divergent for the Jewish and Arab nationalities vying for rule of Jerusalem".[145]

However, it appears that the Waqf's decision to close the sanctuary for Balfour, as well as the other closures of sanctuaries, was motivated by politics, rather than by religion. If access to the sanctuaries was denied to individuals or groups, these measures were temporary and limited to the most prominent sites. Since the denial of access was applied only for a limited period of time, the measures seemed to be symbolic attempts to exclude political opponents, rather than religious groups. Jerusalem was a space that was politically – as opposed to religiously – contested.

This symbolic communication was effective *because* Jerusalem was probably the least religiously fragmented place in Palestine regarding the visiting practices of travellers. Here, the narrative of Palestine as the origin of three world religions was reflected in the visiting practices of tourists. The holy sites of the Jewish, Muslim and Christian faiths were a must-see for tourists: even itineraries for visitors staying in Jerusalem for a few days considered visits to the Church of the Holy Sepulchre, the Haram al-Sharif and the Western Wall mandatory.[146] Likewise, the Dome of the Rock and the Al-Aqsa Mosque not only attracted Muslim and European Christian visitors; Zionist tour programmes also included a visit.[147]

At a time when tourists searching for aesthetic and cultural highlights outnumbered pilgrims, controlling and denying access to monuments was a powerful symbol that would attract international attention. The *Supreme Muslim Council* demonstrated its sovereignty over Jerusalem not by defining access to a Muslim sanctuary, but by defining access to Jerusalem's most prominent tourist attraction – and the visual symbol of the city. The touristic value of the Dome of the Rock for non-Muslims turned its locked doors into a powerful statement.

Above all other monuments, the Dome of the Rock was the visual symbol of Jerusalem. Regardless of the religious adherence of its visitors, the Dome figured prominently both in the photographs tourists took in Jerusalem and on the postcards they sent.[148] In the album of travel photographs depicting P.S.H. Lawrence's journey across the Orient, consisting of both his own photographs and coloured picture postcards, 9 of the 13 representations of Jerusalem showed the Haram al-Sharif (fig. 11).[149] In this regard, Lawrence's album was no exception. In slide shows and travel lectures, as well as in advertisements of tour operators, the Dome of the Rock was a frequently reproduced motif representing Jerusalem,[150] and popular panoramic photographs of the city were often taken from the Mount of Olives, which allowed for a prominent positioning of the iconic cupola of the Dome.[151] The Dome of the Rock was an ideal symbol from an aesthetic point of view. Due to its topographical situation on the Temple Mount and its characteristic shape, it was a clearly recognisable motif to be reproduced in photographs and postcards and could be easily marketed as a tourist attraction. Thus, its prominence reflected the logic of the tourist, rather than of the pilgrim.

It appears that the strong inclination of tourism to visual representation favoured the monumental Dome of the Rock over other meaningful sites such as the Western Wall or the Christian sanctuaries. This is reflected in the comments of European Christian visitors, who admired the aesthetic value of the site over its spiritual significance. The cruise tourist Hélène Vacaresco described the colourful ornaments reflecting the prayers of the worshippers; here, Mary Steele-Maitland found the grace and simplicity that she had been missing in the Church of the Holy Sepulchre.[152] Helena Harrison admired the beauty of the Haram al-Sharif to the degree that she planned her second visit there straight away; and Eunice Holliday, who had moved to Jerusalem recently, was amazed by the beauty of the Dome of the Rock, even during her second visit to the site.[153]

The Muslim sanctuary met the aesthetic expectations and preferences of most tourists to a greater degree than the famed Christian monuments of Jerusalem, such as the Church of the Holy Sepulchre. Indeed, it was for aesthetic reasons that most visitors to the Church of the Holy Sepulchre ended up disillusioned, as the first section of this chapter has shown. The aesthetic qualities and monumental prominence of the sanctuaries, not necessarily the religious piety of the visitors, defined the Jerusalem experience of many tourists.[154]

While Jerusalem may have attracted most European visitors on the basis of its importance in Christian and biblical history, after their arrival the practices of visitors went far beyond religious worship. All of them included sights in their itineraries that fell outside the realms of places related to their own faith. Arab Palestinian communities shaped movements across Jerusalem and defined access to its most prominent monuments in order to demonstrate authority over the space of Palestine. In this context, they presented the Haram al-Sharif not only as a Muslim sanctuary, but as a symbol of Arab Palestine.

Such a presentation of the Haram al-Sharif was a conscious choice made by the *Supreme Muslim Council*. I suggest that the creation of this symbol and the centrality

Figure 11: Memories of a visit to the Haram al-Sharif, Jerusalem (Photographer: P.S.H. Lawrence. Reproduced by kind permission of the National Archives, Kew)

of Jerusalem in the policies of Amin al-Husayni and the *Supreme Muslim Council* not only targeted the international community of Muslims. Rather, the SMC equally addressed European Christians in their capacity as tourists. This attempt to reach out to tourists is manifest in the *Brief Guide* to the Haram al-Sharif, distributed to visitors as an admission ticket upon entering the site.

While the *Supreme Muslim Council* had issued the short brochure of 16 pages, available in English or Arabic, it was mainly Christian contributors who had shaped its contents.[155] The photographers of the American Colony provided the large black-and-white photographs illustrating the brochure, and the author of the guide (cited as "G. A.") was most likely the Christian Arab nationalist George Antonius.[156]

The guide presented the Haram al-Sharif as a monument of national culture rather than as a Muslim sanctuary. After providing a rough overview of the Haram al-Sharif, the author focused on the Dome of the Rock and Al-Aqsa Mosque, now and again introducing other structures and monuments on the site. He described the architectural properties of the monuments, directed the attention of visitors to elements of major interest, and highlighted the features that helped to establish the respective dates of construction. The booklet was written in the neutral, scientific style common to the major guidebooks, and Antonius clearly indicated that he was presenting scientifically validated historical knowledge, as well as information derived from archaeological findings. When he occasionally inserted information about popular beliefs and religious attributions, the information was marked as such.[157]

The guide was a testimony to the significance of Arab art and culture, to the tradition of Arab culture in Palestine, and to the scientific and historical qualifications of its contemporary inhabitants. It was not a political polemic, as it did not contain any reference to recent disputes or the political situation in contemporary Jerusalem. Distributed to all visitors, Antonius' guide pursued a scientific approach to the Dome, making sure that all tourists would understand and appreciate the Haram al-Sharif as a historical monument. In fact, it appears that his objective was fulfilled. Edwin W. Smith mentioned the brochure in his diary, and George H. Williams seems to have deemed the information trustworthy, since he copied information on the monuments from Antonius' text.[158] At the same time, in terms of temporality the guidebook differed from the narrative strategies of Egyptian publications as it created a national past – yet without referring to the modern contemporary society as we have seen, for example, in Khoori's illustration of the Pyramids. Antonius' modernity, though implicit in his scientific approach to the sanctuary, was not a modernity to be exhibited and photographed.

In Jerusalem, then, both Muslim and Christian Palestinian actors used their access to monuments and visitors alike to exhibit proofs of their civilisation and claim sovereignty over the city. While this strategy shaped the space of Jerusalem at a local level, the major function of these policies was to demonstrate Palestinian sovereignty. Although the policies of granting and denying access to the sanctuaries certainly had a spatial dimension, they did not attempt to create a spatial order. Rather, they were part of the Arab Palestinian strategy of symbolic communication aimed at a large, international audience. Such radical and generally temporary measures of control should certainly not be mistaken as indicators of religious strife or division. They articulated Palestinian defiance against the structure of the mandate in symbolic actions, thus constituting a crisis mode of expression.

The Sea of Galilee: Pilgrims into bathers?

> – and so we come to Tiberias & pulled up at the Hotel Tiberias. I had been advised to stay at the Elizabetha – Haven of Rest – opened by Feingold, a Jew. I asked where it was & an [unreadable, jd] young dragoman said it was up the hill. 7 minutes by car. I told him to get a car. But evidently there was objection on the part of the drivers. It didn't need a knowledge of Arabic to divine that evidently they object to taking people to a Jewish establishment. So I gave up the idea & took a room in the Hotel Tiberias. Room No. 17 facing the Lake. Had cup of tea while bath was being heated.[159]

When Edwin W. Smith arrived at Tiberias on the shores of the Sea of Galilee in late 1929, he was immediately exposed to the conflict between Arab Palestinians and Jewish immigrants that had erupted violently in autumn of that year. The scene he described confirmed Zionist complaints that Arab guides and drivers incited tourists to boycott Jewish establishments.[160] Although he was well aware of these developments, Smith did not pay attention to the economic vision for Tiberias that stood behind both ventures, and the change in concessions that separated them.

The *Hotel Tiberias* was by no means an Arab Palestinian venture. It was opened in 1896 by the Protestant Swiss-German Richard Grossmann from the nearby Templar Colony in Haifa, and was soon reputed to be the most comfortable hotel in town.[161] For visitors to the shores of the Sea of Galilee, it offered a luxurious alternative to the more humble hospice of the German Catholic Committee of Palestine at Ain Tabgha.[162] Yet in the 1920s its actual competitor was the *Elizabetha Haven of Rest*, a sanatorium. Both hotels were founded on the expectation of bathing guests: The *Hotel Tiberias* hoped for the Ottoman plans to promote the hot springs, whereas the *Elizabetha* anticipated guests at the Zionist bathing establishment.

With little to no awareness of these ambitions, biblical references remained the major attraction of the Sea of Galilee for Christian visitors like Edwin W. Smith. Travellers like him associated the lake and the surrounding towns and villages such as Capernaum (Kfar Nahum) or Majdal with the life and deeds of Jesus of Nazareth and his disciples. A tour agent specialising in tours for American visitors boasted that he "had taken a lot of them up to Tiberias",[163] which testified to the business side of religious landscape contemplation.

In establishing contact with the land of the Christian revelation, visitors approached the lake either from a scientific or from a romantic, contemplative point of view. Edwin W. Smith chose the first approach. Equipped with archaeological knowledge and accompanied by a German pastor he had met by chance in the *Hotel Tiberias*, he hoped to identify the 'original' places where the wonders attributed to Jesus of Nazareth had occurred.[164] A second group of Christian visitors approached the landscape more intuitively, like Mary Steele-Maitland. Her description of the stormy Sea of Galilee that she contemplated from her hotel window alluded to the legend of Jesus calming the lake before his disciples.[165] For her, the attraction of the Sea of Galilee consisted in an aura of sacredness she attributed to the towns and villages bordering its shores. Landscape contemplation became meaningful once the tourist framed the experience in biblical legends, as the numerous postcards and illustrations in Christian guidebooks showing fishermen in simple wooden boats on the Sea of Galilee implied.[166] They suggested an ongoing continuity linking the disciples to the Palestinian present.[167]

While the ongoing importance of the biblical reference for many European Christian visitors continued to shape perceptions of the Sea of Galilee, a second form of tourism had emerged in late Ottoman times and attracted British interest during the mandate. With leisure tourists in mind, some hotel owners and commentators from Tiberias viewed the shores of the Sea of Galilee as a future bathing resort, modelled on the renowned European spas of Carlsbad or Marienbad. During the mandate, the *Elizabetha Haven of Rest* symbolised the hopes projected at this new form of tourism at the Sea of Galilee. Owned by Solomon Feingold, the *Elizabetha* was inaugurated on Friday 31 January 1929 in the presence of High Commissioner Sir John Chancellor. The event was announced in the *Palestine Bulletin* as the opening of the "most beautiful building ever erected" in Tiberias, to the benefit of Palestinians and tourists alike. Journalists expressed their delight: "The old dream of Tiberians of seeing their city become a first-class health resort, is rapidly materialising at last".[168]

The idea of turning Tiberias into a health resort able to compete with famous European spas was not new. The mineral waters of the hot springs, reputed to cure chronic rheumatism and skin diseases, were the cornerstone of the project.[169] Already in Ottoman times, local representatives pursued an initiative to renovate and modernise the bathing establishment of Tiberias to attract an international bathing public. In 1890, a few years before Grossmann had founded his *Hotel Tiberias*, additions had been made to an existing bathhouse dating back to 1833. In 1912, hoping to further develop the baths, the Ottoman Government granted a concession for their exploitation to two notables from Beirut, Dr Samuel Fakhuri and Amin ʿAbd al-Nur.[170] Yet, with the outbreak of war, the works had not been undertaken.

During the 1920s, the dream of establishing Tiberias as a resort of international renown re-emerged. This time, however, the British mandatory relied on Zionist entrepreneurs to realise the ambition. After the war, the British administration initially upheld and confirmed the Ottoman concession, extending it to two new partners, Sulayman Nassif and Amin Rizq Effendi, for a period of forty years. The concessionaries intended to renovate the baths and to erect a new hotel, as well as a sanatorium.[171] Thus, from the mid- to late 1920s, when tourism expanded and capital was invested in luxurious hotels across the region, the idea of a modern resort in Tiberias started to materialise. These ambitions were obvious to the French traveller Charles Le Gras, who visited the town in 1927: new buildings and avenues lined with palm trees reminded him of modern tourist resorts.[172]

Despite the positive outlook, around this time the concession for the use of the springs was transferred from Sulayman Nassif and his partner to the Zionist *Hamei Tiberia (Tiberias Hot Springs) Company Ltd.* In 1928, the *Hamei Tiberia* acquired 90% of the concession.[173] The background to the transfer is not entirely clear. The historian Jacob Norris assumed political reasons, suggesting that the British Government revised their decision to confirm the Ottoman concession in 1925, when the Jewish concessionary expressed interest. Various local newspapers, by contrast, reported that the Arab concessionaries were unable to invest the sum of 50,000 Palestinian Pounds (£.P.) agreed upon in the concession and therefore had to sell most of their shares, keeping only 10%.[174]

It seems likely that a combination of both elements led to the change. On the one hand, the fact that the Arab investors continued to hold 10% of the shares indicates that the concessionaries sold their interests for financial reasons. Moreover, Sulayman Nassif continued to exploit the baths in the neighbouring village of Al-Hamma. The regional economic crisis of 1925/26 might indeed have caused financial difficulties for the Lebanese concessionaries.[175] On the other hand, a comparison with the economic crisis of 1929/30 suggests a political preference of the British for the Zionist investors: when financial problems did not allow the *Hamei Tiberia* to meet their obligations, the British administration nonetheless extended the term of the lease for the concessionaries. While this renewal of the lease term was not uncontested among the British officials, it seems that ultimately the mandatory power made a political decision in favour of the *Hamei Tiberia*, under the influence of both Zionist networks and their own interests regarding the mandate obligations.

After the first and second terms of the lease had expired in 1931, the British Mandate Government questioned a further renewal of the concession. The officers at the High Commission suggested charging the Municipality of Tiberias, rather than the *Hamei Tiberia*, with the renovation and administration of the baths. Since the initial project of attracting international bathing guests required large amounts of capital for a profound modernisation, they advised the municipality to give up on this vision. They suggested instead that the municipality invest minor sums, which would allow for a resumption of local bathing tourism and at least provide them with these modest revenues.[176] After lengthy debates and negotiations, however, the British administration granted the Zionist concessionary another five-year term of lease, preserving the vision of an international resort.

The factors favouring this prolongation followed political, rather than economic, considerations: Zionist lobbying at the Colonial Office, the ambitions of the local municipality, and the British desire to satisfy the League of Nations. First, Selig Brodetsky, Professor of Mathematics at the University of Leeds and member of the Executive of the Jewish Agency for Palestine, intervened at the British Colonial Office. He argued that the rejection of American Jewish capital to be invested in the development of Palestine would violate the principles of the mandate.[177] Second, the members of the Municipality of Tiberias had a sudden change of heart and refused to take over the concession, thereby leaving the field open to the concessionary. The Mayor of Tiberias, a Sephardic Jew named Zaki Hadef, argued that the baths had the potential to attract an international bathing tourism, and that targeting a local public would understate the potential of the town. As a result, the officials at the Colonial Office decided to follow Brodetsky's argument and recommended a renewal of terms, wary that a decision to decline investments would have to be justified in Geneva.[178]

The British Mandate Government and the *Hamei Tiberia Co.* thus concluded a new agreement in February 1932 and, from March onwards, the directors of the company, Joshua Supraski and Jacob Gesundheit from Tel Aviv, and Bernard A. Rosenblatt from New York, officially administered the baths, promising to develop Tiberias into a bathing resort of international renown.[179] Their networks, not least the connection to Brodetsky, was one factor which contributed to their advantageous position. Brodetsky, a well-regarded professor at a British university, apparently found the right argument when he referred to the mandate obligation of facilitating international investments in Palestine.

In addition, the Zionist vision of a glamorous resort seems to have convinced not only the municipality, but also the Colonial Office. The emergence of an internationally renowned bathing place would be credited to an apt British administration and certainly attract greater prestige than a bathing place for a regional clientele, as envisioned by the Lebanese concessionaries. In their reports to the League of Nations the mandate officials classified the project as a British contribution to local development. Moreover, the tourism advertisement committee promoted the Baths of Tiberias on a set of special stamps advertising tourism to Palestine.[180] Thus, the baths came to represent the successful development of Palestine under the aegis of the British mandate and by means of Jewish capital investments. In confirming the effectiveness of their development policies, tourism had become a resource for the British mandatory. The Ottoman origins of the modern bathing establishment, though the continuity remained obvious (fig. 12), were not mentioned in any of these contexts – in contrast to the Roman example, which served as a major reference point in advertising the beneficial effects of the hot springs.[181] For Zionist guidebook

Figure 12: The bathing establishment at Tiberias, around 1900 and in the 1930s (Photographer: American Colony (Jerusalem), Photo Department. G. Eric and Edith Matson Photograph Collection, Library of Congress, Prints & Photographs Division, [LC-DIG-matpc-07014; LC-DIG-matpc-17113])

authors, the transformation of Tiberias into a health resort with modern amenities marked an exemplary renewal for 'New Palestine'.[182] According to them, the joint efforts of the *Hamei Tiberia* and the mandate administration allowed Palestine to participate in new economic ventures, to establish a modern culture of leisure, and to render it accessible to an international community.

In the case of the Tiberias Baths concession, British political goals regarding the League of Nations outweighed other considerations. It must be noted that the British audience did not play a role in this context. British travellers, ever in search of places, views and sceneries connecting them with revealed "truths",[183] barely noticed the existence of a bathing establishment. Economically speaking – at least from the perspective of Solomon Feingold – the long-term vision of creating an international resort proved to be more of a curse than a promise. By the time the renovation of the Baths of Tiberias had been completed, his *Elizabetha Haven of Rest* no longer existed. Feingold declared bankruptcy in late January 1931.[184]

Moreover, the promised investments did not materialise. In 1935, the French *Guide bleu* still commented on the current state of dilapidation of the public baths, considering the establishments suitable for the local population only.[185] The outbreak of the Palestinian Revolt, which kept tourists away from Palestine, plunged the *Hamei Tiberia Company* into renewed financial difficulties in 1937; even by April 1939 the directors were still seeking the support of the Colonial Office to save the enterprise.[186] Regardless of the actual investments made, the mere potential of creating a luxurious resort had convinced both the municipality of Tiberias and the mandate administration. The promised investments and the prestigious prospects drafted by the concessionary outstripped the anticipated economic revenues of a local or regional bathing establishment in the short term.

While the Lebanese entrepreneurs continued to think within regional frameworks, aiming to attract a regional bathing public, the *Hamei Tiberia* offered a larger perspective. Its directors conceived of a modern bathing resort able to compete with those in European countries and attract international (European and American) leisure tourism. Although the investors could ultimately mobilise neither the required capital nor the international bathing tourists, the promise of substantial American Jewish investments and the vision of a resort of international renown provided them with a decisive advantage over the more modest visions of their Arab competitors. The officers at the British Colonial Office, more eager to convince the international observers in Geneva than the tourists visiting the Sea of Galilee, inferred that this prospect would leave a greater impression on the Permanent Mandates Commission. Thus, the success of the Zionist entrepreneurs in obtaining the Tiberias Baths concession was not only a question of mobilising sufficient capital, but also of mobilising the right networks and the right visions.

The case of bathing tourism at the Sea of Galilee showed that tourism was a resource not only in the sense of local economic development, but also in a political sense. While most European travellers visited Tiberias because of its biblical prominence, the establishment of bathing tourism, originally an Ottoman project, promised the town prestige and international attention. Financed by international investors and attracting wealthy international tourists, the global renown of the bathing place would benefit not only the municipality, but also the reputation of the British mandatory. Such considerations, it seems, motivated the British preference for the Zionist concessionary, even after the economic depression when observers were forced to acknowledge that the anticipated international bathers would never materialise on the shores of the lake.

Redefining the tourist: The Zionist project in the Jezreel Valley

> The tourist to modern Palestine will discover dotted about the country something which the tourist who came some twenty years ago would not have found. The Jewish communal colony, where the element of private property has disappeared, offers a fascinating study to the sociologist and offers an extremely pleasing sight to the visitor.[187]

This recommendation was published in a series of newspaper articles entitled *Notes for Tourists* in the *Palestine Bulletin*. Among tourists, however, the interest in the Jewish settlements was less pronounced than the author suggested. Certainly, they noticed the Zionist settlements when traversing the country. In their diaries they described their effect on the Palestinian landscape, sometimes enthusiastically like Edwin W. Smith, who was delighted at the view of neatly planted trees and the green, healthy and orderly plants around the settlements.[188] Yet only few tourists actually visited these colonies.[189] Some visitors even disapproved of the modernity of 'New Palestine', which they considered inauthentic. During a visit to Ahmya near Mount Carmel in 1936, Naval officer Donovan Roe criticised the "very ugly" houses of the "ultra modern" type characterising the new settlement.[190] Roe understood transformation as a process depriving places of their authenticity, thereby spoiling the experience of immediate access to the unchanged realities of biblical life and Oriental customs. For many European Christian travellers, noteworthy places were defined by past revelations, rather than by their present state.

For the Zionist authors in the *Palestine Bulletin*, however, international tourists were of secondary importance: their messages primarily addressed Jewish visitors, perceived as potential supporters of the Zionist project or even as settlers and investors. In this context, Zionist actors used tourism as a tool to radically redefine the space of Palestine as 'New Palestine'. Taking assumptions and expectations of ordinary sightseeing tourists as a starting point, Zionist institutions and guidebook authors created a Zionist presence by means of tourism. As such, this implied a redefinition of notions of tourist attraction, heritage and the tourist.

First, the unique attraction of Palestine proclaimed in Zionist guidebooks referred to the present, rather than the past. The *Zionist Information Bureau for Tourists*, for example, interpreted travellers as witnesses to the current developments in Palestine, which allegedly brought the Bible to life.[191] This redefinition implied a refocusing of tourists' attention from the past to the present and future. Compared to the messages disseminated by George Antonius and the *Supreme Muslim Council*, the ZIB dedicated less effort to the creation of a national past and the assertion of historical claims. Instead, the ZIB promoted the Zionist vision for the future of the country: "The Past is over: out of their daily labour they must fashion the future; which is the basic sentiment of all Jewish Palestine."[192]

Whereas Zionist guidebook authors redefined the implication of 'sights' in terms of content, form and style, they referred to familiar narrative strategies of tourist publications. *Steimatzky's Palestine Guide* resembled the popular European guidebooks.[193] The narrative structure and a seemingly neutral, fact-based language objectified the description of the settlers' movement, which was paralleled to the biblical movement of the Israelites. The author Zev Vilnay described modern installations such as the Rutenberg electrical works in a way that resembled *Baedeker*'s presentation of archaeological monuments: having described the outer appearance of the site, Vilnay drew attention to characteristic features and explained their meaning. His explanations were based on etymological, archaeological and historical reflections; plans, maps and statistics further contributed to an objective tone.[194] The attention dedicated to industrial installations in the guidebook emphasised their relevance, and the precise accompanying description turned them into monuments. Mechanisms of inventing traditions and creating heritage were attributed to the ongoing process of immigration. The narrative strategies of guidebook authors thereby historicised the contemporary immigration movement, aiming to establish and ultimately legitimise the existence of these sites, and the Yishuvi presence as a whole.

This reinterpretation also implied a processual understanding of 'heritage'. The Zionist 'heritage' advertised in brochures consisted of the transitions occurring at that moment in Palestine. The Jezreel Valley, a particularly fertile plain in central Palestine, is a prime example in this regard, acquiring a prominent position in Zionist guidebooks and brochures. Both its agricultural fertility and its central position had turned the Jezreel Valley (the *Emek* in Zionist terms) into a central region for the Zionist project.[195]

As it was lacking monuments – that is, 'sights' in a *Baedeker* sense – most tourists at best passed the plain without visiting the settlements, contrary to the recommendations of Zionist guidebook authors. Fritz Löwenstein, the director of the *Zionist Information Bureau*, offered a vivid, picturesque description of an imagined ancient past to his guidebook readers. In his account, conquerors and armies passed through the valley across the centuries, shaped by changing civilisations and rulers, while peasant life remained at a standstill. Caravans passed, merchants haggled over prices, and nomads plundered caravans – until Ottoman rule led to its decay:

> Such was the Emek for thousands of years, until neglect of waterworks led to swamps and malarial pests, and a decline in trade, commerce and security was brought about by indifference. It was left to the Keren Kayemeth to redeem the soil (if ever the word "redeem" can be justly used, it is in this connection) and drain it, while the Keren Hayesod made possible the greater part of settlement – these and the toil and blood of thousands of Halutzim [pioneers].[196]

This approach fundamentally altered the understanding of heritage. Rather than well-preserved 'authentic' sites, the ZIB advertised the transformation of the landscape as a major tourist attraction, considering it a 'restoration' of the original state. The attraction consisted in the proclaimed enormity of the task; the almost inhumane efforts necessary to achieve this transformation, and the unique chance to witness the process of transformation before it was fully completed – a heritage-in-progress, so to speak. In a similar case, referring to the so-called Balfour Forest, the *Zionist Information Bureau* asserted that "the opportunity of inspecting the beginning of afforestation on the slopes above the Emek should not be missed."[197] The authors promised that tourists had the unique chance to witness a singular process: the "redemption of the soil", a project of large-scale modernisation clad in religious terms. The concept of eye-witnessing, framed as a variety of sightseeing, not only guaranteed a particularly rare form of authenticity; it also contributed to a re-creation of the understanding of the tourist.

The reordering of the Palestinian lands in terms of a present-as-attraction and a heritage-in-progress has to be understood as a programme, rather than a justification in retrospect. It assigned a new role to the tourist: previously an observer, he had now become a participant. Indeed, many future immigrants seem to have visited Palestine as tourists before they settled in the Mandate. Especially during the 1920s, when the pressure to leave Europe was less acute than in the 1930s, a number of settlers entered the country on tourist visas, hoping to establish business contacts or to buy land before they would emigrate for good.[198] Among those tourists, the British authorities observed in particular Jewish merchants "from Poland and Austria", who bought land and established trade connections while they were in Palestine on tourist visas.[199] Later on, when persecution threatened Jewish Europeans more imminently and British authorities restricted immigration to Palestine in reaction to Arab protests, tourism also provided a means of concealing intentions to migrate. As an economically promising phenomenon, tourist mobilities were harder to question and to prevent than migrant or other movements. This overlap between tourism and migration led to a specific genre in publications related to tourism: the guidebook addressing both tourists and settlers. Their authors assumed a new tourist, whose gaze merged with that of the potential settler.

Such a target group explains the exhaustive enumerations of Yishuvi settlements in guidebooks and brochures. Whereas guidebooks such as the *Baedeker* or *Cook's Handbook* inserted occasional descriptions of picturesque (Arab) villages, Löwenstein's ZIB guidebook gathered detailed information on the Zionist settlements. He described the origin of the inhabitants, the organisational structures, the foundation and history of purchase, its economic potential and the main produce.[200] Thereby, the guidebook not only demonstrated the diversity of the settlement movement and the innovative forms of organising society in the Yishuv;

it also provided future settlers with information that allowed them to assess which settlements corresponded best with their ideals.

Similarly, the map in the ZIB guidebook was directed much more to the needs of the settler than to the desires of the tourist. Information commonly recorded on tourist maps was missing, such as easily identifiable symbols indicating major cities, historical attractions and natural sites. Instead, the map highlighted the major landscape formations, such as the Jezreel Valley, the Plain of Sharon and Galilee, and fertile regions were tinted in light green. Rather than pointing out monuments and excavation sites, the map directed attention towards the Jewish settlements and indicated the respective Jewish landowning organisations. Thus, instead of orienting the spectator towards the major cultural spots, the map directed attention to the most fertile areas of Palestine – all to the benefit of the future settler, rather than the tourist.[201]

Via maps and descriptions, the Zionist guidebook authors added a new layer to the landscape of tourism in Palestine. This vision of a 'New Palestine' implied new attribution of centrality and noteworthiness, while both verbally and visually eradicating the Arab Palestinian past. The tourist space of Palestine outlined in guidebooks did not describe a present state, but a vision for the future of the country – the realisation of which tourism was supposed to play a role in. Since tourist movements were expected to precede settlers' movements, the territorial order drafted by tourist maps and itineraries gradually eclipsed the 'Holy Land', and ultimately threatened the existence of local inhabitants.

Lost hopes at the Dead Sea

At first sight, the Dead Sea appeared to be the ideal spot to illustrate Zionist narratives of conquering a barren land, in which industrial development and tourism went hand in hand. In the nineteenth century, it had attracted the curiosity of geographers and travellers because its geological properties seemed hostile to any form of life. But during the mandate period, the joint efforts of mandate officials and British entrepreneurs as well as Zionist ones, turned the Dead Sea into a centre of industrial production and a popular destination for excursionists.[202] In the early 1920s, however, hardly any visitor foresaw this upsurge in popularity. Certainly, most tourists were attracted to the Dead Sea from the astonishing geological facts reported in guidebooks, but none of them stayed there for long. George H. Williams had "a look" at the Dead Sea and did his best to "photograph the 'town'"[203], before he moved on to visit the River Jordan and the alleged baptism site of Jesus of Nazareth. Less than a decade later, the character of tours to and around the Dead Sea had changed thoroughly. The Dead Sea had become both a site of industrial development and a seaside resort, and it was this combination that turned the

Dead Sea into a colonial outpost during the Palestinian Revolt in the late 1930s. In contrast to the settlements in the Jezreel Valley, however, the promoters of this project did not conceive of it as an exclusivist vision from the outset.

In 1924, neither bathing installations nor any kind of leisure infrastructure existed, and reaching the shores was adventurous: Eunice Holliday and her family had to turn back because the road was flooded and they were unable to proceed.[204] It was up to excursionists to find themselves a proper bathing spot.[205] Around 1930, plans for extracting the minerals of the Dead Sea were still at an experimental stage and the installation of the works was "anxiously awaited."[206] Though at an early stage of development, tourism was already noticeable. Edwin W. Smith reported on "a number of bathing shelters. Drinks sold."[207] A mere two years later, Mary Steele-Maitland observed numerous bathers at the Dead Sea, among them families and tourists:

> We get out – tawdry buildings and shingle down to the water. Children running about and paddling. I taste the water and make a wry face as it is absolute trine[?] – a lady asking in broken English: "Did you find it good" It is so salty that ones [sic] hands are left feeling oily after being wet. The evaporation is so great that though the Jordan flows in there is not an outlet at the other end. We buy post-cards and return to the Hospice.[208]

The popularity of the Dead Sea as a seaside resort was related to the presence of the local factory, which had launched production in early 1932. After the Ottoman defeat, the *Palestine Potash Company*, a business partnership of the Russian Jewish mining engineer Moise Novomeysky and the Scottish engineer Thomas G. Tulloch, had obtained the concession for the exploitation of minerals, aiming to extract bromine and potash – and to exploit the tourist potential. From the outset, the chairman, Major Tulloch, envisaged the creation of a tourist resort.

It was Tulloch's vision that induced Eunice Holliday to undertake a second trip to the Dead Sea in 1932. This time, professional obligations on behalf of her husband Clifford Holliday prompted the journey: as an architect and town planner, he was involved in planning the new pleasure resort. Compared to her first visit, Eunice Holliday experienced a much more comfortable journey: the old track had been replaced by "a marvellous road", new buildings housed the factory workers and a café with bathing cabins had been opened on the shore. During her husband's meeting, Eunice Holliday and their children had a ride on the speedboat of Tulloch's son and visited the hot springs of Callirrhoe on the other side of the Dead Sea – a tour *Kallia* would later offer to guests of the resort.[209]

At the time, Tulloch intended to establish a winter resort able to compete with the Mediterranean resorts and to attract an international public, thus similar to the ambitions of the *Hamei Tiberia* on the Sea of Galilee he hoped to attract a wealthy global clientele. At the same time, Tulloch promoted leisure tourism as an inclusive political project: His leisure centre, open to all, was to contribute to overcoming the "differences between various sects and creeds in Palestine". He hoped that tourism in the Jordan Valley and at the Dead Sea would offer attractions beyond the "spiritual and historical appeals" that so far attracted most visitors.[210] Such leisure attractions for a new clientele would contribute to creating a modern, open-minded and mixed Palestinian society.

The restaurant and the bathing establishment of the resort opened their doors on 4 January 1933. In the opening speech, Tulloch repeated his intention to promote leisure tourism as an example of fruitful Jewish-Arab cooperation. The erection of a hotel was already envisaged, and the company was working on excursion boats for tourists that would transport up to 80 persons to the other seashores. On this occasion, Tulloch explained that the name of his resort, *Kallia*, was inspired by the Arabic word for potassium and stood for his vision of an Arab-Jewish cooperation that had for centuries been the basis for scientific progress.[211] While this interpretation suggested that both *Kallia* and the *Palestine Potash Company* adopted the Arab heritage as a central part of its business identity, the reference seems to have been more of a marketing strategy than a political vision.

Its directors put forward that the company employed 500 workers, both Arab and Jewish, who lived in new buildings close to the factory. However, the historian Jacob Norris revealed that Arab and Jewish workers lived in segregated camps and under very different conditions.[212] Moreover, Arabs were scarce among the higher echelons of the company. On the board of the *Kallia* resort, Ismaʿil al-Husayni, a Palestinian landowning notable and investor from the Jerusalem branch of the al-Husayni family, was the only Arab member, the other members were British, Yishuvi and other international representatives: the aforementioned Major Tulloch, the Zionist and Dutch Consul General Dr Siegfried A. van Vriesland, S. Horowitz, A. Goldwater, L. Green and Miriam Sacher.[213]

Among inhabitants and expat residents of Palestine, the resort became a huge success. The directors had abandoned the idea of promoting the beaches as an international wintering resort for wealthy bathing tourists, and instead focused on a popular, domestic clientele, a strategy that proved successful. Between July 1933 and 1934, the owners of the *Kallia Seaside and Health Resort* continuously offered new attractions, such as an open-air dance floor, special excursions by boat, a café, a restaurant and bathing huts for 250 guests. A bus service transported weekend guests from Jerusalem to the resort, and the number of visitors continued to grow.[214] By the time the hotel was opened in 1935, the resort had obtained permission to sell alcohol. The beach was equipped with fresh-water showers and towels, and the resort had bathing suits and towels for hire. A sanatorium was under construction in 1934, which aimed to "take advantage of the remarkable curative properties of the Dead Sea waters, which are radio-active, and of the mild climate".[215] Even though international guidebooks now included information on the resort and the excursions around the Dead Sea, it still attracted a largely domestic public. Crowds predominantly from Jerusalem, Jaffa and Tel-Aviv visited the resort on weekends and on moonlit nights, when the restaurant offered special entertainment.[216]

From the summer season of 1935 onwards, the *Kallia* Company further extended its business activities. The company started to organise regular car transport services and trips across Palestine, as well as to neighbouring Syria. *Kallia* specifically advertised affordable bus tours to the Jewish settlements in the Jezreel Valley and Galilee, thus clearly targeting Yishuvi excursionists.[217] As the company targeted domestic tourists, its directors increasingly modified its foundational principles.

The movement towards a Yishuvi, and notably Zionist, clientele apparently contradicted Tulloch's initial vision of a leisure space that united Arab Palestinian and Yishuvi workers and excursionists. Yet, economically, the strategy paid off and allowed for further expansion of the business. In October 1936, Tulloch and the board envisioned substantial enlargements of the Dead Sea resort. The owners planned the construction of a new hotel with 35 rooms, including facilities for warm sea-water baths. Moreover, a power plant, another restaurant, and a new motorboat would improve the services offered to tourists.[218] While the opening was initially projected to take place the following spring, it was not until February 1938 that the owners announced the opening of the new *Kallia Hotel* – at the height of the Great Revolt.[219]

By that time, from Tulloch's point of view, the hope of jointly manufactured economic development was shattered. The directorate of *Kallia* readapted its communication strategy, exhibiting a sort of defiance regarding the British and Jewish position in Palestine. In September 1937, Tulloch gave an interview to a British newspaper on "the Life at the Dead Sea", in which he commented on the threat that the Yishuv and British nationals were facing from the Arab population. He explained his strategies of self-defence to readers: a machine gun was mounted on the roof of his house, and an armed guard protected him.[220]

In addition, the domestic situation during the Great Revolt had rendered the initial strategy of attracting international leisure or bathing tourists obsolete. To a certain degree, the company countered assumptions about lacking security through positive publicity. In 1939, a letter of recommendation was published in the *Readers' Letters* section of the *Palestine Post*. An anonymous reader reported on his stay at the *Kallia* resort, attempting to alleviate fears of potential visitors over security issues. He highlighted that the organisers had taken all possible security measures, including an armoured convoy accompanying his car, patrols along the coast and armed guards at the hotel, which were reinforced during the night, "so I should think that even the most nervous of visitors would feel (and be) quite secure, and enjoy his holiday in peace."[221] While it is doubtful whether such a letter, most likely written by managers of the resort themselves, was able to convince local readers, it certainly did not bring back international guests. After the outbreak of the revolt, the number of visitors to Palestine rapidly declined.

The directorate of the *Kallia* company reacted to this decline by focusing on a domestic, Yishuvi target group. To a greater degree than before, they promoted the curative qualities of the Dead Sea, aiming to establish its reputation as a health resort with the assistance of the local press, where experts of international renown and internationally acknowledged papers were cited.[222] Such articles prepared for the somewhat defiant opening of the new luxurious hotel on 16 February 1938.

Parallel to this reorientation concerning target groups, the narratives around *Kallia* began to change. Compared to the tone of the opening speeches and initial articles in 1933, the enterprise was now interpreted in an exclusivist sense. Its promoters adopted narratives of bringing industrial progress to Palestine's underdeveloped landscapes. The correspondent of the *Palestine Post*, for example, interpreted the opening of the new hotel as the "successful transformation of a scene of barrenness and desolation into one of attractiveness and beauty". The act of "carrying through in these disturbed times of a project designed primarily for days of peace and prosperity and an influx of tourists from all over the world" was interpreted as one of heroism. Similarly, the keynote speech of High Commissioner Edward Keith-Roach at the opening framed the hotel as a (luxurious) space of evasion from the harshness and violence of external realities.[223] Major Tulloch's initially envisioned space of contact and cooperation became a space of refuge and evasion. At the same time, the defiant attitude – ranging from armed protection for bathers to claims of heroic stamina in a hostile environment – turned the resort into an outpost of colonial modernity, and tourists and excursionists into markers of the colonial presence.

Economically, the reorientation proved a success – which came at the cost of the initially proclaimed ideals. The new hotel was a Zionist undertaking, both symbolically and in practice. Its manager Harry Levy interpreted the building as a "true Zionist project", and journalists invited to cover the opening represented only Jewish newspapers.[224] The persons and companies involved in the construction of the hotel were exclusively Zionist immigrants. The architect Zeev Rechter from Tel Aviv designed the hotel in the modernist style that came to symbolise the innovative modernity of the Yishuv (fig. 13). The interior design, also in the modernist style, corresponded to these aesthetic ideals. The designers Wachsberger and Leurer produced most of the furniture for the *Kallia Hotel* in their workshops in Tel Aviv.[225] Daniella Ohad Smith demonstrated that the Yishuvi adoption of the modernist style mirrored the general disillusion of Yishuvi immigrants with their initial hopes

Figure 13: The *Kallia Hotel* on the Dead Sea (Photographer: American Colony (Jerusalem),
Photo Department. G. Eric and Edith Matson Photograph Collection, Library of Congress,
Prints & Photographs Division, [LC-DIG-matpc-03722])

and ideals. Especially after the riots of 1929, she argued, Zionist architects stopped
experimenting with local elements intended to create a grounded, 'national' style.
Instead, they adopted the International Style, expressing the ambition to bring
Western progress and, in particular, hygienic conditions, to Palestine.[226] The reinterpretation of *Kallia* as a Zionist enterprise, intended as a symbol of defiance in
difficult times, was part of a larger trend and a conscious choice, expressed in
the telling speech of Major Tulloch at the public opening of the new hotel. On this
occasion, he once again addressed the meaning behind the name of the resort:
Kallia, Tulloch explained, was an acronym and stood for a Hebrew phrase: *"Kam
litehiah yam Hamelah* (The Dead Sea Has Come to Life)."[227]

Conclusion

In the contested tourist space of Palestine under the British mandate, the complex workings of tourism come to the fore in a particularly pronounced manner. Both Palestinian nationalists and Zionists grasped tourism as an occasion to promote their competing visions of Palestine before an international audience. As in other domains, the policies of the British mandate power, which had limited ambitions in shaping the tourist space of Palestine, favoured the Yishuv in this competition. Tourism, considered a sign of the economic and cultural development of Palestine, mattered to the British mandatory in particular as it proved the fulfilment of their mandate obligations. British officials therefore highlighted educational projects such as the Archaeological Museum, but also the disinterested administration of Palestine by the British, allowing foreign investors to realise their business ventures. A British preference for Zionist actors as imperial middlemen – as interlocutors, investors, concessionaries and visionaries – as well as a structural advantage regarding the political representation of the Yishuv provided Zionists in Palestine with more favourable conditions compared to their Arab Palestinian counterparts.

Among Arab Palestinians, mainly members of the middle classes pursued political ambitions in tourism. As in Egypt, their educational and professional background allowed them to present Palestinian history and heritage to visitors, yet they had greater difficulties in reaching out to their potential audiences, as institutions to develop a strategy and voice their perspectives had not emerged. Their collaboration with more powerful actors – the *Supreme Muslim Council* in the case of George Antonius, the *Tourist Development Association* for Tawfiq Canaan – can be understood as an attempt to address larger audiences. The impact of the former on political decision-making was limited though, and the TDA was dominated by Zionist stakeholders, so voices like Canaan's remained marginal. Without institutional support, coherent and encompassing narratives did not emerge. While the Palestinian Arab intellectuals shared the goal of making Palestinian history and culture visible, claiming the historical legitimacy of their presence in Palestine, their voices remained fragmented. Attempts to actively direct tourist movements were restricted to the sanctuaries in Jerusalem, which is why interrupting tourist flows was the main instrument to attract tourists' attention to their political claims. Uses of tourism were thus limited to the creation of powerful symbols, rather than to the active shaping of spaces.

The efforts of Zionist actors, by contrast, were more orchestrated and they had institutional means at their disposal that the mandate treaty granted to the Jewish Agency. In particular the *Zionist Information Bureau for Tourists* (ZIB) adopted a comprehensive, systematic approach to tourism development and seemed to perceive tourism as a resource in shaping spaces before the Yishuv would gain full sovereignty. In order to promote their vision of a modern and progressive 'New Palestine', which allegedly came into being thanks to the efforts of the immigrant community, the ZIB and its affiliated institutions tried to achieve a monopoly on tourist services that would allow them to provide visitors with information on Palestine that was deemed useful to their cause.

In particular, Zionist actors intended to use tourism as a resource in claiming the contested terrain of Palestine. As tourism allowed potential immigrants to establish their first contacts in Palestine, tourism advertisement materials pointed out the most fertile regions and introduced the visitor to settlement structures. The movements of tourists created a presence across the territory that Zionist advocates of tourism development envisioned as preceding actual settlement. Pursuing a spatial rather than a symbolic approach to tourism, the *Zionist Information Bureau for Tourists* stood for a rethinking of the very terms of tourism, as well as notions of 'sights' and 'heritage'. The Jewish tourist was conceived of as a future settler and, consequently, the suggested itineraries, guidebook descriptions and tourist maps catered to the needs of the settler, rather than the temporary visitor. The presentation of the country preceded its occupation.

In the mandate period, European tourists in Palestine seemed to be more open to new interpretations than they were in Egypt. Although they pursued their own agenda, and biblical references still largely shaped their itineraries, the contemporary political transformations and conflicts in Palestine required new explanations. Yet, the advocates of tourism development in Palestine profited unevenly from this window of opportunity. Rather than simply adopting the messages communicated to them, tourists attributed different degrees of credibility to their interlocutors. While they questioned most political remarks uttered by their Arab tour guides, they more readily adopted viewpoints if they read them in printed guidebooks, or if they attributed a certain expertise to their interlocutor – in which attributions of nationality and class played a role. It appears that, in this regard, the strategy of the Zionist associations to establish institutions and disseminate media satisfied the expectations of tourists to a greater extent than the attempts of Arab Palestinian tour guides. The latter were taken seriously as experts on local customs and traditions, much less on the contemporary political situation.

Around the mid-1930s, similar to Egypt, visions of international tourism faded into the background. The activities of Arab Palestinian actors in tourism development seemed to have decreased over time and due to the economic and political crises. Yishuvi entrepreneurs focused on a domestic audience, as the case of the Dead Sea showed. The reorientation towards a more popular Yishuvi tourism went hand in hand with an adoption of Zionist narratives in the later 1930s, which materialised in excursions to Zionist settlements and in the symbol of the *Kallia Hotel*, where Arab workers, collaborators and observers were largely excluded from the planning, construction, and inauguration of the hotel. The reorientation was certainly related to economic considerations, yet the defiant attitude of its directors during the Great Revolt indicated that the leisure resort had also come to establish a both symbolic and material presence of the Yishuv in an increasingly hostile environment.

In this context, the presence of tourists was not only of symbolic relevance; it also created a tangible reality that could be interpreted as a conquest or an occupation. Tourism to 'New Palestine' was thus ideated as a self-fulfilling prophecy: the tourist visited 'New Palestine', which his presence helped to establish.

Notes

[1] BL, Workers' Travel Association, *"What Our Members Think"*, p. 21.

[2] BL, Workers' Travel Association, *W.T.A. Easter Holidays.*

[3] BL, Workers' Travel Association, *Notes on Summer Programme.*

[4] BL, Workers' Travel Association, *Easter and Spring Holidays.* BL, Workers' Travel Association, *Palestine in Spring.*

[5] In 1925/26, a 19-day journey to Palestine and Syria organised by Thomas Cook & Son cost about £ 113, the transfer to the Mediterranean by ship excluded: FDC, Thomas Cook & Son, *L'Egypte, Le Nil, la Palestine*, pp. 70–72. Unfortunately, I did not find prices for *Cook's* tours for the 1930s after the Great Depression.

[6] Jarman, 'Report on the Economic Conditions in Palestine, 1935', V, pp. 406–407.

[7] Urry and Larsen, *The Tourist Gaze 3.0*, pp. 40–41.

[8] BL, Workers' Travel Association, *Travel Log Photograph Competition, 1934.* BL, Workers' Travel Association, *1935 Photograph Competition.* BL, Aitken, *What do you go to see?*

[9] BL, Workers' Travel Association, *Palestine Reunion.*

[10] Jarman, 'Department of Migration, Annual Report 1939'. For the discussion of the transition from 'religious' to 'secular' tourists cf. Cohen-Hattab, 'Zionism, Tourism, and the Battle for Palestine', p. 62. Bar, Doron and Cohen-Hattab, 'A New Kind of Pilgrimage'.

[11] *Yishuv:* The Jewish settler community in Palestine. Reich and Goldberg, 's.v. Yishuv'. On the internal complexity of the Yishuv, the competing visions of Zionism and Palestine cf. Caplan, 'Zionist Visions of Palestine'.

[12] Krämer, *Geschichte Palästinas*, pp. 122–132. Although not all Jewish immigrants were Zionists defending a homogeneous nationalist ideology, Rashid Khalidi has argued that as a settler community the Yishuv "in effect constituted a self-selecting sample", attracting mainly settlers who supported the Zionist ideology, comparatively homogeneous in contrast, for example, to political visions circulating in the Palestinian society: Khalidi, *The Iron Cage*, p. 18.

[13] Cohen-Hattab and Shoval, *Tourism, Religion and Pilgrimage*, p. 8.

[14] Sela, 'The "Wailing Wall" Riots'. Krämer, *Geschichte Palästinas*, pp. 262–271.

[15] Cf. for the „denial of coevalness": Fabian, *Time and the Other*, p. 31.

[16] Norris, *Land of Progress*, pp. 81–91. Norris, 'Development and Disappointment', pp. 286–287. Cf. also Smith, *The Roots of Separatism*, p. 20.

[17] Migdal, *Strong Societies and Weak States*, pp. 149–151.

18 Doumani, 'Rediscovering Ottoman Palestine', pp. 7–9. van Oord, 'The Making of Primitive Palestine', pp. 218–219. Cf. also Osterhammel, '"The Great Work of Uplifting Mankind"', p. 365.

19 Horowitz, 'Remarkable Rather for Its Eloquence', p. 443.

20 Coleman, 'From the Sublime to the Meticulous', p. 282. On landscape as a cultural construction: Urry and Larsen, *The Tourist Gaze 3.0*, pp. 106–113.

21 Howe, 'Revealing the Holy Land', p. 28. Shepherd, *The Zealous Intruders*, pp. 77–84. Coleman, 'From the Sublime to the Meticulous', pp. 277–284.

22 MECA, Harrison, *Journal of Visit to Palestine*, 30/03/1925.

23 Glover, 'Co-Produced Histories', pp. 122–123.

24 Cf. also Mairs and Muratov, *Archaeologists, Tourists, Interpreters*, p. 4.

25 RCSA, Williams, *Travel Diary: Egypt, 1925*, p. 175. RCSA, Williams, *Travel Diary: Palestine and Syria*, p. 117.

26 SOAS, Smith, *Travel Diaries, Diary 1929, II*, 27/11/1929.

27 Robson, 'Becoming a Sectarian Minority', pp. 62–64. NLS, Steele-Maitland, *Diary of Visit to Egypt*, 19/01/1932. Although often attributed to Protestant desires for purity, French Catholic travellers reacted similarly: cf. ANF, Vacaresco, *La Croisière Bleue, in: Croisières*, p. 60.

28 SOAS, Smith, *Travel Diaries, Diary 1929, II*, 27/11/1929. A similar judgement: NLS, Steele-Maitland, *Diary of Visit to Egypt*, 19/11/1932.

29 Research has focused mainly on European reactions: Horowitz, 'Remarkable Rather for Its Eloquence', pp. 453–455, 457–459.

30 MECA, Harrison, *Journal of Visit to Palestine*, 21/03/1925. MECA, Harrison, *Journal of Visit to Palestine*, 22/03/1925. MECA, Harrison, *Journal of Visit to Palestine*, 24/03/1925, 17/04/1925, 18/04/1925, 19/04/1925, 22/04/1925, 29/05/1925.

31 E.g. RCSA, Williams, *Travel Diary: Palestine and Syria*, pp. 68, 111.

32 RCSA, Williams, *Travel Diary: Palestine and Syria*, p. 111.

33 RCSA, Williams, *Travel Diary: Palestine and Syria*, p. 41.

34 RCSA, Williams, *Travel Diary: Palestine and Syria*, p. 68.

35 SOAS, Smith, *Travel Diaries, Diary 1929, II*, 28/11/1929.

36 SOAS, Smith, *Travel Diaries, Diary 1929, II*, 09/12/1929.

37 Cohen-Hattab and Shoval, *Tourism, Religion and Pilgrimage*, p. 64. Cf. also Berkowitz, *Western Jewry*, p. 128.

38 The Jewish Agency as a quasi-Government: Krämer, *Geschichte Palästinas*, p. 225.

39 Gelvin, *The Modern Middle East*, pp. 233–234. Lockman, *Comrades and Enemies*, pp. 47–53.

40 Cohen-Hattab and Shoval, *Tourism, Religion and Pilgrimage*, p. 42. Berkowitz, *Western Jewry*, p. 125.

41 Jarman, 'Report on Palestine Administration 1923', I, p. 407. NLI, *Tourist Traffic Decreasing, 02/01/1934*.

42 Jarman, 'Report on the Economic and Financial Situation, 1927', II, p. 353. Especially low-budget tour agencies such as the WTA profited from lower prices in summer and preferably organised summer tours: eg BL, Workers' Travel Association, *Notes on Summer Programme*.

43 Jarman, 'Report on the Administration of Palestine 1924', I, p. 539. Jarman, 'Report on the Administration of Palestine 1926', II, p. 283. Andrea Stanton established that "special trains" in Palestine generally transported Egyptian and Iraqi summer guests to Lebanon: Stanton, 'Locating Palestine's Summer Residence', p. 55.

44 IFP, Chamber of Commerce of Jaffa & District, *Annual Bulletin 1927–1928*, p. 1. IFP, Chamber of Commerce of Jaffa & District, *Annual Bulletin 1928–1929*, p. 27. Schulze, *Geschichte der islamischen Welt*, pp. 121–123.

45   IfP, Stead, *Economic Conditions in Palestine*, pp. 8, 27. IfP, Chamber of Commerce of Jaffa & District, *Annual Bulletin 1929–1930*, p. 7. IfP, Chamber of Commerce of Jaffa & District, *Annual Bulletin 1931–1932*, p. 25. Cohen-Hattab and Shoval, *Tourism, Religion and Pilgrimage*, p. 42.

46   IfP, Chamber of Commerce of Jaffa & District, *Annual Bulletin 1928–1929*, p. 27.

47   Cohen-Hattab and Shoval, *Tourism, Religion and Pilgrimage*, pp. 57–58.

48   Ohad Smith, 'Hotel Design in British Mandate', pp. 102–106. Berkowitz, *Western Jewry*, pp. 134–135.

49   MECA, *Inauguration Dinner King David Hotel*. Holliday, *Letters from Jerusalem*, p. 123.

50   Cohen-Hattab and Shoval, *Tourism, Religion and Pilgrimage*, p. 60.

51   Cohen-Hattab and Shoval, *Tourism, Religion and Pilgrimage*, p. 61.

52   Berkowitz, *Western Jewry*, p. 134.

53   IfP, Government of Palestine, Department of Customs, Excise and Trade, *First Census of Industries 1928*, pp. 18, 62. Smith, *The Roots of Separatism*, p. 161. Interestingly, a large part of the women in the tourism sector were widows, unmarried, or divorced, thereby confirming similar findings from tourism research stating that for solitary women, accommodating tourists was a preferred means of income: Cohen-Hattab and Katz, 'The Attraction of Palestine', p. 174. Cf. for example Battilani and Fauri, 'The Rise of a Service-Based Economy'.

54   IfP, Government of Palestine, Department of Customs, Excise and Trade, *First Census of Industries 1928*, p. 3.

55   Cohen-Hattab and Shoval, *Tourism, Religion and Pilgrimage*, p. 43.

56   IfP, Chamber of Commerce of Jaffa & District, *Annual Bulletin 1932–1933*, p. 1. IfP, Stead, *Economic Conditions in Palestine*, p. 27. IfP, *Annual Report of the Department of Customs, 1935*, p. 43. E.g. the *King David* and the *Palace Hotel*. IfP, Matson, *The American Colony Palestine Guide*.

57   Cohen-Hattab and Shoval, *Tourism, Religion and Pilgrimage*, p. 43.

58   Cohen-Hattab and Shoval, *Tourism, Religion and Pilgrimage*, p. 44.

59   Cohen-Hattab and Katz, 'The Attraction of Palestine', pp. 169–170. Cohen-Hattab and Shoval, *Tourism, Religion and Pilgrimage*, p. 75. Jarman, 'Herbert Samuel to Earl Curzon, 1920', I, p. 134. Jarman, 'Report on Palestine Administration 1920–1921', I, p. 326. Jarman, 'Report of the High Commissioner, 1920–1925', II, pp. 9, 21.

60   Yale Law School, 'The Covenant of the League', Art. 22. Krämer, *Geschichte Palästinas*, pp. 191–192.

61   Council of the League of Nations, 'The Palestine Mandate'.

62   Krämer, *Geschichte Palästinas*, pp. 193–206. Khalidi, *The Iron Cage*, pp. 31–48.

63   Pedersen, 'The Impact of League Oversight', pp. 40–41.

64   Jarman, 'Report on the Administration of Palestine 1927', II, p. 366. IfP, Vilnay, *Steimatzky's Palestine Guide*, p. 179.

65   St. Laurent and Taşkömür, 'The Imperial Museum of Antiquities', p. 37. Cohen-Hattab and Shoval, *Tourism, Religion and Pilgrimage*, pp. 53–54.

66   Jarman, 'Report on the Administration of Palestine 1927', p. 435. On the particular interest of imperial powers in the exhibition of antiquities cf. Anderson, *Imagined Communities*, pp. 179–183.

67   St. Laurent and Taşkömür, 'The Imperial Museum of Antiquities', pp. 6, 16–21, 32, 37–38.

68   For the French case cf. the following chapter. Jarman, 'An Interim Report, 1920–1921', I, p. 182. Jarman, 'Report on Palestine Administration 1920–1921', I, 235. Jarman, 'Report on the Administration of Palestine 1924', I, p. 537. Jarman, 'Report of the High Commissioner, 1920–1925', II, p. 21.

69   Jarman, 'Report on the Economic and Financial Situation, 1927', II, p. 354.

70   NA, *J.A. Seiler, Palestine Hotels*. NA, *Palestine Hotels Ltd., 1931–1932*.

71   NA, *Young to Parkinson, 07/02/1933*.

72   NA, *E. Mills to Parkinson, 28/04/1933*.

73   NA, *Wauchope to the President of the JNF, 09/02/1933*.

74 Migdal, *Strong Societies and Weak States*, p. 151.

75 NA, *Travel facilities for tourists, 1931*.

76 Jarman, 'Report on the Administration of Palestine 1933', IV, p. 301.

77 NA, *Regulations regarding travellers visas, 14/03/1034*.

78 Jarman, 'Report on the Administration of Palestine 1933', IV, p. 312.

79 NA, *£60 Deposit requirement, 1934*. NA, *Werner, to the Commissioner for Migration, 08/03/1936*. NA, *Wauchope to Thomas, 30/03/1936*.

80 NA, *£60 Deposit requirement, 1934*.

81 NA, *The British Passport Control Officer, The Hague, 19/03/1934*, p. 2.

82 NA, *The British Passport Control Officer, The Hague, 19/03/1934*, pp. 2–3.

83 NA, *Spencer, Visas for Palestine, 02/06/1933*.

84 NA, *Memorandum, to Wauchope, 23/01/1937*.

85 NLI, Robinson, *Palestine, the Tourists' Paradise, 09/1926*.

86 Council of the League of Nations, 'The Palestine Mandate', Art. 4. Krämer, *Geschichte Palästinas*, pp. 198–200, 222–226. Reich and Goldberg, 's.v. Jewish Agency'.

87 Next to Robinson's statement, these aims were expressed by the *Zionist Information Bureau for Tourists*: cf. Cohen-Hattab and Shoval, *Tourism, Religion and Pilgrimage*, p. 66. Cf. also Cohen-Hattab, 'Zionism, Tourism, and the Battle for Palestine', pp. 62–63.

88 Berkowitz, *Western Jewry*, pp. 132–133.

89 NLI, *News About Jews Everywhere, 04/1922*. Cohen-Hattab and Shoval, *Tourism, Religion and Pilgrimage*, pp. 64–71.

90 Cohen-Hattab and Shoval, *Tourism, Religion and Pilgrimage*, p. 70. Cohen-Hattab, 'Zionism, Tourism, and the Battle for Palestine', p. 73.

91 NLI, Robinson, *Palestine, the Tourists' Paradise, 09/1926*. On "dirty Arab boatmen" carrying tourists from the ships to the harbour: NLI, *An Advertiser Sees Palestine, 05/1931*.

92 Cohen-Hattab and Shoval, *Tourism, Religion and Pilgrimage*, p. 70. Cohen-Hattab, 'Zionism, Tourism, and the Battle for Palestine', pp. 71–72. Berkowitz, *Western Jewry*, pp. 129–132.

93 NA, *Despatch No. 8 of the High Commissioner, 03/01/1927*. NA, *Guides Ordinance No. 18, 01/07/1927*. Cohen-Hattab and Shoval, *Tourism, Religion and Pilgrimage*, p. 71.

94 Pappé, *A History of Modern Palestine*, pp. 93–97. Sela, 'The "Wailing Wall" Riots'.

95 SOAS, Smith, *Travel Diaries, Diary 1929*, II, 28/11/1929. Cohen-Hattab, 'Zionism, Tourism, and the Battle for Palestine', p. 75.

96 Cohen-Hattab and Shoval, *Tourism, Religion and Pilgrimage*, pp. 72–73. Berkowitz, *Western Jewry*, p. 132. IfP, Vilnay, *Steimatzky's Palestine Guide*, pp. 150–151. IfP, Zionist Information Bureau for Tourists in Palestine, *Guide to New Palestine 1934*, advertisement section. FDC, Thomas Cook & Son, *L'Egypte, Le Nil, la Palestine*, p. 73.

97 Cohen-Hattab and Katz, 'The Attraction of Palestine', p. 173. Khalidi, *The Iron Cage*, pp. 11–13.

98 Cohen-Hattab and Katz, 'The Attraction of Palestine', p. 170. Klein, *The Second Million*.

99 Mammon, 'Pioneers of Modern Tourism', pp. 30–35.

100 Mammon, 'Pioneers of Modern Tourism', pp. 36–37.

101 It had been founded in 1923: NLI, Grunhut, *The Zionist Information Bureau for Tourists, 18/11/1931*. Already before, some Zionist tourist guidebooks had been published in the early 1920s: Cohen-Hattab, 'Zionism, Tourism, and the Battle for Palestine', p. 67.

102 IfP, Zionist Information Bureau for Tourists in Palestine, *Guide to New Palestine 1934*, p. 9.

103 BL, Zionist Organization in Palestine, *Report on the Work of the Zionist Organization, 11/08/1929*, pp. 11–12.

104 BL, Zionist Organization in Palestine, *Report on the Work of the Zionist Organization, 11/08/1929*, p. 172.

105 For the previous publications cf. Cohen-Hattab and Shoval, *Tourism, Religion and Pilgrimage*, pp. 66–67. IfP, Zionist Information Bureau for Tourists in Palestine, *Guide to New Palestine 1934*. BL, Zionist Organization in Palestine, *Report on the Work of the Zionist Organization, 11/08/1929*, p. 12. BL, Löwenstein, *A Guide to Jewish Palestine, 1927*.

106 IfP, Zionist Information Bureau for Tourists in Palestine, *Guide to New Palestine 1934*, p. 7.

107 IfP, Zionist Information Bureau for Tourists in Palestine, *Guide to New Palestine 1934*.

108 BL, Zionist Organization in Palestine, *Report on the Work of the Zionist Organization, 11/08/1929*, p. 174.

109 Cohen-Hattab and Shoval, *Tourism, Religion and Pilgrimage*, p. 55.

110 Krämer, *Geschichte Palästinas*, pp. 198–200, 254–260.

111 Krämer, *Geschichte Palästinas*, pp. 254, 259–260.

112 Kupferschmidt, *The Supreme Muslim Council*, pp. 133, 237–240. Cohen-Hattab and Shoval, *Tourism, Religion and Pilgrimage*, p. 51. Khalidi, *The Iron Cage*, pp. 55–64.

113 The Waqf (pl. Auqaf) were Islamic religious endowments, among other obligations they funded educational or medical institutions. After the end of the Ottoman Empire, the British assigned the administration of these endowments in Palestine to the *Supreme Muslim Council.*

114 Wallach, *A City in Fragments*, p. 199. Wallach also pointed out that the hotel hosted the Arab Exhibitions of 1933 and 1934, similarly conveying a message of Arab modernity: Wallach, *A City in Fragments*, p. 201.

115 Kupferschmidt, *The Supreme Muslim Council*, p. 136.

116 Ohad Smith, 'Hotel Design in British Mandate', p. 102. The architect is identified as Nihad bey in the Palestine Bulletin: NLI, *Jerusalem's New Hotel.* Uri Kupferschmidt transliterates his name as Nahhas bey: Kupferschmidt, *The Supreme Muslim Council*, p. 136. In other secondary sources, he is also known as Mehmet Nihat Nigisberk.

117 Kupferschmidt, *The Supreme Muslim Council*, fig. 4.

118 Tom Segev interpreted the relation as a pragmatic stance on behalf of Amin al-Husayni: Segev, *Es war einmal ein Palästina*, pp. 303–304. The striking imbalance in access to capital between Palestinian and Zionist actors might have been a reason. Khalidi, *The Iron Cage*, pp. 12–14. Smith, *The Roots of Separatism*. Norris, *Land of Progress*, pp. 64–68, 91–98. Certainly, the cooperation was no exception. Cyrus Schayegh analysed the economic cooperation and the interdependence between the Yishuv and Syria: Schayegh, *The Middle East and the Making of the Modern World*, pp. 226–242. // Wallach mentioned that Qatinqa later installed secret microphones in the building, in the service of the Haganah: Wallach, *A City in Fragments*, p. 199.

119 NLI, *Jerusalem's New Hotel.* Kupferschmidt, *The Supreme Muslim Council*, pp. 136–137.

120 NLI, *The New Jerusalem Palace Hotel.* NLI, *Jerusalem's New Hotel.* Cohen-Hattab and Shoval, *Tourism, Religion and Pilgrimage*, pp. 58–59. Kupferschmidt, *The Supreme Muslim Council*, pp. 136–137.

121 NA, *Despatch No. 985, 23/09/1932.*

122 Birzeit University, Khoori, *Jerusalem: How to see it*, p. 36. In the advertisement for his Jerusalem guide, Khoori stated that the author was from the Holy Land: BL, Khoori, *Cairo: How to See It*, p. 182.

123 Birzeit University, Khoori, *Jerusalem: How to see it*, pp. 34–36.

124 Mammon, 'Pioneers of Modern Tourism', p. 33.

125 NA, *Young to Parkinson, 07/02/1933.* NLI, *Advertise Palestine, 05/1932.*

126 Jarman, 'Report on Palestine Administration 1922', I, pp. 336–337, 348.

[127] NLI, *Quidnunc, In a Few Lines.* NLI, *In Brief.* The Jamal Brothers were tour agents in Jerusalem: IfP, Vilnay, *Steimatzky's Palestine Guide*, p. 426. JD PRIVATE COLLECTION, Tourist Development Association of Palestine, *Ancient and Modern Palestine*, p. 8.

[128] NLI, *Social and Personal, 01/1933.* In 1935, Sargent was no longer mentioned as a member, he seems to have been replaced by Cecil R. Webb: JD PRIVATE COLLECTION, Tourist Development Association of Palestine, *Ancient and Modern Palestine*, p. 5.

[129] NLI, *Advertise Palestine, 05/1932.* NLI, *Tourist Association Meeting, 11/1933.* The Press Bureau of the Executive in Jerusalem was abolished in the spring of 1927 for economic reasons. G. Agronsky, however, volunteered to act in an honorary capacity as Director of Information, meet local and foreign press respresevatives as well as distinguished visitors. BL, Zionist Organization in Palestine, *Report on the Work of the Zionist Organization, 11/08/1929*, p. 11.

[130] NLI, Settel, *Selling Egypt, 04/1937.* Andrea Stanton highlighted that the *Palestine Post* regulaly reported on Lebanese activities in tourism advertisement as well. Stanton, 'Locating Palestine's Summer Residence', p. 55.

[131] NLI, *Advertise Palestine, 05/1932.*

[132] NLI, *Tourist Association Meeting, 11/1933.*

[133] The Zionists well understood their advantage over the Palestinians in demography and capital in the 1930s: Khalidi, *The Iron Cage*, p. 12.

[134] NLI, *Tourist Association Meeting, 11/1933.*

[135] JD PRIVATE COLLECTION, Tourist Development Association of Palestine, *Ancient and Modern Palestine*, pp. 27–33. Tamari, 'Lepers, Lunatics, and Saints'.

[136] Eg JD PRIVATE COLLECTION, Tourist Development Association of Palestine, *Ancient and Modern Palestine*, pp. 20–26.

[137] On the arrival of the film in Palestine: NLI, *Social and Personal. NA, Letter by Stewart Perowne, 16/12/1938.*

[138] NA, *Letter by Stewart Perowne, 16/12/1938.*

[139] Cf. Fabian, *Time and the Other*, p. 31

[140] NA, *J. C. Lamont, 14/06/1939.*

[141] RCSA, Williams, *Travel Diary: Palestine and Syria*, p. 124. Eunice Holliday reported another version of the event: According to her, Balfour was able to visit the Dome of the Rock thanks to the disguised gendarmerie guiding him there: Holliday, *Letters from Jerusalem*, p. 52. On the visibility of tourists in the streets of Jerusalem (1924) cf. also Holliday, *Letters from Jerusalem*, p. 43.

[142] Cohen-Hattab, 'Zionism, Tourism, and the Battle for Palestine', p. 75.

[143] Steimatzky's guide for Jewish travellers mentioned the restrictions regarding tour guides, yet the author described the church in detail, including a precise plan of the Church's interior: IfP, Vilnay, *Steimatzky's Palestine Guide*, pp. 169–174.

[144] SOAS, Smith, *Travel Diaries, Diary 1929, II,* 28/11/1929. Cohen-Hattab, 'Zionism, Tourism, and the Battle for Palestine', p. 75.

[145] Cohen-Hattab and Shoval, *Tourism, Religion and Pilgrimage*, pp. 6–11. The edited volume by Dalachanis and Lemire introduces many of the Jerusalemite communities: Dalachanis and Lemire, *Ordinary Jerusalem.*

[146] E.g. „Proposed itinerary for a visit of three days to Jerusalem and vicinity": IfP, Matson, *The American Colony Palestine Guide*, p. vii.

[147] Berkowitz, *Western Jewry*, p. 142.

[148] In the photograph album of Rowland Christopher Jerram, five photographs represented Jerusalem, three of which showed the inner city. The clearest and best shot was certainly a clearly structured black-and-white photograph showing the cupolas of the Dome of the Rock and the smaller Dome

of the Chain (Kubbet al-Silsileh) next to it: NMM, Jerram, *Photograph Album, 1922–1925*. They date from 1924.

149   NA, Lawrence, *Travel photographs: Voyage to Egypt*. Moreover, Edwin W. Smith and George H. Williams mentioned in their diaries that they had taken photographs of the Dome: SOAS, Smith, *Travel Diaries, Diary 1929, II*, 28/11/1929. RCSA, Williams, *Travel Diary: Palestine and Syria*, p. 113.

150   BL, Messinger Murdoch, *The World in Its True Colours*, p. 101. FRENCH LINES, Compagnie des Messageries Maritimes, *Souvenirs d'un voyage*.

151   Cf. for example the postcard in Dora Hilda Southgate's Album: SOAS, Southgate, *Postcard Album*. IfP, Matson, *The American Colony Palestine Guide*, pp. 54–55, 60.

152   ANF, Vacaresco, *La Croisière Bleue, in: Croisières*, pp. 63–64. NLS, Steele-Maitland, *Diary of Visit to Egypt*, 19/01/1932.

153   MECA, Harrison, *Journal of Visit to Palestine*, 24/05/1925. Holliday, *Letters from Jerusalem*, p. 66.

154   Robson, 'Becoming a Sectarian Minority', pp. 62–67.

155   Smith noted that he received the brochure as an admission ticket: SOAS, Smith, *Travel Diaries, Diary 1929, II*, 28/11/1929. George H. Williams copied information on the site's history from the booklet: RCSA, Williams, *Travel Diary: Egypt, 1925*, pp. 175–178. IfP, Vilnay, *Steimatzky's Palestine Guide*, p. 148.

156   IfP, *A Brief Guide*, p. 3. Irving, '"This is Palestine"', p. 2.

157   IfP, *A Brief Guide*.

158   SOAS, Smith, *Travel Diaries, Diary 1929, II*, Nov 28, 1929. RCSA, Williams, *Travel Diary: Egypt, 1925*, pp. 175–178.

159   SOAS, Smith, *Travel Diaries, Diary 1929, II*, 05/12/1929.

160   Cohen-Hattab and Shoval, *Tourism, Religion and Pilgrimage*, p. 70. IfP, Vilnay, *Steimatzky's Palestine Guide*, p. 473.

161   IfP, Matson, *The American Colony Palestine Guide*, advertisement section. IfP, Vilnay, *Steimatzky's Palestine Guide*, p. 472. Monmarché, *Syrie – Palestine – Iraq – Transjordanie*, p. 519. Hope, *Hotel Tiberias*, pp. 190, 196.

162   IfP, Matson, *The American Colony Palestine Guide*, p. 312.

163   RCSA, Williams, *Travel Diary: Palestine and Syria*, p. 70.

164   SOAS, Smith, *Travel Diaries, Diary 1929, II*, 06/12/1929. Cf. also Thomas Cook & Son et al., *The Traveller's Handbook for Palestine and Syria, 1929*, pp. 256–259.

165   NLS, Steele-Maitland, *Diary of Visit to Egypt*, 20/01/1932.

166   Monmarché, *Syrie – Palestine – Iraq – Transjordanie*, p. 520. Coleman, 'From the Sublime to the Meticulous', pp. 282–283.

167   E.g. SOAS, Southgate, *Postcard Album*. IfP, Matson, *The American Colony Palestine Guide*, p. 318. Coleman, 'From the Sublime to the Meticulous', p. 284.

168   NLI, *Palestine from Day to Day*.

169   NA, *Palestine as a Health Resort*. JD PRIVATE COLLECTION, Vilnay, *Steimatzky's Palästina-Führer*, pp. 75–78.

170   NA, *Palestine as a Health Resort*.

171   NLI, *Syrian Company to Lease Tiberias*. NA, *Palestine as a Health Resort*.

172   Le Gras, *Le Lac Méditerranéen*, p. 112.

173   NA, *Memorandum on Tiberias Baths Concession*.

174   Norris, *Land of Progress*, pp. 93–94. NLI, *Palestine from Day to Day*. NLI, *Jewish Company Acquires Tiberias Hot Springs*. NLI, *Hamamat Tabariyya yubaʿu imtiyazha li-l-yahud*.

175   Khalidi, *All That Remains*, pp. 518–520. According to Khalidi, Sulayman Nassif received the concession for the exploitation of the springs of al-Hamma in 1936 only: Khalidi, *All That Remains*, p. 519.

176   NA, *Despatch No. 931, 24/10/1931*.

[177] Norris highlighted the significant impact of scientific credentials in attracting support from British political circles (sometimes outweighing financial difficulties of the Zionist competitors) in the example of the Dead Sea concession: Norris, *Land of Progress*, pp. 154–156. Cf. also Krämer, *Geschichte Palästinas*, pp. 203–206.

[178] NA, *Tiberias Baths Concession 1931–1932*. NA, *Meeting No. 12, 08/10/1931*. Pedersen, 'The Impact of League Oversight'.

[179] NA, *The Hot Springs of Tiberias*.

[180] NLI, *Activities among the Zionists*. Jarman, 'Report on the Administration of Palestine 1933', pp. 318–319. The committee suggested that the stamps advertised visits to the following sites: the baths of Tiberias and Al-Hamma, the citrus groves of Jaffa, as well as the harbour of Haifa: NA, *Tourist Traffic 1933*.

[181] JD PRIVATE COLLECTION, Vilnay, *Steimatzky's Palästina-Führer*, pp. 75–77.

[182] IfP, Vilnay, *Steimatzky's Palestine Guide*, p. 91.

[183] NLS, Steele-Maitland, *Diary of Visit to Egypt*, 20/01/1932. IfP, Matson, *The American Colony Palestine Guide*, pp. 311–313.

[184] NLI, *Notice, 01/1931*.

[185] JD PRIVATE COLLECTION, Chardon, *Méditerranée Orientale – Egypte*, p. 162.

[186] NA, *Tiberias Baths Concession 1938*. For the competition from al Hamma cf.: NA, *Despatch No. 931, 24/10/1931*.

[187] NLI, *Notes for Tourists VII*.

[188] SOAS, Smith, *Travel Diaries, Diary 1929, II*, 27/11/1929, 01/12/1929, 02/12/1929. RCSA, Williams, *Travel Diary: Palestine and Syria*, p. 7.

[189] Cf. RCSA, Williams, *Travel Diary: Palestine and Syria*, p. 7.

[190] NMM, Roe, *Journal of Donovan C. Roe*, 24/08/1936. NLI, *Palestine Tourist Agencies Overlook Jewish Settlements*. MECA, Harrison, *Journal of Visit to Palestine*, 18/03/1925. An exception was Eunice Holliday, who had studied architecture and apparently enjoyed the modernist buildings of Tel Aviv and Nathanyah: Holliday, *Letters from Jerusalem*, pp. 61, 158–159.

[191] IfP, Zionist Information Bureau for Tourists in Palestine, *Guide to New Palestine 1934*, front section advertisements, p. 9.

[192] IfP, Zionist Information Bureau for Tourists in Palestine, *Guide to New Palestine 1934*, p. 53. Cohen-Hattab, 'Zionism, Tourism, and the Battle for Palestine', p. 69.

[193] IfP, Vilnay, *Steimatzky's Palestine Guide*, f.

[194] IfP, Vilnay, *Steimatzky's Palestine Guide*, pp. 65–67. Similarly, the Shemen factory, producing olive oil and soaps, offered guided tours around the factory to tourists: IfP, Zionist Information Bureau for Tourists in Palestine, *Guide to New Palestine 1934*, advertisement section.

[195] Krämer, *Geschichte Palästinas*, pp. 136–137, 226, 357. Norris, *Land of Progress*, pp. 47–48.

[196] IfP, Zionist Information Bureau for Tourists in Palestine, *Guide to New Palestine 1934*, pp. 82–83. The Keren Hayesod (Palestine Foundation Fund) was the major fund-raising and financial institution of the Zionist Organization. The Fund provided financial aid to finance immigration to and settlement in Palestine: cf. Reich and Goldberg, 's.v. Keren Hayesod'. On the Keren Kayemeth Le Israel (Jewish National Fund): Reich and Goldberg, 's.v. Jewish National Fund'. Cf. also Pappé, *A History of Modern Palestine*, p. 97.

[197] IfP, Zionist Information Bureau for Tourists in Palestine, *Guide to New Palestine 1934*, pp. 83–87.

[198] IfP, Zionist Information Bureau for Tourists in Palestine, *Guide to New Palestine 1934*, pp. 31–49.

[199] Jarman, 'Report on the Administration of Palestine 1924', I, p. 539.

[200] IfP, Zionist Information Bureau for Tourists in Palestine, *Guide to New Palestine 1934*, e.g. pp. 73–77. IfP, Vilnay, *Steimatzky's Palestine Guide*, pp. 41–42.

201 IfP, Zionist Information Bureau for Tourists in Palestine, *Guide to New Palestine 1934*. Map signed Zink. M. Pikovsky, Jerusalem.

202 On the development and nineteenth-century discourses on the Dead Sea: Norris, *Land of Progress*, pp. 142–150.

203 RCSA, Williams, *Travel Diary: Palestine and Syria*, pp. 189–190. Norris argued that for that reason it was the ideal site to exhibit the positive effects of the British colonial presence: Norris, *Land of Progress*, pp. 139–140.

204 Holliday, *Letters from Jerusalem*, p. 38.

205 Thomas Cook & Son, Luke and Garstang, *The Traveller's Handbook for Palestine and Syria, 1924*, pp. 167–169.

206 IfP, Matson, *The American Colony Palestine Guide*, p. 249. Norris, *Land of Progress*, pp. 150–160.

207 SOAS, Smith, *Travel Diaries, Diary 1929, II*, 29/11/1929.

208 NLS, Steele-Maitland, *Diary of Visit to Egypt*, 18/01/1932. The buildings she mentioned belonged most likely to the town of Jdeideh, founded by Jamal Pasha in 1915 as a place of distributing grain and salt for transport via railway to different parts of the empire (the salt was extracted by the Ottoman company *Djebel Usdum*, continued during the mandate as a Zionist company, *Palestine Salt*). IfP, Matson, *The American Colony Palestine Guide*, p. 247. Monmarché, *Syrie – Palestine – Iraq – Transjordanie*, p. 603.

209 Monmarché, *Syrie – Palestine – Iraq – Transjordanie*, p. 603. Holliday, *Letters from Jerusalem*, pp. 129–130. IfP, Matson, *The American Colony Palestine Guide*, p. 247. Seikaly, *Men of Capital*, p. 7. Vilnay mentions the Novomeysky-Tulloch Company as concessionaries for the exploitation of the Dead Sea minerals: IfP, Vilnay, *Steimatzky's Palestine Guide*, p. LXXXIV.

210 NLI, Tulloch, *Letter to the Editor*.

211 NLI, *Dead Sea Resort*.

212 IfP, Zionist Information Bureau for Tourists in Palestine, *Guide to New Palestine 1934*, p. 63. IfP, Vilnay, *Steimatzky's Palestine Guide*, p. 267. Norris, *Land of Progress*, p. 201.

213 NLI, *Quidnunc, In a Few Lines*. NLI, *Dead Sea Resort*. Al-Husayni: While his position towards Zionism is not entirely clear, he did not support Arab nationalism and cooperated with Jewish entrepreneurs for business purposes. Pappé assumes that his distance towards Arab nationalism was partly due to a generational gap, the nationalist movement appearing alien to him and his pro-Ottoman leanings. Pappé, *The Rise and Fall of a Palestinian Dynasty*, pp. 104–106, 145, 176, 320.

214 NLI, *Social and personal, 07/1933*. NLI, *Sunny Christmas Holidays*. NLI, *Visit "Kallia"*. NLI, *Come to Kallia's Moonlight Dance*. NLI, *Kallia on the Dead Sea*. The destinations of the excursions were the Zarqa River, Masada, and the Hot Springs of Callirhoe: Thomas Cook & Son, Lumby and Garstang, *Cook's Traveller's Handbook to Palestine, Syria & Iraq, 1934*, pp. 179–180.

215 IfP, Vilnay, *Steimatzky's Palestine Guide*, pp. 257, 265. Thomas Cook & Son, Lumby and Garstang, *Cook's Traveller's Handbook to Palestine, Syria & Iraq, 1934*, p. 179. NLI, *Drinks till Midnight at Kallia*.

216 IfP, Zionist Information Bureau for Tourists in Palestine, *Guide to New Palestine 1934*, p. 63. Thomas Cook & Son, Lumby and Garstang, *Cook's Traveller's Handbook to Palestine, Syria & Iraq, 1934*, p. 179. Eunice Holliday and the children visited the Dead Sea several times: E.g. Holliday, *Letters from Jerusalem*, pp. 65, 148.

217 NLI, *Kallia Taxi Office*. NLI, *"Kallia" Ltd., Bus Service*.

218 NLI, *Palestine Potash 1935*.

219 NLI, *Social and Personal, 02/1938*.

220 NLI, *Major Tulloch's Description*.

221 NLI, *Readers' Letters: Dead Sea Holiday*.

222 NLI, Tulloch, *Kallia-on-the-Dead-Sea*.

223 NLI, *New Life.*

224 Ohad Smith, 'Hotel Design in British Mandate Palestine', p. 109.

225 NLI, *New Life.*

226 Ohad Smith, 'Hotel Design in British Mandate Palestine', pp. 107–109.

227 NLI, *In Good Humour.*

References

Sources

Archives nationales (France), Pierrefitte-sur-Seine

VACARESCO, HÉLÈNE, *La Croisière Bleue, in: Croisières du Yacht Alphée*, ed. Yvonne Cotnaréanu, Fonds Lucien Romier, Voyages, ANF/408 AP/2.

Birzeit University, Digital Library

KHOORI, ALEXANDER R., *Jerusalem: How to see it: Including Palestine, Syria & Lebanon*, London, 4th edn., 1929 <https://fada.birzeit.edu/handle/20.500.11889/6006>, accessed 30 Apr 2020, Birzeit University.

British Library, London

AITKEN, D. F., *What do you go to see?*, The Travel Log, 12.1 (1937) p. 6–8, BL.

KHOORI, ALEXANDER R., *Cairo: How to See It*, Alexandria: The Anglo-Egyptian Supply Association, 8th edn., 1928, BL.

LÖWENSTEIN, FRITZ, *A Guide to Jewish Palestine*, ed. by Keren Kayemeth Leisrael (Jewish National Fund), Keren Hayesod (Palestine Foundation Fund), Jerusalem 1927, BL/X.708 42882.

MESSINGER MURDOCH, HELEN, *The World in Its True Colours*, The Photographic Journal, LXX (1930), pp. 100–102, BL/Ac.5051.

WORKERS' TRAVEL ASSOCIATION, *1935 Photograph Competition*, The Travel Log 3.18 (1935), p. 8, BL.

WORKERS' TRAVEL ASSOCIATION, *Easter and Spring Holidays*, The Travel Log 3.18 (1935), p. 9, BL.

WORKERS' TRAVEL ASSOCIATION, *Notes on Summer Programme*, The Travel Log 3.15 (1934), p. 11, BL.

WORKERS' TRAVEL ASSOCIATION, *Palestine in Spring*, The Travel Log 4.22 (1936), p. 11, BL.

WORKERS' TRAVEL ASSOCIATION, *Palestine Reunion*, The Travel Log 3.13 (1934), p. 17, BL.

WORKERS' TRAVEL ASSOCIATION, *Travel Log Photograph Competition, 1934*, The Travel Log 3.18 (1935), pp. 6–8, BL.

WORKERS' TRAVEL ASSOCIATION, *W.T.A. Easter Holidays – Abroad and at Home*, The Travel Log 3.14 (1934), pp. 10–11, BL.

WORKERS' TRAVEL ASSOCIATION, *"What Our Members Think"*, The Travel Log 3.17 (1934), pp. 20–21, BL.

ZIONIST ORGANIZATION IN PALESTINE, *Report on the Work of the Zionist Organization in Palestine Submitted to the Council of the Jewish Agency Zurich, 11/08/1929*, BL.

Association French Lines, Le Havre

COMPAGNIE DES MESSAGERIES MARITIMES, *Souvenirs d'un voyage en Orient, 1926–1927, photographs*, French Lines/1997 002 5012.

Fouad Debbas Collection, Sursock Museum, Beirut

THOMAS COOK & SON, *L'Egypte, le Nil, la Palestine et la Syrie, Services du Nil, Saison 1925–26.*, Programme des Arrangements pour la visite de l'Egypte, du Nil, du Soudan, de la Palestine et de la Syrie. Avec cartes, illustrations et plans des bateaux de touristes, FDC.

Institute for Palestine Studies, Beirut

A Brief Guide to al-Haram al-Sharif, Jerusalem, 1924, published by the Supreme Moslem Council, Jerusalem: Moslem Orphanage Printing Press, 1924, signed G.A. (George Antonius), IfP/CA 297.35 B853.

Annual Report of the Department of Customs, Excise, and Trade, 1935, Special Supplement to the Palestine Commercial Bulletin, 03/1936, IfP/CA 336.26 P157a.

CHAMBER OF COMMERCE OF JAFFA & DISTRICT, *Annual Bulletin 1927–1928*, IfP/CA 380.956 C443a.

CHAMBER OF COMMERCE OF JAFFA & DISTRICT, *Annual Bulletin 1928–1929*, IfP/CA 380.956 C443a.

CHAMBER OF COMMERCE OF JAFFA & DISTRICT, *Annual Bulletin 1929–1930*, IfP/CA 380.956 C443a.

CHAMBER OF COMMERCE OF JAFFA & DISTRICT, *Annual Bulletin 1931–1932*, IfP/CA 380.956 C443a.

CHAMBER OF COMMERCE OF JAFFA & DISTRICT, *Annual Bulletin 1932–1933*, IfP/CA 380.956 C443a.

GOVERNMENT OF PALESTINE, DEPARTMENT OF CUSTOMS, EXCISE AND TRADE, *First Census of Industries 1928*, 1929, IfP.

MATSON, G. OLAF, *The American Colony Palestine Guide*, Jerusalem: The American Colony Stores, 3rd edn., 1930, IfP/CA 915.69404 M 434a.

STEAD, K. W., *Economic Conditions in Palestine*, Department of Overseas Trade, London 1931, IfP.

VILNAY, ZEV, *Steimatzky's Palestine Guide*, General Survey in Collaboration with Dr A. Bonne, Steimatzky Publishing, Jerusalem 1935, IfP/CA 915.694 V762s.

ZIONIST INFORMATION BUREAU FOR TOURISTS IN PALESTINE, *Guide to New Palestine 1934–1935*, Jerusalem: Asriel Press, 8th edn., 1934, IfP.

jd private collection

CHARDON, J., *Méditerranée Orientale – Egypte: Guide du passager pour les escales des croisières et pour les excursions en Syrie et en Égypte*, Paris: Hachette, 1935, jd private collection.

TOURIST DEVELOPMENT ASSOCIATION OF PALESTINE, *Ancient and Modern Palestine: Composed by Various Local Authorities*, Jerusalem: Golderg's Printing Press, 1935, jd private collection.

VILNAY, ZEV, *Steimatzky's Palästina-Führer*, Jerusalem: Steimatzky Publishing, 1935, jd private collection.

Middle East Centre Archive, University of Oxford

HARRISON, HELENA, *Journal of Visit to Palestine, 1925*, Helena Harrison Papers, MECA/GB 165-0136.

Inauguration Dinner King David Hotel Jerusalem, 19/01/1931, Cecil R. Webb Papers, MECA/GB 165-0297.

National Archives, Kew

£60 Deposit requirement, 1934, Colonial Office, Palestine Original Correspondence, NA/CO 733/260/1.

Commissioner for Migration and Statistics, Memorandum, enclosed to the despatch of High Commissioner Arthur Wauchope, 23/01/1937, Colonial Office, Palestine Original Correspondence, NA/CO 733/300/8.

Despatch No. 8 of the High Commissioner, Jerusalem, to L. C. Amery, Secretary of State for the Colonies, 03/01/1927, Colonial Office, Palestine Original Correspondence, NA/CO 733/134/1.

Despatch No. 931, Young to J.H. Thomas, 24/10/1931, Colonial Office, Middle East, Tiberias Baths Concession 1931–1932, NA/CO 733/210/10.

Despatch No. 985 from High Commissioner Arthur Wauchope to Sir Philip Cunliffe-Lister, Secretary of State for the Colonies, 23/09/1932, Colonial Office, Palestine Original Correspondence, NA/CO 733-225-11.

E. Mills, Chief Secretary's Office, Jerusalem, to Parkinson, Colonial Office, 28/04/1933, Colonial Office, Palestine Original Correspondence, NA/CO 733/242/3.

Guides Ordinance No. 18 of 1927, An Ordinance to Provide for the Licensing and Control of Guides, 01/07/1927, Colonial Office, Palestine Original Correspondence, NA/CO 733/134/1.

J. C. Lamont, Under Secretary of State, Colonial Office, 14/06/1939, Colonial Office, Palestine Original Correspondence, NA/CO 733/406/4.

J.A. Seiler, Palestine Hotels Ltd., Letter to the Chief Secretary, Government of Palestine, Jerusalem, 09/04/1931, Colonial Office, Palestine Original Correspondence, NA/CO 733/225/11.

LAWRENCE, *Travel photographs: Voyage to Egypt, 1937*, NA/CO1069/797.

Letter by Stewart Perowne, BBC, to H.F. Downie, Colonial Office, 16/12/1938, Colonial Office, Palestine Original Correspondence, NA/CO 733/386/3.

Memorandum on Tiberias Baths Concession, Colonial Office, Middle East Dept., Tiberias Baths Concession 1932, NA/CO 733/227/1.

Memorandum: Regulations regarding travellers visas for Palestine, attached by Passport Officer Parkin to a letter adressed at British Palestine & Eastern Tours, 14/03/1034, Colonial Office, Palestine Original Correspondence, NA/CO 733/260/1.

Minutes of Meeting No. 12 of Tiberias Municipal Council held on 08/10/1931, Colonial Office, Middle East Dept., Tiberias Baths Concession 1931–1932, NA/CO 733/210/10.

Palestine as a Health Resort: Tiberias Hot Springs Concession to Jewish Company Signed, Jewish Telegraphic Agency Bulletin, 22/02/1932, Colonial Office, Middle East Dept., Tiberias Baths Concession 1932, NA/CO 733/227/1.

Palestine Hotels Ltd., 1931–1932, Colonial Office, Palestine Original Correspondence, NA/CO 733/225/11.

Spencer, Passport Control Department, Foreign Office, to His Majesty's Consular and Passport Control Officers, Visas for Palestine, 02/06/1933, Colonial Office, Palestine Original Correspondence, NA/CO 733/260/1.

The British Passport Control Officer, The Hague, to H. M. Minister, The Hague, 19/03/1934, Colonial Office, Palestine Original Correspondence, NA/CO 733/260/1.

The High Commissioner Arthur G. Wauchope, to the President of the Jewish National Fund, 09/02/1933, Colonial Office, Palestine Original Correspondence, NA/CO 733/242/3.

The High Commissioner Arthur Wauchope to J.H. Thomas, His Majesty's Principal Secretary of State for the Colonies, 30/03/1936, Colonial Office, Palestine Original Correspondence, NA/CO 733/300/7.

The Hot Springs of Tiberias, Palestine (Hamei-Tiberia), The Tiberias Hot Springs Co. Ltd., 2nd edn., July 1933, Colonial Office, Middle East Dept., Tiberias Baths Concession 1938, NA/CO 733/386/12.

Tiberias Baths Concession 1931–1932, Colonial Office, Middle East Dept., NA/CO 733/210/10.

Tiberias Baths Concession 1938, Colonial Office, Middle East Dept., NA/CO 733/386/12.

Tourist Traffic 1933, Colonial Office, Palestine Original Correspondence, NA/CO 733/242/3.

Travel facilities for tourists, 1931: Communication between Spencer, Clauson, Williams, 17/12/1931 – 23/01/1932, Colonial Office, Palestine Original Correspondence, NA/CO 733/196/2.

Werner Senator, Executive of the Jewish Agency for Palestine, to the Commissioner for Migration and Statistics, 08/03/1936, Colonial Office, Palestine Original Correspondence, NA/CO 733/300/7.

Young, Chief Secretary's Office, Jerusalem, to Parkinson, Colonial Office, London, 07/02/1933, Colonial Office, Palestine Original Correspondence, Tourist Traffic 1933, NA/CO 733/242/3.

National Library of Israel, Jerusalem

Activities among the Zionists, B'nai B'rith Messenger, 01/06/1922, p. 5. Historical Jewish Press, NLI.

Advertise Palestine: Tourist Association's First Function, The Palestine Bulletin, 08/05/1932, p. 4. Historical Jewish Press, NLI.

An Advertiser Sees Palestine, The Palestine Bulletin, 21/05/1931, p. 3. Historical Jewish Press, NLI.

Come to Kallia's Moonlight Dance [advertisement], The Palestine Post, 18/10/1934, p. 6. Historical Jewish Press, NLI.

Dead Sea Resort Springs to Life: Formal Opening of Kallia, The Palestine Post, 01/05/1933, p. 5. Historical Jewish Press, NLI.

Drinks till Midnight at Kallia, The Palestine Post, 05/11/1935, p. 5. Historical Jewish Press, NLI.

GRUNHUT, M., *The Zionist Information Bureau for Tourists*, The Palestine Bulletin, 18/11/1931, p. 2. Historical Jewish Press, NLI.

Hamamat Tabariyya yuba῾u imtiyazha li-l-yahud, [The concession of the Tiberias Baths is sold to the Jews], Filastin, 25/02/1932, p. 2. Arabic Newspaper Archive of Ottoman and Mandatory Palestine, NLI.

In Brief, The Palestine Post, 06/07/1934, p. 4. Historical Jewish Press, NLI.

In Good Humour, The Palestine Post, 18/02/1938, p. 4. Historical Jewish Press, NLI.

"Kallia" Ltd., Bus Service: A Three Days' Popular Trip to the Emek and Galilee, The Palestine Post, 20/09/1935, p. 7. Historical Jewish Press, NLI.

Jerusalem's New Hotel, The Palestine Bulletin, 23/12/1929, p. 4. Historical Jewish Press, NLI.

Jewish Company Acquires Tiberias Hot Springs, The Sentinel, 17/02/1928, p. 4. Historical Jewish Press, NLI.

Kallia on the Dead Sea [advertisement], The Palestine Post, 23/12/1934, p. 6. Historical Jewish Press, NLI.

Kallia Taxi Office [advertisement], The Palestine Post, 20/06/1935, p. 3. Historical Jewish Press, NLI.

Major Tulloch's Description: Life at the Dead Sea, The Palestine Post, 22/09/1937, p. 3. Historical Jewish Press, NLI.

New Life at the Dead Sea: District Commissioner opens Kallia Hotel, The Palestine Post, 17/02/1938, p. 1. Historical Jewish Press, NLI.

News About Jews Everywhere: Keep Tourists off from Jewish Colonies, The American Jewish World, 21/04/1922, p. 4. Historical Jewish Press, NLI.

Notes for Tourists VII: A Jewish Communal Settlement, The Palestine Bulletin, 12/02/1930, p. 2. Historical Jewish Press, NLI.

Notice, The Palestine Bulletin, 31/01/1931, p. 4. Historical Jewish Press, NLI.

Palestine from Day to Day, The Palestine Bulletin, 29/01/1928, p. 3. Historical Jewish Press, NLI.

Palestine from Day to Day, The Palestine Bulletin, 15/01/1929, p. 3. Historical Jewish Press, NLI.

Palestine Potash 1935: Considerable Increase of Sales, The Palestine Post, 09/10/1936, p. 15. Historical Jewish Press, NLI.

Palestine Tourist Agencies Overlook Jewish Settlements, The Sentinel, 12/05/1938, p. 20. Historical Jewish Press, NLI.

Quidnunc, In a Few Lines, The Palestine Bulletin, 04/05/1932, p. 3. Historical Jewish Press, NLI.

Quidnunc, In a Few Lines, The Palestine Bulletin, 17/11/1932, p. 4. Historical Jewish Press, NLI.

Readers' Letters: Dead Sea Holiday, The Palestine Post, 16/02/1939, p. 6. Historical Jewish Press, NLI.

ROBINSON, M., *Palestine, the Tourists' Paradise*, The Sentinel, 24/09/1926, p. 2. Historical Jewish Press, NLI.

SETTEL, ARTHUR, *Selling Egypt: How Tourists Are Attracted to the Land of the Nile. Setting an Example for the Holy Land*, The Palestine Post, 25/04/1937, p. 5. Historical Jewish Press, NLI.

Social and Personal, The Palestine Post, 06/07/1933, p. 5. Historical Jewish Press, NLI.

Social and Personal, The Palestine Post, 11/02/1938, p. 3. Historical Jewish Press, NLI.

Social and Personal, The Palestine Post, 06/03/1938, p. 3. Historical Jewish Press, NLI.

Social and Personal, The Palestine Post, 18/01/1933, p. 5. Historical Jewish Press, NLI.

Sunny Christmas Holidays on the Dead Sea: Special Excursions by Twin-Screw Motorboat "Kallirhoe", The Palestine Post, 22/12/1933, p. 5. Historical Jewish Press, NLI.

Syrian Company to Lease Tiberias Baths, The Palestine Bulletin, 19/10/1926, p. 1. Historical Jewish Press, NLI.

The New Jerusalem Palace Hotel, The Palestine Bulletin, 28/02/1929, p. 3. Historical Jewish Press, NLI.

Tourist Association Meeting: Palestine Folder 60.000 copies, The Palestine Post, 15/11/1933, p. 2. Historical Jewish Press, NLI.

Tourist Traffic Decreasing: Adverse Effects of Government Measures Feared, The Palestine Post, 02/01/1934, p. 1. Historical Jewish Press, NLI.

TULLOCH, T. G., *Kallia-on-the-Dead Sea: A Great Health Resort*, The Palestine Post, 14/02/1938, p. 2. Historical Jewish Press, NLI.

TULLOCH, T. G., *Letter to the Editor: Kallia on the Dead Sea*, The Palestine Bulletin, 22/11/1932, p. 2. Historical Jewish Press, NLI.

Visit "Kallia" on the Dead Sea [advertisement], The Palestine Post, 18/02/1934, p. 7. Historical Jewish Press, NLI.

National Library of Scotland, Edinburgh

STEELE-MAITLAND, MARY, *Diary of Visit to Egypt, Palestine and Turkey, 24/12/1931 – 29/01/1932*, NLS/ACC 5553.

National Maritime Museum, Greenwich

JERRAM, ROWLAND CHRISTOPHER, *Photograph Album, 1922–1925*, NMM/JRR/26.

ROE, DONOVAN C., *Journal of Donovan C. Roe, during his Naval Service in the Far East, 1930–1937*, NMM/JOD/192/5.

Royal Commonwealth Society Archives, Cambridge University Library

WILLIAMS, GEORGE H., *Travel Diary: Egypt, 1925*, RCSA/RCMS 53/3/1.

WILLIAMS, GEORGE H., *Travel Diary: Palestine and Syria, 1925*, RCSA/RCMS 53/3/2.

School of Oriental and African Studies, London

SMITH, EDWIN W., *Travel Diaries. Diary. 1929. Vol. II.*, SOAS/MMS/17/02/05/02/09/02/02.

SOUTHGATE, DORA HILDA, *Postcard Album*, London Missionary Society, Personal Collections: India, SOAS/CWM-LMS-09-10-12-178.

Other published sources

COUNCIL OF THE LEAGUE OF NATIONS, 'The Palestine Mandate', signed 24/07/1922, The Avalon Project, Lillian Goldman Law Library, 2008 <https://avalon.law.yale.edu/20th_century/palmanda.asp>, updated 2008, accessed 9 Jun 2020.

HOLLIDAY, EUNICE, *Letters from Jerusalem: During the Palestine Mandate* (London, New York: Radcliffe, 1997).

JARMAN, ROBERT L., 'An Interim Report on the Civil Administration of Palestine during the period 1st July 1920–30th June 1921, London 1921', in Robert L. Jarman (ed.), *Palestine and Transjordan Administration Reports 1918–1948*, 16 vols. (Archive Editions, 1995), I, pp. 169–197.

JARMAN, ROBERT L., 'Colonial Office, Department of Migration, Annual Report 1939, London 1940', in Robert L. Jarman (ed.), *Palestine and Transjordan Administration Reports 1918–1948*, 16 vols. (Archive Editions, 1995), VIII, pp. 645–700.

JARMAN, ROBERT L., 'Colonial Office, Report by His Britannic Majesty's Government on the Administration under Mandate of Palestine and Transjordan for the Year 1924, London 1925', in Robert L. Jarman (ed.), *Palestine and Transjordan Administration Reports 1918–1948*, 16 vols. (Archive Editions, 1995), I, pp. 529–598.

JARMAN, ROBERT L., 'Colonial Office, Report by His Britannic Majesty's Government to the Council of the League of Nations on the Administration of Palestine and Trans-Jordan for the Year 1926, London 1927', in Robert L. Jarman (ed.), *Palestine and Transjordan Administration Reports 1918–1948*, 16 vols. (Archive Editions, 1995), II, pp. 227–329.

JARMAN, ROBERT L., 'Colonial Office, Report by His Britannic Majesty's Government to the Council of the League of Nations on the Administration of Palestine and Trans-Jordan for the Year 1927, London 1928', in Robert L. Jarman (ed.), *Palestine and Transjordan Administration Reports 1918–1948*, 16 vols. (Archive Editions, 1995), II, pp. 363–492.

JARMAN, ROBERT L., 'Colonial Office, Report by His Majesty's Government in the United Kingdom of Great Britain and Northern Ireland to the Council of the League of Nations on the Administration of Palestine and Trans-Jordan for the Year 1933, issued by the Colonial Office, London 1934', in Robert L. Jarman (ed.), *Palestine and Transjordan Administration Reports 1918–1948*, 16 vols. (Archive Editions, 1995), IV, pp. 285–621.

JARMAN, ROBERT L., 'Colonial Office, Report on Palestine Administration 1923, London 1924', in Robert L. Jarman (ed.), *Palestine and Transjordan Administration Reports 1918–1948*, 16 vols. (Archive Editions, 1995), I, pp. 391–442.

JARMAN, ROBERT L., 'Empson, C., Department of Overseas Trade, Report on the Economic Conditions in Palestine, July 1935, London 1935', in Robert L. Jarman (ed.), *Palestine and Transjordan Administration Reports 1918–1948*, 16 vols. (Archive Editions, 1995), V, pp. 325–429.

JARMAN, ROBERT L., 'Government of Palestine Report, Report on Palestine Administration, July 1920 – December, 1921, London 1922', in Robert L. Jarman (ed.), *Palestine and Transjordan Administration Reports 1918–1948*, 16 vols. (Archive Editions, 1995), I, pp. 199–332.

JARMAN, ROBERT L., 'Government of Palestine, Report on Palestine Administration, 1922, London 1923', in Robert L. Jarman (ed.), *Palestine and Transjordan Administration Reports 1918–1948*, 16 vols. (Archive Editions, 1995), I, pp. 333–390.

JARMAN, ROBERT L., 'Report of the High Commissioner on the Administration of Palestine, 1920–1925: London 1925', in Robert L. Jarman (ed.), *Palestine and Transjordan Administration Reports 1918–1948*, 16 vols. (Archive Editions, 1995), II, pp. 1–59.

JARMAN, ROBERT L., 'Stead, K. W., Report on the Economic and Financial Situation of Palestine: Department of Overseas Trade, London 1927', in Robert L. Jarman (ed.), *Palestine and Transjordan Administration Reports 1918–1948*, 16 vols. (Archive Editions, 1995), II, pp. 331–361.

JARMAN, ROBERT L., 'The Right Hon. Herbert Samuel to Earl Curzon, Jerusalem, April 2, 1920. Enclosure 2, Communication by the right Honourable Herbert Samuel to the Press, March 25, 1920', in Robert L. Jarman (ed.), *Palestine and Transjordan Administration Reports 1918–1948*, 16 vols. (Archive Editions, 1995), I, p. 134.

LE GRAS, CHARLES, *Le Lac Méditerranéen: Pèlerinage à Jérusalem et croisière en Méditerranée orientale*, du 5 Avril au 19 Mai 1927 (Avignon: Aubanel Frères, 1927).

MONMARCHÉ, MARCEL, *Syrie – Palestine – Iraq – Transjordanie* (Les Guides bleus, Paris: Hachette, 1932).

THOMAS COOK & SON, ELSTON, ROY, LUKE, HARRY CHARLES et al., *The Traveller's Handbook for Palestine and Syria* (2nd edn., London: Simpkin Marshall, 1929).

THOMAS COOK & SON, LUKE, HARRY CHARLES, and GARSTANG, JOHN, *The Traveller's Handbook for Palestine and Syria* (London: Simpkin Marshall, 1924).

THOMAS COOK & SON, LUMBY, CHRISTOPHER, and GARSTANG, JOHN, *Cook's Traveller's Handbook to Palestine, Syria & Iraq* (6th edn., London: Simpkin Marshall, 1934).

YALE LAW SCHOOL, 'The Covenant of the League of Nations', signed 28/06/1919, effective 10/01/1920, The Avalon Project, Lillian Goldman Law Library, 2008 <https://avalon.law.yale.edu/20th_century/leagcov.asp>, updated 2008, accessed 9 Jun 2020.

Secondary Literature

ANDERSON, BENEDICT R., *Imagined Communities: Reflections on the Origin and Spread of Nationalism* (2nd edn., London, New York: Verso, 2016).

BAR, DORON, and COHEN-HATTAB, KOBI, 'A New Kind of Pilgrimage: The Modern Tourist Pilgrim of Nineteenth-Century and Early Twentieth-Century Palestine', *Middle Eastern Studies*, 39/2 (2003), pp. 131–148.

BATTILANI, PATRIZIA, and FAURI, FRANCESCA, 'The Rise of a Service-Based Economy and its Transformation: Seaside Tourism and the Case of Rimini', *Journal of Tourism History*, 1/1 (2009), pp. 27–48.

BERKOWITZ, MICHAEL, *Western Jewry and the Zionist Project, 1914–1933* (Cambridge, New York: Cambridge University Press, 1997).

CAPLAN, NEIL, 'Zionist Visions of Palestine, 1917–1936', *Muslim World*, 84/1–2 (1994), pp. 19–35.

COHEN-HATTAB, KOBI, 'Zionism, Tourism, and the Battle for Palestine: Tourism as a Political Propaganda Tool', *Israel Studies*, 9/1 (2004), pp. 61–85.

COHEN-HATTAB, KOBI, and KATZ, YOSSI, 'The Attraction of Palestine: Tourism in the Years 1850–1948', *Journal of Historical Geography*, 27/2 (2001), pp. 166–177.

COHEN-HATTAB, KOBI, and SHOVAL, NOAM, *Tourism, Religion and Pilgrimage in Jerusalem* (London: Routledge, 2015).

COLEMAN, SIMON, 'From the Sublime to the Meticulous: Art, Anthropology and Victorian Pilgrimage to Palestine', *History & Anthropology*, 13/4 (2002), pp. 275–290.

DALACHANIS, ANGELOS, and LEMIRE, VINCENT (eds.), *Ordinary Jerusalem, 1840–1940: Opening New Archives, Revisiting a Global City* (Leiden, Boston: Brill, 2018).

DOUMANI, BESHARA, 'Rediscovering Ottoman Palestine: Writing Palestinians into History', *Journal of Palestine Studies*, 21/2 (1992), pp. 5–28.

FABIAN, JOHANNES, *Time and the Other: How Anthropology Makes its Object* (New York: Columbia University Press, 2014).

GELVIN, JAMES L., *The Modern Middle East: A History* (4th edn., New York, Oxford: Oxford University Press, 2016).

GLOVER, NIKOLAS, 'Co-Produced Histories: Mapping the Uses and Narratives of History in the Tourist Age', *Public Historian*, 30/1 (2008), pp. 105–124.

HOPE, SEBASTIAN, *Hotel Tiberias: A Tale of Two Grandfathers* (New York: Harper Collins, 2005).

HOROWITZ, ELLIOTT, "Remarkable Rather for Its Eloquence than Its Truth": Modern Travelers Encounter the Holy Land – and Each Other's Accounts Thereof', *The Jewish Quarterly Review*, 99/4 (2009), pp. 439–464.

HOWE, KATHLEEN STEWART, 'Revealing the Holy Land: Nineteenth Century Photographs of Palestine', in Kathleen Stewart Howe (ed.), *Revealing the Holy Land. The Photographic Exploration of Palestine* (Santa Barbara: Santa Barbara Museum of Art, 1997), pp. 16–46.

IRVING, SARAH, "This is Palestine": History and Modernity in Guidebooks to Mandate Palestine', *Contemporary Levant*, 2019, pp. 1–11.

KHALIDI, RASHID ISMAIL, *The Iron Cage: The Story of the Palestinian Struggle for Statehood* (Oxford: Oneworld, 2007).

KHALIDI, WALID (ed.), *All That Remains: The Palestinian Villages Occupied and Depopulated by Israel in 1948* (Washington, D.C.: Institute for Palestine Studies, 1992).

KLEIN, CHAIM H. (ed.), *The Second Million: Israel Tourist Industry Past-Present-Future* (Tel Aviv: Amir, 1973).

KRÄMER, GUDRUN, *Geschichte Palästinas: Von der osmanischen Eroberung bis zur Gründung des Staates Israel* (6th edn., München: C.H. Beck, 2015).

KUPFERSCHMIDT, URI M., *The Supreme Muslim Council: Islam under the British Mandate for Palestine* (Leiden, New York: Brill, 1987).

LOCKMAN, ZACHARY, *Comrades and Enemies: Arab and Jewish Workers in Palestine, 1906–1948* (Berkeley: University of California Press, 1996).

MAIRS, RACHEL, and MURATOV, MAYA, *Archaeologists, Tourists, Interpreters: Exploring Egypt and the Near East in the Late Nineteenth and Early Twentieth Centuries* (London: Bloomsbury, 2015).

MAMMON, BENJAMIN S., 'Pioneers of Modern Tourism', in Chaim H. Klein (ed.), *The Second Million. Israel Tourist Industry Past-Present-Future* (Tel Aviv: Amir, 1973), pp. 29–37.

MIGDAL, JOEL S., *Strong Societies and Weak States: State-Society Relations and State Capabilities in the Third World* (Princeton: Princeton University Press, 1988).

NORRIS, JACOB, *Land of Progress: Palestine in the Age of Colonial Development, 1905–1948* (Oxford, New York: Oxford University Press, 2013).

NORRIS, JACOB, 'Development and Disappointment: Arab Approaches to Economic Modernization in Mandate Palestine', in Cyrus Schayegh and Andrew Arsan (eds.), *The Routledge Handbook of the History of the Middle East Mandates* (London: Routledge, 2015), pp. 275–290.

OHAD SMITH, DANIELLA, 'Hotel Design in British Mandate Palestine: Modernism and the Zionist Vision', *Journal of Israeli History*, 29/1 (2010), pp. 99–123.

OSTERHAMMEL, JÜRGEN, "The Great Work of Uplifting Mankind": Zivilisierungsmissionen und Moderne', in Boris Barth and Jürgen Osterhammel (eds.), *Zivilisierungsmissionen. Imperiale Weltverbesserung seit dem 18. Jahrhundert* (Konstanz: UVK Verlagsgesellschaft, 2005), pp. 363–425.

PAPPÉ, ILAN, *A History of Modern Palestine: One Land, Two Peoples* (2nd edn., Cambridge, New York: Cambridge University Press, 2006).

PAPPÉ, ILAN, *The Rise and Fall of a Palestinian Dynasty: The Husaynis, 1700–1948* (Berkeley: University of California Press, 2010).

PEDERSEN, SUSAN, 'The Impact of League Oversight on British Policy in Palestine', in Rory Miller (ed.), *Britain, Palestine, and Empire. The Mandate Years* (Farnham, Burlington: Ashgate, 2010), pp. 39–65.

REICH, BERNARD, and GOLDBERG, DAVID H., 's.v. Jewish Agency', in Bernard Reich and David H. Goldberg (eds.), *Historical Dictionary of Israel* (2nd edn., Lanham, Toronto, Plymouth: Scarecrow Press, 2008), pp. 257–259.

REICH, BERNARD, and GOLDBERG, DAVID H., 's.v. Jewish National Fund (JNF; Keren Kayemet Le Israel)', in Bernard Reich and David H. Goldberg (eds.), *Historical Dictionary of Israel* (2nd edn., Lanham, Toronto, Plymouth: Scarecrow Press, 2008), p. 260.

REICH, BERNARD, and GOLDBERG, DAVID H., 's.v. Keren Hayesod (Palestine Foundation Fund)', in Bernard Reich and David H. Goldberg (eds.), *Historical Dictionary of Israel* (2nd edn., Lanham, Toronto, Plymouth: Scarecrow Press, 2008), p. 273.

REICH, BERNARD, and GOLDBERG, DAVID H., 's.v. Yishuv', in Bernard Reich and David H. Goldberg (eds.), *Historical Dictionary of Israel* (2nd edn., Lanham, Toronto, Plymouth: Scarecrow Press, 2008), p. 554.

ROBSON, LAURA, 'Becoming a Sectarian Minority: Arab Christians in Twentieth-Century Palestine', in Laura Robson (ed.), *Minorities and the Modern Arab World. New Perspectives* (Syracuse: Syracuse University Press, 2016), pp. 61–76.

SCHAYEGH, CYRUS, *The Middle East and the Making of the Modern World* (Cambridge, Mass.: Harvard University Press, 2017).

SCHULZE, REINHARD, *Geschichte der islamischen Welt: Von 1900 bis zur Gegenwart* (München: C.H. Beck, 2016).

SEGEV, TOM, *Es war einmal ein Palästina: Juden und Araber vor der Staatsgründung Israels* (München: Pantheon, 2006).

SEIKALY, SHERENE, *Men of Capital: Scarcity and Economy in Mandate Palestine* (Stanford: Stanford University Press, 2016).

SELA, AVRAHAM, 'The "Wailing Wall" Riots (1929) as a Watershed in the Palestine Conflict', *The Muslim World*, 84/1-2 (1994), pp. 60–94.

SHEPHERD, NAOMI, *The Zealous Intruders: The Western Rediscovery of Palestine* (San Francisco: Harper & Row, 1988).

SMITH, BARBARA J., *The Roots of Separatism in Palestine: British Economic Policy, 1920–1929* (Syracuse: Syracuse University Press, 1993).

ST. LAURENT, BEATRICE, and TAŞKÖMÜR, HIMMET, 'The Imperial Museum of Antiquities in Jerusalem, 1890–1930: An Alternative Narrative', *Jerusalem Quarterly*, 55 (2013), pp. 6–45.

STANTON, ANDREA L., 'Locating Palestine's Summer Residence: Mandate Tourism and National Identity', *Journal of Palestine Studies*, 47/2 (2018), pp. 44–62.

TAMARI, SALIM, 'Lepers, Lunatics, and Saints: The Nativist Ethnography of Tawfiq Canaan and His Circle', in Salim Tamari (ed.), *Mountain Against the Sea. Essays on Palestinian Society and Culture* (Berkeley: University of California Press, 2009), pp. 93–112.

URRY, JOHN, and LARSEN, JONAS, *The Tourist Gaze 3.0* (3rd edn., Los Angeles, London: Sage, 2011).

VAN OORD, LODEWIJK, 'The Making of Primitive Palestine: Intellectual Origins of the Palestine-Israel Conflict', *History & Anthropology*, 19/3 (2008), pp. 209–228.

WALLACH, YAIR, *A City in Fragments: Urban Text in Modern Jerusalem* (Stanford: Stanford University Press, 2020).

CHAPTER 4

Contested rule and fragmented space in French Mandate Syria

Abstract

In Syrian tourism development, the French Mandate Government was the dominant actor. Tourism was not only advertised in order to disseminate propagandistic messages, but the presence of tourists was part of French strategies of imperial rule. Local urban initiatives aiming at tourism development were circumvented by the French authorities, but also by Syrian nationalist notables, who began to use tourism in a political sense only in the late 1930s. Due to the weakness of Syrian political institutions and the exclusion of the middle classes, a coherent vision of a Syrian nation did not emerge, and tourists continued to perceive the country in either Orientalist or geopolitical terms.

Keywords: tourism; French occupation of Syria; social classes; modernity; nationalism; Syrian history

The regions of the Near East united under the name of Syria and Lebanon are politically divided into two States, the Lebanese Republic and the State of Syria, and two 'governorates': those of Lattaquieh and Djebel-Druze. [...] They are bounded by Turkey on the north, Irak on the east, Transjordania and Palestine on the south, and the Mediterranean on the west.[1]

This is how the guidebook *Syria and Lebanon: Holidays off the beaten track*, published around 1930 in both English and French under the patronage of the French High Commission, introduced the French Mandates of Syria and Lebanon to its readers.[2] The clumsy introduction describing the administrative divisions of the mandate territory and its borders exemplifies the territorial approach of the French mandate power towards the region under their tutelage. The editors presented an overview of the new post-Ottoman political and geographical structure to potential visitors, which in fact was still in the making. After all, the spatial order described above was far less evident than the introduction suggested. A variety of actors nourished different and often competing hopes regarding the future order of the region. Anti-colonial conflicts and imperial competition entered into the debate, just as inner-Syrian conflicts and rivalries did.

The French mandate for Syria was harshly contested from the outset. An overwhelming majority of the Syrian population rejected the idea of mandate rule per se, and French rule in particular. The British, whose troops far outnumbered the French military presence in the Middle East, had installed the Hashemite Prince Faysal as King of Greater Syria, but accepted the French mandate to avoid threatening the British-French alliance. That the French overthrew Faysal's Government at the battle of Maysalun in 1920 and expelled him into exile caused consternation among the British.[3]

The British and French regimes installed in Syria revealed their respective views of Syria and strategies of governance. Both mandate powers preferred to collaborate with allegedly aristocratic elites rather than with the urban middle classes, often broadening the power base of the former compared to the pre-mandate era.[4] While the British had backed Faysal to reward an ally, the French mandatory based their administration on divide-and-rule tactics, a strategy which they had previously implemented in Morocco. They had been granted the mandate for Syria against Syrian resistance and realised their ambitions by means of diplomatic and military pressure. Whereas the notion of the mandate comprised promises of economic and political development, French rule in Syria was marked instead by a lack of investment, violent campaigns of counterinsurgency, the hampering of democratic institutions, and attempts to play some groups off against others. Generally suspicious of the urban bourgeoisie and their claims for independence, the French mandatory attempted to find allies among religious 'minorities' as well as rural leaders and played groups of notables off against each other.[5] The geographic division of the French Mandate along alleged religious or ethnic lines was part of this strategy. With the separation of Syria and Lebanon being the most stable division, there was a constant reshuffling of administrative boundaries, particularly during the 1920s.[6]

The political system of the Mandate, based on the French model of governing Morocco, made it difficult for Syrians to participate in processes of decision-making. Demands for a liberal parliamentary constitution were overheard by the mandate power and, once a chamber was introduced, French manipulations of elections, as well as suspensions of the chamber, more often than not hampered its functionality.[7] The highest authority in the French Mandate was the High Commissioner for Syria and Lebanon, seated in Beirut. Collaboration with local officials was avoided as far as possible, and the mandatory rather relied on French officials (often with limited understanding of local society) and intelligence services. On the lower levels of the administration, Syrian bureaucrats such as district governors were 'assisted' by French advisors overseeing political decisions.[8] Only after the Syrian Revolt from 1925–1927, when rebels in the hinterlands and supporters among the urban notables had aimed to overthrow French rule, the French modified their position to some degree and established cooperation with Syrian urban notables, who had wielded significant political influence under the Ottomans. These notables, they hoped, would serve as an intermediary between the mandate power and the potentially insurrectionist lower classes. The notables, in turn, although striving for an end of French rule in Syria, opted for cooperation with the French in order to build a power base that would allow them to take over after independence. From 1930 onwards, a French-imposed constitution defined offices such as a Syrian president, prime minister and a legislative assembly, but French vetoes, dissolutions of the chamber, and the manipulation of elections did not allow Syrian representatives to effectively participate in political decision-making.[9]

Despite fundamental and violent opposition against French rule since the inception of the mandate, colonial officials and supporters of French imperialism started to discuss options for tourism development early in the 1920s. In this chapter, I argue that for the French mandate power, tourism development was part of their military strategy, as tourist movements served as a tool to define, order and control space, thereby resembling the Zionist approach to tourism and French policies of tourism development in North Africa.[10] Yet, in contrast to both the Yishuv and French North Africa, in Syria the French officials could not build on the support of a local settler community. As in the Yishuv, settlers in Morocco, for example, contributed to realise tourism development policies by establishing both infrastructures and disseminating information, but they also constituted a reservoir of potential travellers.[11] In Syria, by contrast, French imperial tourism development often built on military officials as proponents of tourism development, and the number of tourists did not create the spatial presence the French High Commission might have intended. Still, the French mandatory maintained its ambitions for political reasons: in contrast to other means of exercising territorial control, tourism development was an economic and cultural undertaking in accordance with the mandate regulations, and even testifying to the good administration of the territory.

Syrian approaches to tourism development were less systematic and less coherent. The Syrian notables in the Chamber of Deputies had a marginal interest in tourism and seem to have used tourism development policies mainly as a commitment to secure loyalties in the Syrian *estivage* regions. As in Egypt and Palestine, the middle classes were the main advocates of tourism development in Syria, which was supposed to support their claims for self-governance and assist in laying the grounds for a historically rooted national consciousness.[12] For these bourgeoning Syrian middle classes though, the French approach to governing Syria limited the opportunities to participate in processes of political and administrative decision-making. Their access to both Syrian national politics and the French mandate administration was limited, and the nationwide implementation of tourism development policies did not occur. As they lacked a forum for debate in Syria, narratives of Syrian history, emancipation, historicity and modernity remained particularistic, often limited to urban contexts rather than acquiring national relevance, and passed largely unnoticed by travellers. The European tourists themselves either questioned or defended the legitimacy of the French presence, yet without questioning the necessity of imperial rule as such. Both critics and advocates among the tourists derived their arguments from debates led in their home countries and their national press, rather than from considerations of Syrian claims.

Competing narratives presented to tourists thus emerged in Syria, which were tied to different spatial scales. Syrian middle-class actors in the urban centres largely argued for Syrian modernity on a local level, while the French mandate power and its British imperial rival assumed Syrian backwardness in need of imperial assistance, perceiving Syria as part of a larger Middle Eastern geopolitical framework. On a national level, Syrian notables shaped politics within the limits of French rule, though without pursuing the aim of creating a national idea.

Since the French principles of governing had, to a large extent, deprived the Syrian urban middle classes of political representation, parliamentary debates hardly represented their positions. The first section on picture postcards is therefore an attempt to grasp their perceptions of Syria as opposed to the views of European visitors. The second section considers potential actors in tourism development more systematically. While the room for manoeuvre of the Syrian middle classes was limited, the Syrian notables shaping national politics were barely interested in reaching out to tourists. Tourism development in Syria was therefore a largely French imperial ambition and, as such, connected with French attempts to establish territorial control. The four case studies in the final section show that in Syria, a coherent national Syrian tourist space did not emerge. The competing and often contradictory dynamics in tourism development impeded the existence of both a coherent, overarching narrative, and unanimously acknowledged tourist places.

The two sides of the same postcard

In the summer of 1927, Lucie Dufour, the fiancée of André, a French soldier stationed in the Mandate of Syria, received two picture postcards showing "general views" of Hama, a town in northern Syria. Despite similar captions identifying the dense urban structure in the photographs as *Panorama de Hama*, the two illustrations gave very different impressions of the town (fig. 14, 15).[13] In the foreground of the black-and-white postcard, the minaret of a mosque directed Dufour's gaze further

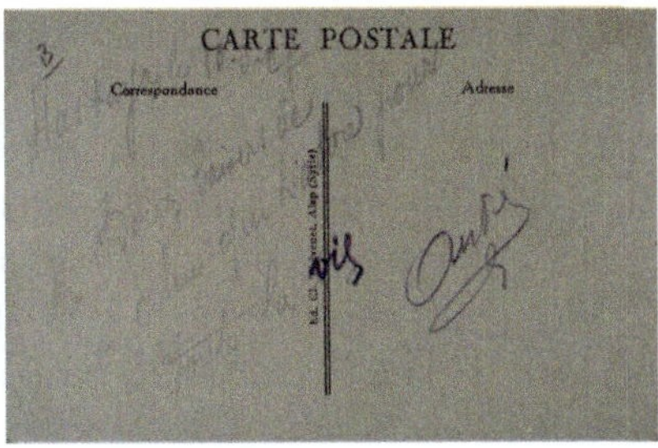

Figure 14: Clovis Thévenet's panoramic view of Hama (Photographer: Clovis Thévenet. Reproduced by kind permission of The Fouad Debbas Collection / Sursock Museum, Beirut)

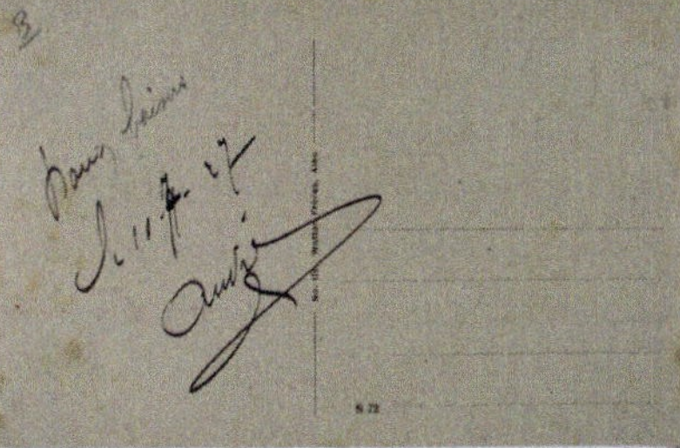

Figure 15: Wattar Frères, Panoramic view of Hama (Photographer: Wattar Frères Credits: Reproduced by kind permission of The Fouad Debbas Collection / Sursock Museum, Beirut)

towards several striped stone buildings with half-dome roofs, thereby highlighting the Oriental aspect of the city. The second postcard showed a hill with brightly coloured buildings. The minaret in this photograph, though in central position, was hardly recognisable among the numerous rectangular buildings, and the landmark attracting the most attention was a large white building of late-Ottoman shape, the white vaulted arches of which formed a sort of arcade. The most exotic sign in this picture, from Lucie Dufour's point of view, must have been the words in Arabic script, a translation of the French caption. We may assume that, for her, the most important message of both cards consisted in the tender greetings expressed by André on the verso side. But these personal messages were complemented by a second one: the broad ideas the picture side spread about Syria.

As both postcards were sent to France by the same person to the same addressee within a few weeks, neither the mindset of the sender nor the context of communication accounted for the different representations of Hama. In this case, the editors made the difference. Both had their studios in nearby Aleppo, yet the first postcard was edited by French citizen Clovis Thévenet (who had his studio between c. 1905–1927), and the second by the Syrian photographers *Wattar Frères* (c. 1921–1930).[14] The presence of foreign visitors and the popularity of postcards provided these photographers and postcard editors with the opportunity to shape visions of Syria among faraway audiences, and to inform the expectations of future travellers.

Writing postcards used to be a popular tourist practice in the late nineteenth and early twentieth centuries. Low fares, increasingly well-connected postal systems, and the collectability of images contributed to the rise of postcards as a means of communication across long distances. Whereas in the late nineteenth century mostly bourgeois women collected picture postcards, they turned into a means of mass and non-elite communication during and after World War I.[15] As postcards were affordable for large parts of the population and circulated widely, both in a geographical sense and among different social classes, they provide access to social strata which otherwise barely leave written traces in history. In Mandate Syria, the authors were not only bourgeois tourists, but also soldiers and officers from various social backgrounds who were stationed in the country.[16] However, often only bits and pieces of communication remain: the traces of communication rarely ended up in public archives and while postcard collections offer an opportunity to access them, most collectors tend to focus on the picture sides, rather than the senders' stories.

Many of the studies based on postcards have contoured these ambiguities and uncertainties by opting for an understanding of postcards as a form of "visual mass media".[17] Such an approach led Malek Alloula in his study on North African postcards to suggest that the often erotic representations of Algerian women allow critical observers to draw conclusions about the French colonial imaginary. Tellingly, Alloula's analysis ends in 1930, when erotic motifs stopped circulating. Alloula explains that "their mission is accomplished. Colonial cinema and tourism will take their place."[18]

In my opinion, this explanation is hardly convincing, as a growing number of tourists, whose Orientalist visions had been nourished by colonial cinema, was likely to result in an increasing circulation of Orientalist postcards. Moreover, the approach overlooks the practices attributing meaning to the individual postcard.[19] The mere existence of a postcard neither testifies to the availability of the motif in bookshops, kiosks, and hotels, nor does it inform us about the number of persons purchasing certain images or about contexts of use. Therefore, it is necessary to consider both producers and distributors of images; to reconstruct their audiences and ask how postcards were used, commented on and sent. That is, to do justice to their function not only as visual media, but also as a means of private communication.[20]

For analyses of postcards, the historian depends to a large extent on collectors' preferences, interests and sources of acquisition. The reflections in the present chapter are mostly based on postcards from the extensive Fouad Debbas Collection; in addition, Wolf-Dieter Lemke kindly showed me a significant part of his collection. I also included major websites offering postcards for sale in my analyses.[21] The sample comprised 524 postcards written and sent during the 1920s and 1930s, with 416 of them depicting motifs from Lebanon and Syria. Due to the dependence on the preferences of individual collectors, a bias in the sample cannot be ruled out. In my particular sample, the overwhelming majority of postcards from Lebanon and Syria were written by male French soldiers or military officials. Fouad Debbas collected his cards predominantly in a francophone context, so most of his cards were sent by French writers to France. Similarly, on German websites for collectors, cards authored by French senders prevailed, whereas on British websites, very few cards represented Syria and Lebanon. The high percentage of postcards written by members of the military may indicate actual proportions, but it may also be attributed to practices of conservation among the receivers. In my sample, soldiers or army officials stationed in Syria or Lebanon wrote around eight times as many cards as tourists, mostly to their parents, wives and fiancées.[22] It is likely that the families of military personnel attributed great value to the letters and postcards they received and thus preserved the correspondence. Given the number of postcards sent annually from the British Mandate of Palestine (more than three million in the 1930s), it has to be assumed that a large number of postcards sent by civilian visitors have not been preserved.[23]

The majority of the recipients were female. Soldiers sent cards to their wives, fiancées, mothers and sisters, and more rarely to brothers, fathers and former comrades. Travellers addressed couples, female friends and relatives.[24] Very often, the cards were sent in envelopes, so the recipients cannot be identified. In such cases, the date of expedition is lacking as well. The photographs displayed on the picture cards are not dated. Many photographers reprinted their images various times or sold the rights to reprint to other photographers and editors.[25]

With these circumstances in mind, I gathered two different corpora of postcards for the analysis. In general, I considered only postcards that had been posted: that is, that showed traces such as written messages or addresses and stamps. For the analysis of senders' reflections, I consider postcards sent by tourists. Postcards authored by military staff members or expatriate residents are only included in the event that they describe the author's touristic activities. They often reported upon excursions to famous monuments or trips around the region during their leave; during these activities, I assume they were part of the general tourist movement. In the second part on postcard producers, I include cards of both tourists and soldiers that were sent between 1920 and 1940. It must be assumed that, regardless of the military background of the senders, these pictures shaped the idea of the metropolitan audience of the region, as well as the expectations of potential visitors.

Recto: "They are of absolute indifference –"

By far the largest number of postcards tourists or visitors sent from Syria (excluding Greater Lebanon/ the Lebanese Republic) showed Damascus on the picture side (17 postcards). Few tourists sent postcards from Hama (6); even fewer from Aleppo (4), Homs (2) and Palmyra (2). This corresponds roughly to tourists' itineraries and the number of visitors to the respective cities. While soldiers tended to report more exhaustively on their short trips or excursions, most authors used the cards to send brief greetings or to announce the next stops of their planned trip, rather than to comment on their experiences. These few general remarks turned the photograph into the main element of communication. The following text, part of a longer message spread over a number of different cards, which are now missing, is a rare example in which a visitor to Damascus commented on the picture side, explaining his choice:

> 2) I am not satisfied at all with this series of cards. Nothing good exists here in this regard, and one has to stick to this poor selection. What is most fashionable are the persons, Arab types, more or less dressed women, etc. ... This is not what one sees in the streets and this one [referring to the postcard] looks like a ready-made painting, at any rate "conventional". If I find a better series, I'll send it to you, but it is not likely.[26]

The author had opted for a coloured photograph of the municipality building in Damascus: in the background is a mountain range, and in the foreground Marjah square and a partly visible wagon, possibly part of the tramway. Though the author was not convinced by the picture, he stated that he preferred it over the seemingly ubiquitous postcards representing "Arab types", among them erotic photographs of women.

Apparently, he expected the postcard to provide the addressee with an impression of Syria that corresponded to his observations on contemporary daily life, thereby contradicting arguments of De Roo and Alloula, who argued that the attraction of erotic postcards consisted in showing the inaccessible places visitors assumed to exist behind the doors of local houses. In fact, in the postcard collections of Fouad Debbas and of second-hand postcard distributors, around the turn of the century Syria seems to have been represented particularly by so-called 'type photographs' of the erotic or the exotic kind.[27] However, after World War I, few tourists actually sent such exotic images to friends or relatives. Among the 98 Syrian postcards in my sample, only 14 cards belonged to the 'ethnographic' or 'type' category of photography. Moreover, the authors of all such postcards were members of the military stationed in Syria rather than civilian tourists, and only one had been sent by a soldier on a touristic excursion.

The absence of Orientalist 'types' did not imply that tourists did not share the condescending feelings towards the local population associated with Orientalist thinking. Visitors easily combined denigrating remarks about locals with postcards illustrating the modernity of the region, as another postcard demonstrates: the sender commented on a photograph of the *Serail*, the residence of the Governor of Syria in Damascus, taken by the local photographers *Soubhi & Aïta*.[28] The author noted that he saw the *Serail* illuminated the day the Syrian Federation was enacted: "The illumination was intense and the ensemble was magnificent. No need to add that no Muslim deigned to turn the head to see it. They are of absolute indifference –"[29] From his point of view, the late Ottoman, specifically Hamidian, modernity these pictures illustrated did not necessarily presuppose the modernity of the Syrians. As observed in comments on Cairo's modernity, the coexistence of assumed Oriental backwardness and the witnessed 'modern' standards did not perplex tourists. The author adopted colonialist narratives attributing modernity to the successful efforts of the colonisers, and the assumed ignorance of the local population confirmed the necessity of continued European presence.[30]

In contrast to the trends observed in Williams' travel photographs from Egypt, which portrayed local inhabitants and scenes of daily life, most senders of postcards preferred images of cities and monuments to illustrate their travel experiences. Monuments, sights, public buildings and general views were more frequent among the cards written and sent. Apparently, at least in the 1920s and 1930s, the Orientalism on the picture sides was not as pronounced as the works by Said and Alloula lead one to expect. The disappearance of 'type' photographs must partly be attributed to preferences of the senders, such as that of the author who did not consider such views sufficiently authentic. As such views of a local population enjoyed continued popularity in travel photography, however, an alternative explanation for the changing views will be discussed in the following section: changes among the suppliers.

Verso: Showcasing urban modernity

On the supply side, a major variable in postcard production changed. In the first half of the twentieth century, a number of local photographers and postcard producers in the Arab East took over the business from their European competitors, most of whom left the region after World War I. This process was particularly pronounced in Beirut and Jerusalem.[31] In Syria, notably in Aleppo, several European photographers kept their businesses running and older postcard motifs continued to be sold (available from stock or as reprints), bought and collected in the 1920s and 1930s. The new images produced at the time differed significantly from the preceding motifs. Both the shift in production and new Syrian self-conceptions, inspired by Ottoman ideas of modernity as well as nationalism, were responsible for this change in the imagery of postcards.

In the nineteenth century, European photographers had been producing most of the picture postcards available in the Arab East. The photographs were edited in Europe, printed, and reimported to travellers' hubs such as Jerusalem, Cairo, or Beirut.[32] The pictorial programmes produced by studios such as *Lehnert & Landrock, Bonfils*, or *Neurdein Frères* typically represented local societies as 'backward' societies. The activities of some photographers were directly linked to the French colonial expansion, as the example of *Neurdein Frères* shows, a postcard editor partially funded by the French Government. DeRoo argued that *Neurdein Frères*, famous for their views of the French North African territories in particular, disseminated postcards putting forward the civilisational distance between the French colonists and the local population.[33]

After World War I, type photographs from the region continued to be produced by the remaining European photographic studios, such as *Clovis Thévenet* in Aleppo. *Thévenet*'s photographs showed entirely veiled women, peasant women or members of the Bedouin population in gloomy light and dark, shaded tones, thereby reproducing conceptions of a mysterious, threatening and archaic Orient.[34] Similar representations distributed by local photographers were rare in the 1920s. I came across one example resembling the type photographs at first sight, but it was framed very differently: a postcard by *Wattar Frères* depicting a group of heavily armed fighters. However, *Wattar Frères* did not label their sitters as 'types', but rather named and individualised the protagonist in the caption: *Aleppo. The Emir Moudchem with his followers* (fig. 16).[35] The caption of *Wattar Frères* thus individualised the sitters and turned the postcard motif into a portrait. Rather than representing a generic *La Syrie*, the photographers shifted attention to the represented individual and the way these men wished to be seen.

It is not entirely clear whether this shift was part of a larger trend. In Williams' amateur photography, after all, we have seen a similar consciousness among the sitters about how to pose in front of the camera – both among the tour guide, Mahmud, but also among the young woman carrying the jar (cf. fig. 5). Regarding professional photographs, Egyptian deputies indeed attempted to control and reduce the circulation of erotic views. Laws prohibited the importation and circulation of erotic postcards, and in 1927, parliament members suggested that the derogatory 'type' photographs should be banned as well.[36] Similar legislation was not introduced in the French-governed mandates, and the remaining European studios continued to offer familiar Orientalist views. In Syria, the shift in representing the population seems to have been a choice of the photographers rather than a reaction to legislation, but the arguments put forward in the Egyptian debate might have well been reflected in the photographs of *Wattar Frères*.

Figure 16: Portrait rather than type photograph: *Wattar Frères*, Aleppo (Photographer: Wattar Frères. jd private collection)

As shown in the previous section, most tourists preferred cards representing urban settings, panoramic views of cities, as well as monuments. This way of representing places, was, according to Wolf-Dieter Lemke, due to a general shift in taste around the turn of the century, which can be observed in European photography and postcards too.[37] In the Ottoman Empire, it emerged during Hamidian times and remained popular even after World War I, at least in Syria, where it was adopted by an increasing number of local photographers. *Wattar Frères* and *Chouha Frères* produced in Aleppo, and *Terazi & Churbadji* and *Soubhi S. & Munir Aïta* were popular postcard producers in Damascus. Moreover, the portfolio of some photographic studios from Beirut, such as *André Terzis & Fils* and *Dimitri Tarazi & Fils*, comprised similar motifs from the Syrian part of the Mandate.[38] These locally produced postcards portrayed the major Syrian cities and drew attention mainly to the built environment, complementing the exoticism of postcards produced by European photographers by exhibiting symbols of progress.

Some of these postcard motifs date back to Hamidian times. Sultan Abdülhamid II had consciously fostered the technology of photography in order to promote his politics of modernisation and to demonstrate the achieved progress to Western great powers.[39] He gifted a famous photograph collection to European and American representatives and libraries, which captured buildings of new administrative or military institutions, or insignia of modern infrastructures such as tramways, steamships, or bridges of steel, but also formations of athletes or military parades.[40]

The Syrian postcards of the 1920s presented similar visions of modernity. They showed panoramic views of major cities, institutions such as the Municipality of Damascus or the *Serail.* Others captured major historical monuments, for example the ruins of Palmyra, the watermills (Noriahs) at the Orontes River, or the Ummayad Mosque in Damascus.[41] It was the vision of a well-administered urban Syria with a long historical tradition, described best in its urban landscapes, its institutions and its historical sites. The postcards hence perpetuated the idea of Syria as a polycentric structure, where the noteworthy sights were to be found in urban centres.

Such views of Syria as a country characterised by its urban centres contrasted with representations of the neighbouring Mandate of Lebanon. Not only was Lebanon covered more extensively on postcards than the much larger Syrian territory, but landscape photographs of Mount Lebanon far outnumbered representations of the Lebanese cities. The production of one of the earliest and most prolific local photographers, *Sarrafian Brothers,* shows that local fashions of picture postcards changed in the 1920s. These pioneers of local photography, based in Beirut, were among the first local photographers who started to commercialise their photographs as postcards. The three brothers Abraham, Boghos and Samuel Sarrafian had been born into a Protestant Armenian family in Diyarbekir, leaving the region as a consequence of the 1895 massacres. In 1897, they settled down in Beirut, where they opened a photographic studio. In the following thirty years, *Sarrafian Bros.* would publish more than a quarter of all available postcards depicting Lebanon.[42] The various series of 'type' photographs in their repertoire, showing inhabitants in their supposedly authentic costumes, dated back to Ottoman times. Whereas these cards can be read as an expression of 'Ottoman Orientalism' and a condescending view of rural populations by urban dwellers, the visual programmes of the *Sarrafian Brothers* changed in the post-war period.[43] Series of landscapes and monuments began to largely dominate their production.

In the Mount Lebanon area particularly, the Sarrafian brothers covered seemingly every mountain village, and many of them from numerous angles (fig. 17). This is significant insofar as the hinterland of Mount Lebanon constituted a key element in the emerging national narrative; it was a central element in the distinction from neighbouring countries, as well as a major asset in the advertising strategy for tourism to Lebanon. The high mountain chains, snow-capped in winter, caused a relative abundance of water in comparison with the arid lands of *Bilad al-Sham* and nationalist narratives referred to the 'nature' of the nation, literally and metaphorically, as well as to its mythic origin and its landscapes.[44] Waterfalls and rivers, shady valleys and forests characterised Mount Lebanon in the perception of *Shami* people and postcards representing these landscapes thus spread its key visual symbols.

Figure 17: Views of ʿAley (Photographer: Sarrafian Brothers. Reproduced by kind permission of The Fouad Debbas Collection / Sursock Museum, Beirut)

In the French Mandate of Syria, the Syrian visual self-representation vis-à-vis outsiders exhibited its urban civilisation and a well-ordered state in (post-)Ottoman terms. The representations of urban centres on postcards, in particular Damascus, Aleppo, Hama and, to a lesser extent, Palmyra and Homs, illustrated local notions of belonging of the urban population. These representations suggested that Syria continued to be perceived as a network of cities. The urban middle-class photographers and editors did not include the hinterland in the visual representation of the country, nor did they claim a territorial rootedness as postcard editors did in neighbouring Lebanon.

The postcards of Syrian photographers, exhibiting post-Ottoman culture and civilisation while insisting on the centrality and pioneering role of the urban spaces, thus not only claimed the noteworthiness of an urban middle-class modernity consciously ignored by the French mandate power. Compared to trends in neighbouring Lebanon, they also indicate that the Syrian urban middle-class actors, of which photographers and editors were a part, did not attempt to define Syria as a national entity. Rather, their postcards suggest that notions of belonging continued to be imagined in Ottoman categories, referring in particular to towns or cities.[45] Photographers expressed their pride in the Syrian urban modernity, but visual representations of a territorially integrated Syria and a sense of an overarching national belonging were not discernible.

Urban pride and the imperial politics of tourism

During the 1920s, the French High Commission and commercial entrepreneurs seemed to have been the major actors in developing and establishing tourism in Syria – where it often legitimised and implemented French rule, specifically in regions of anti-French resistance. Among Syrian actors, the interest in tourism was restricted to a small area in the Anti-Lebanon Mountains throughout the 1920s. Syrian notables rather focused on diplomacy in order to present their claims about the future of Syria on an international level and it was not until the late 1930s that some Syrian nationalist politicians created an information bureau addressing tourists, among others. The way tourists appraoched Syria bears similarities to visitors in Palestine: while both British and French visitors initially perceived Syria as a rather unspecific part of an Oriental continuum, after the Syrian Revolt their perceptions and judgements were much more influenced by international debates about contemporary conflicts. At the same time, the diaries and travel notes of my sample did not reflect particular interest in local perspectives.

The High Commission: 'Pacification' and touristification

Suwayda was a small town in southern Syria, in the area of the Jabal al-Druze. Despite its small size and remote location, a postcard presented the place to a French audience (fig. 18).[46] It was published by L. Ferid, a bookseller and photographer with studios in Istanbul and Beirut, who maintained connections to the French mandate power.[47] On the picture side, it showed the photograph of four officers standing next to a car in a desert landscape, discussing the situation. The caption informed the viewer that the man at the centre of the group was General Andréa, who led a military column to Suwayda in September 1925.[48] Similar motifs showed French military activities in Musayfrah, a village and military camp located between Suwayda and Darʿa in the Jabal al-Druze.[49] These postcards presented the Jabal al-Druze as a territory under the control of French officers and generals. Mostly sent by soldiers during the revolt, the dissemination of images from southern Syria accompanied military conquest and marked places on the mental maps of the metropolitan French. Actual tourist visits to the area remained low, but a guidebook to the region authored by a French officer indicates that the anticipated tourism development was a core element of the French imperial presence (cf. p. 209).

Three hopes were placed in tourism that explained the interest of the French mandate power in the phenomenon. First, tourism was expected to contribute to regional economic development; second, it created an opportunity to address both the French metropolitan population and an international audience; and third, tourism as a method of controlling space was part of the strategies to establish

Figure 18: Greetings from the Jabal al-Druze (Photographer: L. Férid. Reproduced by kind permission of The Fouad Debbas Collection / Sursock Museum, Beirut)

French rule.[50] The most obvious benefit of tourism was its contribution to economic recovery after the Great War. Around the time France was confirmed as the mandate power of Syria, Albert Sarraut, at the time Minister of Colonies, published his tract on the *mise en valeur* of the colonies. Sarraut, the former Governor General of Indochina, proclaimed the idea that France as a colonial power should pursue a colonial policy beneficial to both the colonies and the colonial power, an important pillar of which was 'economic development'. The French overseas territories should generate sufficient profit in order to financially sustain themselves and lighten French expenditure on the empire.[51] For the metropolitan economy, tourism was considered to have a doubly positive impact: a stronger local economy and a higher level of taxation would lessen the expenses necessary to administer the colony; moreover, observers expected beneficial effects for French companies.[52]

Moreover, as economic development was considered a valid argument in favour of French foreign rule, the importance of strengthening local tourism was evident to the High Commission. Already in the early reports to the League of Nations, the High Commission identified tourism as a significant future source of revenue. The potential was particularly evident in the Lebanon and Anti-Lebanon mountain chains, where an accommodation sector attracted *estiveurs* mainly from the cities of *Bilad al-Sham*, Egypt and Iraq. Whereas the larger part of the *estivage* area belonged to the state of Greater Lebanon after the division of the Mandate, a small section remained with Syria, and the High Commission registered a growing interest in the branch in the Syrian part of the Mandate.[53]

In the early years of the mandate, the High Commission generally approved of initiatives to strengthen tourism, yet its functionaries did not actively foster the business. Instead, the Mediterranean shipping companies seem to have played a key role. The *Messageries Maritimes*, a French company with close relations to the Parisian Foreign Office and the mandate administration, was one such active proponent of tourism development. The first tourist roundtrips organised with French official support targeted the Lebanese and Syrian youth: agents of the company planned a trip to France for school pupils and adolescents during the summer holidays of 1923, which would introduce the major French cities and some famous historical sites to "young people who have absorbed the French culture".[54] In a letter to the High Commission, the organisational team expressed their hopes to enhance the good reputation of the mandate power and its prestige in Syria.

Originally, the programme had been drafted for Egyptian school pupils. For the *Messageries Maritimes,* the extension of the programme to Syrian (including Lebanese) pupils and the collaboration with the High Commission was a pragmatic choice. The company presumably intended to fill cruise ships on their return trips, after having transported Arab summer guests to the Eastern Mediterranean. Supporting the mandate appears to have been rather an intended side effect.[55] By contrast, the advisor for the Department of Education considered the suggestion to be of "political" rather than educational significance and strongly recommended that the High Commission support the project.[56] The idea behind these tours was to create bonds with France among the future Syrian and Lebanese generations.

In the context of the Great Syrian Revolt, the High Commission began to more comprehensively exploit the propagandistic potential of attracting tourists to Syria, as if to re-establish their reputation after the bombings of the old town of Damascus and many civilian casualties had caused an international outcry.[57] In the French reports to the League of Nations, tourism, in particular the protection and promotion of Syria's historical heritage, was now presented as a contribution to the economic and "civilisational" development of Syria.[58]

Additionally, tourism was expected to popularise the overseas possessions among the metropolitan population, parts of which had started to question the expenses necessary to maintain the empire. Increasing resistance in the colonies necessitated both monetary funds and military force, neither of which was available in abundance in the aftermath of the Great War and in the context of the economic depression.[59] Letters such as the one addressed to High Commissioner Henri Ponsot on 17 October 1930 were therefore more than welcome: the industrialist and adventurer Georges-Marie Haardt, a director of the *Citroën* company and leader of popular car expeditions across the Sahara and the African continent,[60] was planning an expedition through Asia, including a visit to Syria. Haardt asked the High Commissioner to establish a programme for their intended visit, reasoning that the expedition would be an opportunity to draw public attention to the important humanist and scientific projects of Frenchmen in Syria.[61]

Not only among the High Commission, but also among the highest echelons of French metropolitan politics, reaching up to the President of the Republic, such arguments were met with interest. Philippe Berthelot, General Secretary at the Foreign Office in Paris, recommended a visit to the ruins of Palmyra, Tyre, Acre and Baalbek to Haardt.[62] From a French imperial perspective, these destinations stood for the historical presence of France in the region: in Acre and Tyre, Crusaders' monuments were among the major attractions, whereas Palmyra and Baalbek were famous for their impressive Roman ruins. Such references were put forward as historical truths testifying to the long-standing French-Syrian connections, simultaneously reaffirming the legitimacy of French rule and passing over the Islamic heritage of the region.[63]

Interestingly, the energy French actors devoted to excavating antiquities and promoting tourism focused in particular on unruly regions. This illustrates a further function of tourism from a French imperial point of view: tourism served the implementation and stabilisation of colonial rule after the conquest. Both the ʿAlawite state and the Jabal al-Druze were advertised as tourist destinations shortly after the suppression of the insurrections. Authors of guidebooks and brochures, often French officers, promoted the archaeological sites in both areas as attractive destinations to the French public.[64] As in the ʿAlawite region, tourism development measures ranging from archaeological excavations to public promotion of the destination were applied in the Jabal al-Druze. In December 1930, Abbé Joseph Mascle, a military chaplain based in Suwayda, drafted a plan for the tourism development of the area.[65] Mascle stated that after the 'pacification' of the region, the French administration had brought new life to the area, in economic as well as cultural terms:

> The Jabal al-Druze, pacified by our arms and resurrecting to life under the aegis of the French Administration, today attracts not only those who are interested in the economic evolution of a country, but all those who, due to professional obligation or cultured spirit, dedicate themselves to the study of ancient monuments.[66]

The author highlighted in particular the modern infrastructures, guaranteeing the easy accessibility of the Jabal al-Druze, and the comfort awaiting the visitor in *Hôtel Royal* in Suwayda. The greatly improved road network he mentioned, in fact constructed to facilitate French warfare, granted easy access – the French military had literally paved the way for tourists.[67] Terms such as "pacify", "resurrect to life", and "economic evolution" were understood by Mascle in a colonial sense. According to him, the ancient – Roman, thus pre-Islamic – heritage of the Jabal al-Druze was not only of historical interest, but also allowed tourists to comprehend the legitimacy of imperial rule, evident – from his point of view – in the ruins of the past and the contemporary beneficial effects of French colonialism.

Colonial officials of the Third Republic, such as Louis Hubert Lyautey and Joseph Gallieni, had firmly anchored the idea in French colonial thought that after the end of military conquest, the so-called pacification should secure long-term rule. Lyautey in particular had acquired a reputation as a benevolent governor of Morocco, whose colonial regime was based on cooperation, respect and civilian governance, rather than outright military force. This reputation resulted from proclamations that military conquest should be accompanied by measures of reconstruction – that is, by rebuilding infrastructures, developing the local economy and establishing institutions. Yet, such measures were part of a more encompassing strategy to reshape colonised societies in a way to facilitate colonial rule, which implied harsh measures of colonial force.[68] Tourism development as an imperial strategy bore striking similarities to the general concept of pacification, and the encompassing tourism development activities of French settlers in Morocco as well as their praise for Lyautey and his successors support the assumption of a close connection.[69] Tourism was supposed to bear the potential of arousing sympathies for and identification with the French imperial project.

On the ground, however, it served the French administration not only as an occasion to convince and persuade, but also as a resource of domination. The number of military staff members involved in touristic publications is considerable: guidebook author Noël Maestracci, Joseph Mascle and the author of the *Guide bleu* on the Mediterranean, J. Chardon,[70] were all affiliated with the French military. Similar observations regarding tourism in the Algerian Sahara during the interwar period have led historian Arnaud Berthonnet to attribute a pioneering role to the French military in making regions accessible to a larger French audience. Yet this success story of the military and strategic exploration to the benefit of the masses needs to be questioned, as it overlooks the negative side of the functional alliance of tourism and territorial control.[71]

Tourism implied bringing civilians to newly 'pacified' places. Daniel Neep argued that one of the main goals of French counterinsurgency strategies consisted in evoking the illusion of densely controlling the territory: the French military aimed to create mobility, rather than establish a static presence.[72] Hence, mobile tourists, whose main activity was to discover and see, may be conceived as civilian allies of the military. They circulated on the same roads, accessed towns, villages and buildings; they demonstrated the presence of Europeans and thereby supported, voluntarily or not, the project of penetrating the spaces of the insurgents. Similar to Zionist tourism development, the interest of tourism did not only lie in the dissemination of imperial propaganda, but also in its contribution to establishing effective colonial rule in the long term.[73]

Tourism thus provided a civilian continuation of military conquest, and within French attempts to control the insurrectionist regions effectively, the vision of developing tourism was linked to the idea of firmly anchoring colonial rule. From an imperial perspective, tourists ideally adopted and spread positive views about mandate rule and, more importantly, they reinforced the French presence in the country. Through defining noteworthy sights, places, and monuments; through framing history, and through shaping territories, borders and belongings, concepts of tourism development served the French administration in stabilising, anchoring and spreading their own vision of the mandate territory. Beyond that, tourists created a physical European presence in the area; they justified military presence and thereby contributed to controlling and supervising the spaces of insurgents. As a resource of colonial dominance, tourism was hoped to establish effective rule after military conquest.

Syrian initiatives: Local initiatives or national efforts?

In contrast, tourism development on a local Syrian level seems to have been the concern of individual entrepreneurs, such as the bankers Asfar and Sara from Damascus, who had conceived the idea of building a hotel to international standards already in May 1926. They had acquired a large plot of land next to the railway station and asked the High Commission for a credit to facilitate the realisation of the project.[74] Although hotel credits were a widespread instrument to support tourism development in the French Empire, in the case of Asfar and Sara – the Syrian Revolt meanwhile intervening – the authorities did not respond to the request.[75] After the end of the revolt in November 1927, the investors therefore approached the High Commissioner's delegate once again. Joined by the experienced hotel entrepreneurs Joseph and Elias Khaouam, they renewed their solicitation.[76] The Syrian Government under Ahmad Nami, appointed by High Commissioner de Jouvenel during the Great Revolt, had apparently expressed their support for the project, regretting the lack of a modern and spacious hotel in Damascus, expected to be a "future tourist centre and transit station for convoys to Iraq and Persia", yet having refrained from making any financial promises.[77] Likewise, the delegate of the French High Commission, Desloges, approved of the project in principle, but refrained from making any financial promises. That the initiative succeeded and the hotel was constructed resulted from the insistence of the respective signatories, Asfar and Sara, as well as the Khaouam brothers.[78]

The hotel project exemplified a general pattern in Syrian tourism development. While neither the French High Commission nor the Government of Syrian notables dedicated more than occasional attention to the matter, some entrepreneurs or other interested individuals pursued tourism development in a more energetic way. These middle-class actors pursued their entrepreneurial, cultural, or political interests largely without support from official institutions or deputies. The initiatives themselves were centred around concrete, local projects, rather than drafting a comprehensive plan for tourism development.

It is difficult to assess the capabilities the Syrian Government had to back local projects such as the hotel in Damascus, or even to draft larger plans, as the reality of decision-making in the French Mandate of Syria remains somewhat nebulous. Nominally, the Syrian Government was to oversee legislation and implement measures concerning internal affairs, such as tourism. In practice, Syrian ministers and administrations were 'assisted' by French 'advisors' who are considered to have had a significant influence on Syrian policies.[79] Even when a parliament was established, it was hardly an efficient institution. The French regularly manipulated the election process and suspended parliament meetings, while Syrian nationalists, in turn, boycotted elections and parliament sessions, all of which impeded a continuous debate. As a result, the major programme of economic development introduced in Syria, which included measures to develop tourism, was – more or less – dictated to the Syrian Government installed by High Commissioner Damien de Martel in 1934.[80] The authorship of political measures is therefore not always evident, and other agents might have been involved than the ones who signed laws or decrees.

For the deputies, tourism development was no political priority, and, according to the *Journaux officiels*, it played a subordinate role in legislation – in contrast to neighbouring Lebanon. There, as the following chapter will demonstrate, support for tourism generally united the assembled deputies, regardless of their backgrounds, and a variety of actors, often organised in committees and commissions, placed tourism high on the political agenda. In Syria, where tourism was of lesser economic significance and the tension between Syrian representatives and the French much more pronounced, deputies gave priority to other projects.

Two commissions concerned with tourism development existed in Syria, though their impact remained limited. The so-called *Commission of Tourism*, established by the French High Commission in 1923, covered both Syria and Lebanon, but the tourist brochure *Holidays Off the Beaten Track* mentioned above (cf. p. 193) appears to have been its only output.[81]

The second attempt to anchor tourism at an administrative level occurred in 1933, shortly after a new, moderate government came to power.[82] This *Commission of Tourism* was established as part of the Agricultural and Economic Services, an agency of the Syrian Ministry of Finance. It was charged mainly with executing a decree established in 1928, the realisation of which had presumably been postponed due to the economic crisis: the commission would decide the tax exemptions to be granted to ten new hotels and one hundred 'villas' erected in the Syrian *estivage* centres. The composition of the institution reflected the modest nature of these responsibilities. Its four members were administrative staff members without any particular interest in tourism: namely, the General Director of Hygiene and Public Health, the Chief Engineer of the Ministry of Public Works, the *Chef de bureau* of the Ministry for Agriculture and Commerce, and a financial auditor.[83] The commission was obviously designed to guarantee the functioning of administrative procedures, but it would not actively shape tourism development.

In contrast to Zionist and Lebanese tourism development policies, the programmes of the Syrian Government did not aim to use tourism as a means of realising a political agenda, or of spreading national narratives. Measures and regulations concerning tourism development focused almost exclusively on the region around Bludan and the Zabadani River, an area where tourism was already well established.[84] Here, *estivage* played an important role in the local economy and interested spokespersons managed to attract attention, investments, and subsidies, whereas in other areas the Syrian notables did not envision tourism as an encompassing development strategy for the entire country. In Syria, local actors, rather than national institutions or representatives promoted tourism development.

Not before the late 1930s did local initiatives and tourist development associations did receive political and financial support in Syria. From April 1936 onwards, thus after the election of a new leftist government in France (the 'Popular Front') and the opening of negotiations for a Franco-Syrian treaty, the budget regularly comprised funds for projects related to tourism, mostly advertisements.[85] The Syrian Government now granted subsidies to institutions which edited brochures advertising tourism to Syria and Lebanon, among them the *Revue archéologique d'Alep* (cf. p. 227), Georges Samné's *Information and Tourist Office for Syria and Lebanon* in Paris, and the *Société d'encouragement au tourisme* (SET), based in Beirut.[86] Moreover, the Government created incentives to improve the accommodation experience of travellers. A new hotel was planned in Bludan, and facilities such as a permanent tourist office should enhance the comfort of visitors.[87] Thus, the Syrian subsidies benefited both the well-established *estivage* regions within Syria and various Lebanese associations, as in the cases of Samné and the SET. The limited number of projects, the fragmentary measures and the coincidence of French-installed governments and tourism development policies applied outside Syria suggest that the measures were inspired by French 'advisors'.

Only around 1938 did some Syrian notables begin to promote tourism more systematically, apparently intending to disseminate their own spatio-political visions. According to the Italian orientalist and journalist Virginia Vacca,[88] who regularly reported on Syrian economic development, the Damascene *Arab National Office for Public Enlightenment* gathered information and spread Syrian nationalist perspectives among institutions and media inside and outside the country, in particular in France and England. While informing tourists was not its primary concern, the office also worked on 'multilingual' tourist guides and published books on Arab countries targeting a tourist audience. The office itself had been founded in 1934 by Fakhri al-Barudi, a former ally of King Faysal and known Arab nationalist, but only in 1938 do sources mention its activity in tourism advertisement, thus in a time when the conflict with France over Syrian independence and debates about reunification of Syria and Lebanon entered a new stage.[89] It appears that the *Arab National Office* hoped to profit from the growing number of foreign visitors to the region in order to spread its viewpoints. Officially, the *Arab National Office* was an apolitical organisation. Yet Vacca argued that, given the influence al-Barudi wielded in Syrian politics, it had to be understood as a quasi-governmental institution, "which, in the long transitory period from mandate to independence, is about to take actions that the government cannot yet perform directly."[90]

In this context, it is significant that al-Barudi's *Office* continued to defend the idea of a united Syrian space, which comprised the Lebanese territory. The development of tourism and *estivage* al-Barudi envisioned included Lebanon, although Vacca commented that, to her knowledge, Lebanese representatives were not involved in the office's activities.[91] This observation is telling. By the late 1930s, the Lebanese administration had established its own institutions of tourism development, including a Ministry of Tourism, which advertised tourism to a Lebanese nation-state, and excluded Syrian destinations. Al-Barudi's initiative did not resonate beyond the Anti-Lebanon Mountains.

While al-Barudi continued to claim Syro-Lebanese unity, protectionist tendencies became increasingly discernible on a practical level, assuming and perpetuating the order of nation-states in the region. In political measures and regulations, the Syrian Government aimed in particular to protect Syrian labour, while it paid lip service to the idea of a united *Bilad al-Sham*. In 1938, tourist guides and dragomans addressed a petition to the Syrian Government, demanding the prioritising of Syrian guides over their foreign colleagues. As justification, they pointed at their superior experience and knowledge of the country, adding that Syrian guides were not allowed to guide tourists in neighbouring Lebanon and Palestine either.[92] Until then, foreign guides who possessed an official certificate issued by their respective governments were allowed to perform the function of guide in Syria, in the case that they accompanied a tourist group on its entire trip across the *Bilad al-Sham*.[93]

Although the issue was closed without further modification, as the accusations could not be verified,[94] the petition of the guides and dragomans testified to an ongoing spatial fragmentation of the region in two regards. First, by the late 1930s, the guides obviously assumed that Palestine, Lebanon and Syria were distinct countries competing for tourists, as well as jobs and income linked to tourism.

Second, their argument relied on the assumption that historical and archaeological knowledge specific to the nation was required in order to present these states to their visitors. This assumption indicated that history and heritage in the three countries under mandate had come to be conceived of as separate, nationally distinct trajectories. Even though the disentanglement of narratives was driven by business interests rather than manifestations of belonging, the claim as such was remarkable. As the guides argued that understanding each country required unique expertise in its respective national past, by this point, the larger unit of *Bilad al-Sham* as a historical and cultural entity had apparently lost its relevance.

Tourists: Mouthpieces of the French High Commission?

By the 1920s, travelling to Egypt was a routine practice; guided trips to Palestine were becoming increasingly popular, and the extension and improvement of the Lebanese accommodation and transport network were in full swing. In Syria, the expansion of organised tourism was less pronounced than in the neighbouring countries. First, the infrastructures for large-scale tourist accommodation remained to be constructed. In contrast to the well-established Nile tourism or Palestinian experiences in accommodating pilgrims, in many parts of Syria tourism and travel had barely played a role. Second, the armed resistance against French rule in the early 1920s, as well as the Syrian Revolt between 1925 and 1927, kept many potential visitors from travelling to the country.[95] However, the French High Commission identified tourists as potential targets for political messages: in their places of origin, they confirmed the state of security reigning in Syria as eyewitnesses.

As argued above, the French High Commission paid attention to improving the experiences of visitors in order to defend its presence and policies in Syria. This ambition was facilitated by the fact that in the early years of the mandate, it was rather common for travellers to contact the French Foreign Office, respectively the mandate authorities, before departure.[96] The habit diminished from the late 1920s onwards; it is likely that at the time tourism to the mandate was well established and better organised, so information was more easily available and the support and protection of the French authorities were no longer necessary. Consequently, most persons who did leave traces in the archives in the 1930s were travellers whom the mandate authorities considered potentially troublesome, or those pursuing special projects, such as André Citroën and Georges Marie Haardt during their *"Croisière Jaune"* across Asia (cf. p. 208).[97]

Especially in politically difficult times such as during the Great Revolt, tourists and visitors were assisted by the High Commission. Receptions with the High Commissioner or assistance in the preparation of travel routes were provided not only to French nationals such as Charles Le Gras and his group of pilgrims, but also to British visitors, such as Humphrey Bowman, the Director of Education in the British Mandate of Palestine, or Norah Rowan-Hamilton, who had joined a delegation of the *British Archaeological Society* for a visit to Syria.

The Foreign Ministry in Paris cared particularly about tourists who were likely to report on their impressions during the journey. Since the French mandatory was concerned about the public image of the mandate both within France and on an international level, writers and journalists travelling across Syria were important assets, all the more so when they were not French citizens and thus considered neutral observers. When the Belgian journalist Louis Piérard planned a trip of two weeks across the Syrian Mandate in early February 1927, which he would report on in the Belgian papers *Peuple* and *Soir*, the expenses during his stay in Syria were covered by the High Commission.[98] As the French authorities reasoned that an international audience would accept the reports of journalists, authors, publishers and moviemakers as independent eyewitness accounts, they endeavoured to provide these travellers with positive experiences.[99]

Alongside the international press, the French High Commission also targeted individuals supposed to have close ties with the League of Nations, such as Henry Vallotton-Warnery from Geneva, a member of the Swiss National Council and of the *Swiss Touring Club*. In 1926, he undertook a trip by car from Paris to Cairo via Istanbul, Alexandretta, Aleppo, Baghdad, Beirut and Jerusalem. At the time, the French were harshly criticised for their measures to crush the Syrian Rebellion – not least in Geneva, the seat of the League of Nations and the place where the Permanent Mandates Commission gathered. Given that Henry Vallotton-Warnery not only published his *Impressions of Syria* in a local newspaper, the *Journal de Genève*, but also wielded a certain influence in his hometown and in the Swiss National Council, the French High Commission hoped that his positive testimony would back the French mandate power.[100] On the visit to the Jabal al-Druze, Vallotton-Warnery was accompanied in person by the delegate of the High Commission in Damascus, de Reffye. While there is no evidence on the nature of their exchange, the reports Vallotton-Warnery published after his return confirmed official French explanations of the Syrian uprising, which suggested that the French military intervention had set an end to Syrian armed robbery, restoring peace and stability to the benefit of the local population. Moreover, it was probably more than welcome that Vallotton-Warnery dispelled potential worries of future visitors and recommended a visit to his readers: In late October, when the article was published, the French strategy of controlling space to quell the insurgence began to take effect.[101] An increased presence of European tourists was thus in the interest of the French mandate power aiming to establish territorial control.

It seems as well that the French High Commission encouraged in particular French travel groups to spend substantial time in the French Mandate. At least the itineraries they suggested to French groups included more stops in Syria and recommended longer stays than the advice they gave to other nationals, such as groups of Romanian and Bulgarian pupils, Italian university students, and Polish pilgrims. The French High Commission seemed to accept that for these groups, the Mandate was a mere stop on the way to their final destination: Palestine, and their recommendations pointed out the main highlights.[102] In contrast, it was assumed that for French groups, Syria was at the core of their journey and even minor sights should be visited.[103]

While in the 1920s, numerous visitors to the Mandate – school classes, as well as individual travellers – seem to have exchanged information with the French authorities, in the 1930s, the communication between individual travellers, clubs and organisations and the mandate authorities declined, presumably due to a lack of requests from travellers. By that time, travel agencies had set up well-organised trips, and tourists did not require the assistance of the French High Commission any longer. Paradoxically, the expansion of organised tourism increased the possibilities for actors other than the High Commission to shape travellers' ideas of Syria.

No common ground: The divided landscapes of tourism in Syria

Throughout the mandate period, divergences in perceptions of tourism, and of Syria, among travellers, the French administration and Syrian actors persisted. Four case studies exemplify different approaches to the Syrian territory and demonstrate that neither of these groups managed to impose a coherent interpretation, nor did consistent collaboration between some of the actors lead to a compromise in developing an overarching narrative for a space that was considered temporary, while practices rendered it permanent.

The case of Damascus illustrates the diverging views of Syria among British and French tourists. After the French bombardments of the city during the Great Revolt, Damascus (and Syria) turned into a focal point of Anglo-French imperial rivalry in the region, which led tourists to perceive Damascus in terms of this geopolitical ambition. This perception was framed as a debate about the legitimacy of French rule, while visitors ignored Syrian voices.

Nationalists from Syria instead perceived tourism and *estivage* in terms of mobility and sociability. The section on the *estivage* region of Bludan and Zabadani shows that the mountainous area, marginal on French strategic maps, was a centre of pan-Arab interaction and nationalist agitation. While the mountain villages were convenient places for exchanges within elite circles, they were not attributed meaning in communication – neither towards visitors nor other sections of Syrian society.

By contrast, communication was the key strategy of middle-class actors such as Kamil al-Ghazzi and the *Archaeological Society* in Aleppo, who suggested alternative mechanisms and principles of claiming (national) sovereignty. Notions of historical lineage bore the potential for turning the preservation and presentation of historically relevant sights into narratives of national belonging. In theory, the interest in the history and traditions of a place could have served as a basis for communication between the Syrian middle-class actors and their European visitors. Yet the latter failed to hear the voices of al-Ghazzi and the *Archaeological Society*, which did not have the necessary networks at their disposal to establish regular communication with international tourists.

Finally, the example of Palmyra represents a place that was promoted as an attraction for tourists – to the detriment of its inhabitants. In Palmyra, an educated urban middle class, as in Aleppo or Damascus, barely existed, and the French administration, backed by French media, promoted both the ancient ruins and the supposedly archaic way of life in the desert as potential tourist attractions. The exhibition of the ruins, however, required the expulsion of the inhabitants. Thus, the integration of Palmyra into the Syrian tourist space was fostered by actors alien to the place and often perceived as a threat by locals.

Damascus: Tourists on the battlefield

> It is a big and very ancient street, it has a roof of iron riddled with holes from machine gun-fire either during a civil war or during wars with foreign countries. [signature illegible][104]

After the end of the Syrian Revolt, the historical *Street called straight* in Damascus, the name of which had provided puns and anecdotes for numerous guidebook authors and travellers, turned into a site for 'dark tourism'. The photograph on the postcard (fig. 19) showed a street scene, even though, due to the underexposure of the photograph, the passers-by in the street were hardly discernible. Light entered the roof by some rectangular windows as well as by the holes that the author mentioned, which dotted the ground with tiny spots of light. Although the photograph ("a real photograph", as the editors stated expressly on the recto side) testified to French acts of warfare in the inner city of Damascus, the author had apparently heard contradictory explanations about the origin of the holes and remained vague in his explanation. While narratives still had to be created, the images documenting the violence of the recent events were quickly edited into postcards.[105] What is therefore certain is that the revolt, as cynical as it may sound, provided the visitors with new and unseen sights. The destruction challenged prevailing narratives of an Oriental Damascus and tourists had to situate the city in a new, geostrategic context.

After all, in the imagination of tourists searching for the Orient, Damascus had been one of the places where these expectations crystallised. French visitors

Figure 19: Postcard showing the destruction in Damascus, similar to the motif described above (Photographer: Luigi Stironi. jd private collection)

in particular continued to frame Damascus and Syria as archetypes of the Orient throughout the mandate period, picturesque in their daily life and markets, and culturally defined by Islamic traditions and architecture, while biblical stories connected them to familiar traditions and histories.[106] In British narratives, however, a marked change occurred in the autumn of 1925, when the French bombardment of Damascus turned the city from an Oriental oasis into the capital of a war zone.[107]

Before the event, British travellers perceived Damascus as a place of picturesqueness and ancient tradition, similar to descriptions of French tourists and in accordance with Orientalist representations. George H. Williams, who commented extensively on the long history of the city, probably quoting passages from his guidebook, described the flourishing Damascene civilisation in antiquity and the medieval period, which declined after the arrival of the Mongols and subsequently the Turks.[108] Typically, the travellers expressed disappointment given the transformation of Damascus to a modern city, assuming that the daily life and living conditions had lost their allegedly authentic Oriental flavour. Williams' father had visited Damascus "30 or 40 years ago" and recommended a visit to Damascus as the perfect place to discover Oriental life in the bazaars, but Williams junior found them much less picturesque than the bazaars in North Africa: "the Holy Land is a mass of delusions."[109]

During the Syrian Revolt, the descriptions of Damascus in British travel accounts lost the overtones of Oriental picturesqueness that had prevailed in

Williams' diary. From now on, they tended to narrate their travel experiences as illustrations of the recent political events, thereby transforming the story of Syrian picturesqueness into the narrative of an oppressed population. Although the French mandate power had not been held accountable for the attacks, which had caused a large number of civilian casualties, the British press had debated the events and questioned the French capability for governing Arab populations.[110] Even the *Mediterranean Guide* of the shipping company *Blue Star Line* denounced the illegitimacy of French rule and regretted that Syria had suffered from bad rule and worse rulers for centuries. Unlike Arab nationalists, however, the authors and travellers did not come to the conclusion that Syrian self-government might be a solution. Instead, they insinuated the necessity of a disinterested rule of experts in imperial administration: the British.[111] Rather than empathy for the population of Damascus, it was the Franco-British rivalry which set the tone.

British travelogues came to resemble investigation reports of wartime correspondents. Norah Rowan-Hamilton, the widow of an Irish lieutenant killed in 1915, who visited Palestine, Jordan and Syria with a group of British archaeologists, was particularly harsh in her judgement of the French presence in Syria. In her travel report, published in 1928, the title of her chapter on Damascus read "War". Her travelogue, in which descriptions and dialogues often conveyed political messages, read like a eulogy of the British Empire at the expense of the rival French colonial system. Although her narrative comprised the standard tropes about Damascus and her illustrations showed the sights any tourist would visit there such as the "Street called straight", the story she told barely referred to the Damascus of the tourist.[112] Instead, Rowan-Hamilton described a gloomy atmosphere reigning in the city. She accused the French rulers of having wrapped up the city in barbed wire, dotted its buildings with bullet holes, and sent in colonial troops accused of pillage and rape. She reported that, as a consequence of French action, the economy in general and tourism in particular had broken down, and that the formerly colourful Oriental street scenes had given way to scenes of warfare. Her ironic comments showed that she considered the French notoriously incapable of handling Syrians who "just want a firm but kind hand".[113] In her account, the Syrian Revolt thus served as an argument justifying British claims in Syria. The contested nature of the newly created state of Syria was not attributed to the demands of Syrian nationalists and their rejection of colonial rule, but to the inadequate French approaches, experiences and capacities in governing an Arab state.

The diaries of Humphrey Bowman, Director of Education in the British Mandate of Palestine, show that such critical judgements were directly related to the bombardment of Damascus. While Bowman had described Damascus as a picturesque Arab town during his first visit in 1922,[114] Bowman's view of the Syrian capital changed during his second visit in 1926. Invited by the French High Commissioner de Jouvenel, his descriptions were dominated by reports on the inadequate behaviour of his French travel companions and on the destruction resulting from French warfare.[115]

The events not only left their traces in travelogues written during the revolt, but durably changed perceptions of Damascus. Even years after the end of the revolt, observations on French colonial policies had become a part of the visiting programme of British tourists.[116] Since the mid-1920s, the violent repression of the Syrian Revolt framed the region in British travel accounts as a battleground, and Damascus became a symbol of French incapacity in handling the Arab population. Ultimately, from their perspective, Damascus had come to represent the legitimate British ambitions to play a leading role in the Middle East.

In terms of geographical conceptions and mental maps, the revolt and French warfare in Syria led to an increasing differentiation between the two major sections of the French Mandate: the State of Greater Lebanon (from 1926 onwards, the Republic of Lebanon) and the State of Syria. When Humphrey Bowman informed the Director of the American University of Beirut, Bayard Dodge, about the invitation to visit Damascus in the spring of 1926, the latter was somewhat concerned. Dodge warned him that "the Americans still regard '*La Syrie*' as the Eastern side is called as in a state of war [...]". So far, Bowman had subsumed both Syria and Lebanon under the name of Syria, which was not unusual. After all, at the time, the separation was a mere administrative divide, without further implications for the imaginaries of travellers and with a limited impact on their practices. Other tourists had a similar comprehensive vision of Syria. The revolt, however, and specifically the French reaction to it, contributed to the formation of separate, distinctive visions of Syria and Lebanon in the minds of travellers. From that time, the Mandate was divided into a troublesome warzone to the East, and a more stable Lebanon to the West.

For French travel authors instead, Damascus remained an Oriental city, the architectural and cultural character of which was preserved thanks to the efforts of French administrators. Again, Morocco served as the blueprint for a preservation discourse advertised by the mandate power and confirmed by French visitors. While in Norah Rowan-Hamilton's account the famous sights of Damascus were silent testimonies to the destructive consequences of French rule, Charles Le Gras, who visited the city in the same year on his pilgrimage to Jerusalem, painted a very different picture. In familiar Orientalist tropes, he described the typical activities of any tourist in Damascus: Le Gras bought souvenirs in bazaars swarming with people, strolled in the streets, and observed the coexistence of an archaic past and a modern present: veiled women, donkeys and other animals shared the streets with bourgeois Syrian women in European-style dresses.[117]

Like the British visitors, Le Gras saw and recorded traces of war: he described destroyed quarters and houses in ruins, the traces of bombardments, machine guns and barbed wire, and the tight French regime in the city. Yet whereas for Rowan-Hamilton or Bowman such measures had been reason to question the legitimacy of French mandate rule, Le Gras' report did not suggest French responsibility. His descriptions of the destroyed quarters were written either in the passive voice or chose impersonal constructions. Rather than humans, he chose weapons or occurrences such as "bombardments" as the semantic agents in his sentences. Consequently, the syntax of his sentences did not allow for the identification of actors. He never mentioned French perpetrators in the context of the military campaigns.[118] Detailed descriptions of destruction and human suffering, in particular of innocent victims such as civilians, women and children, were omnipresent in Rowan-Hamilton's story, yet marginal in the account of Charles Le Gras. In his report, the events appeared as a quasi-natural consequence of warfare against insurgents.

Instead, Le Gras inserted signs and symbols creating the impression of a legitimate French presence. The French flag at the border to Palestine marked the empire's outpost in the Eastern Mediterranean, welcoming the visitors on seemingly familiar ground.[119] On the ten pages covering his visit to Syria, three anecdotes identified bonds connecting France and Syria: the author perceived similarities between Syria and the French South, analysed the characteristic shape of Roman architecture "in their numerous colonies in the Orient", and lauded the preservation policies of General Gouraud aiming to safeguard the Arab-Islamic cultural heritage.[120] This strategy of justifying the mandate by referring to the shared Mediterranean geographical and cultural heritage, as well as the beneficial effects of French cultural policies, was a common trope. Not only the French High Commission emphasised these connections in reports to Geneva; they were also confirmed by various – notably French or francophone – visitors, who displayed a fundamental loyalty to French mandate rule.[121]

While both British and French narratives clearly marked Syria as French territory, after the Great Syrian Revolt they did so with opposing overtones. French authors assumed the legitimacy of French rule. Even if they did not approve of the violence during the suppression of the revolt, they accepted it as a necessity in the struggle against illegitimate insurgencies. From a British point of view, the French were illegitimate rulers over Syria, who had usurped power. Their condescending paternalism with regard to "the Arabs", deemed unable to govern themselves, led them to conclude that a British mandate would have been preferable.

After the Great Syrian Revolt, Damascus became a symbol of the geopolitical rivalry of the great powers, even for visitors without any apparent interest in politics. Although the actual visits centred on well-known historical sights and places, the stories tourists narrated in diaries and travelogues celebrated the imperial governance of the British and the French respectively. Apparently, the literary creation of authenticity relied on the consideration of current debates and events. Like the

Palestinian case, in a time of intense press coverage of foreign affairs, assuming timelessness was no option. Instead, both British and French tourists agreed that the challenge of administering mandates consisted in establishing "good rule" over potentially "unruly" subjects, thereby reproducing the colonial paternalism of their respective colonial administrations. Sharing similar perceptions of the population of Damascus, they differed in their judgement of French imperial policies. The efforts of the French High Commission to reach out to tourists, accompany them on their journeys and circulate certain visions of Syria, as we have seen in the previous section, were only partly successful. Whereas the British tourists, whose expectations had been shaped by critical British press coverage, were not convinced; francophone visitors were more likely to accept the interpretation of the mandate power. In the end, both approaches consolidated the position of Syria as a discursive battlefield for Anglo-French geopolitical rivalry.

The Syrian summer resorts: Hotbeds of agitation?

After many years of absence, the Palestinian Grand Mufti of Jerusalem, Amin al-Husayni, was back in Lebanon and Syria in the summer of 1935. The French authorities had finally granted him a visa, after they had denied him access to the Mandate for political reasons in the previous years. The journalist Laura Veccia Vaglieri, who reported on the journey in her overview of news from the local press, quoted a source close to the Mufti who confirmed that he was looking forward to a pleasurable, relaxing summer holiday. In fact, the journey was very likely intended to strengthen the standing of the Mufti and to forge political alliances.

Estivage seems to have served as a pretext to al-Husayni for a trip that did have political implications. In Ehden, a mountain village in Northern Lebanon, Amin al-Husayni met the famous anti-French rebel Ibrahim Hananu "and other Syrian nationalists". In Tripoli, a Sunni stronghold, the (equally Sunni) Mufti mediated in a fight between rivalling families of 'his' religious community, an act corroborating his reputation as a leader, and in his visit to Homs, he was accompanied by the anti-French Syrian notable Hashim al-Atasi. The Jewish press in Palestine guessed that the actual purpose of al-Husayni's visit to Syria was to meet with various Arab leaders and find a common position in the event of an Italian-Ethiopian war.[122] Whatever the content of the discussions, it is indeed obvious that al-Husayni's journey was far from being an apolitical one. Yet, the example shows that even though the French authorities were deeply suspicious of pan-Arab agitation, the reference to the habitual holiday practice of *estivage* served notables as a convincing pretext for travelling and provided 'natural' gathering places for exchanges with other Arab leaders.

After the French Mandate had been divided into the states of Greater Lebanon and Syria, the *estivage* region around Bludan and the Zabadani district in the

Anti-Lebanon mountain chain remained a part of Syria. Politically, the *estivage* practice of the notables turned the region during the summer months into a meeting point for members of pan-Arabist and anti-imperial networks.[123] It appears that the *estivage* villages were ideal meeting places for pan-Arab activities, providing opportunities for communication among actors whom the French authorities considered potential insurgents, or at least suspect political leaders. In the mountain villages, the French network of spies and informants was not as dense as in the urban centres.

Moreover, ambitions of the High Commission to develop tourism as an economic branch often translated into strengthening the Syrian *estivage* resorts and travel restrictions almost inevitably led to protests among the local population. Initially, political incentives strengthening the tourism and *estivage* business had been limited in Syria, but the economic programmes of the 1930s comprised measures to support this endeavour. Many adopted measures suggested a significant French influence. Initiatives such as hotel credits, tax exemptions, or the renovation of villages had been employed by French colonial administrations in various parts of the empire (and in metropolitan France).[124] It is therefore likely that the French High Commission, considering tourism as a somewhat universally applicable pill to cure all economic illness, had inspired a great number of the decrees adopted by the Syrian Government.

Despite the widely accepted notion that tourism would strengthen the economy, particularly during moments of political and/or social crisis, a certain ambivalence with regard to its potential persisted among the French mandate officials. In 1936, a large general strike launched in the early months of the year across the major cities of Syria paralysed the country and demonstrated to the French not only the limits of their own control, but also the limits of the Syrian notables whom they had chosen as their primary collaborators. While it is commonly accepted that the new, leftist government in France modified its politics about Syria and that the French approach in general became more favourable towards concluding a treaty, on the ground, the protests seemed to confirm a prevalent French mistrust of the Syrian population.[125]

Political processes thus continued to be strictly controlled, including incentives to strengthen economic development.[126] When the Syrian entrepreneurs Osman Zayn Beyhum, ˤAbbud ˤAoun and Chéhadé Ghattasse, for example, launched an initiative to recreate the village of Mount Kassioun as an *estivage* centre, the French administration tightly supervised the decision-making process. A commission was formed comprising the Minister of the Economy, Mustafa al-Qusayri, representatives of his ministry and the Ministry of Interior, in addition to three French 'advisors'. The fact that even a commission of reduced impact was made up of an equal number of French and Syrian representatives shows the determination of the High Commission to supervise any political processes. At the same time, the active support of the Syrian Government and political representatives remained limited. To the representatives in the Chamber, the *estivage* business was neither economically vital nor did it play a significant role in their plans for the future of the country. Support for the measures aimed at attracting a greater number of tourists was expressed once more by members of the middle classes: the journalists in the acknowledged nationalist paper of Damascus *Alif Ba* approved of the measures.[127]

Whereas the French High Commission tried to strengthen the *estivage* centres for economic reasons, they were suspect meeting places for political – anti-French – activity. Indeed, they were an ideal opportunity for gathering. In contrast to the rivalling urban centres such as Aleppo, Homs and Damascus, the smaller towns and villages in the Anti-Lebanon were, in a way, "neutral ground". They were not strongholds associated with major families and, as various notable families from the entire Eastern Arab region used to spend their summer there anyway, they facilitated meetings and exchange among notables across mandate borders.

While the Syrian notables remained the main actors in Syrian politics, a camp for scouts taking place in the summer of 1936 indicated that the notables adopted new patterns of political mobilisation. Scout camps, taking place in the summer resort of Bludan, were subsidised by the Syrian Government.[128] The boys were welcomed by the Syrian prime minister and other representatives of the Government, and the Damascene nationalist newspaper *Alif Ba* covered the camp as an event of national relevance, arguing that it pioneered in establishing a true Syrian scout movement.[129] The presence of prominent politicians and the coverage by one of the most important nationalist newspapers reflect the relevance and the pronounced political implications of youth movements. Anti-imperial mobilisation included, as in many countries in the 1930s, youth organisations. The continuity in choosing the *estivage* centre of Bludan as a meeting place and the fact that politicians and representatives of the notable families uttered the introductory words reaffirms that new forms of political action and new groups of actors in Syria did not emerge in opposition to the spatial formations and political structures associated with the Syrian notables. Rather, old patterns of legitimacy and new forms of mobilisation joined forces in opposing the mandate power.[130] As rivalries between the urban centres persisted, the choice of Bludan as an enclave for the Syrian anti-French youth may have been not only a pragmatic choice, but also a consciously set symbol for the reinvigorated resistance against French rule.

To the French secret service (*Sûreté générale*), the mountain villages were suspect, as their remoteness and comparatively difficult access made them hard to survey.[131] These fears seem justified, given that 'suspect individuals' such as Amin al-Husayni exchanged information and ideas there, shielded comparatively well from observation of the authorities. Yet, the villages should not be mistaken for carefully selected hiding places – from the Arab notables' point of view, they were a rather natural choice. From their perspective, the mountain resorts were centres of sociability, rather than remote villages. They had already become political centres during summer in Ottoman times, and under the mandate regime the *estivage* practice provided the additional advantage of justifying meetings between Arab notables from the entire region. Whereas the French High Commission applied the criteria of military access and intelligence-gathering, the notables read the space in terms of sociability and exchange. From their point of view, the importance of the villages was three-fold: first, in comparison with the cities, French supervision was less pronounced there; second, they brought together notables from entire *Bilad al-Sham*; and third, they were not defined in national or factional terms. This kind of neutrality distinguished them from the urban centres such as Hama, Aleppo and Damascus. In the urban centres, 'the street' expressed resistance against the French in demonstrations of force, which often had strong local references and overtones. Thus, the *estivage* resorts served – on purpose – as transnational centres for meetings and as pan-Arab platforms for the expression of big ideas and political visions that aimed at a larger restructuring of the region in its present state.

Aleppo's isolated middle classes

Great divergences in reading and interpreting spaces not only occurred between Syrian notables and the French mandate administration, but also between the middle-class inhabitants of the Syrian cities and European visitors. Even though both groups spoke a similar idiom of modernity: many city-dwellers had been educated in European missionary schools or in Europe, and worked in liberal professions such as journalism, bureaucracy, law or medicine.[132] In order to advocate their interests and to establish social networks, they typically formed circles and clubs defined by the topics they aimed to advance. Often, these associations facilitated their members to establish contact with similar circles in other countries, such as the tourist development associations in Palestine and Egypt, for example. Whereas a comparable body did not exist in Syria, a group of middle-class members in Aleppo founded the *Société Archéologique d'Alep* (Archaeological Society of Aleppo).[133] Beyond pursuing their archaeological interests, the group also provided analyses of tourist movements and lobbied in favour of better advertisement campaigns as well as the improvement of accommodation facilities.[134]

The case of the *Société Archéologique d'Alep* not only shows the significance these local middle classes attributed to tourism, but also indicates their position in political decision-making. For their project, they needed to win over the French High Commission as well as the local upper and middle classes. Aleppo suffered from the new spatio-political realities in terms of tourism, since the mandate borders had deprived Aleppo of important connections.[135] In late 1930, an average of 4 tourists visited the city per day. Such a low number could not only be attributed to the economic depression – after all, Damascus was frequented in the same year by 15 visitors per day.[136] In fact, most travellers on trips around the Eastern Mediterranean skipped Aleppo. George H. Williams did not visit the city, and Mary Steele-Maitland, whose travel route led her and her husband from Egypt via Palestine and Syria on to Turkey and the Balkans, passed Aleppo on the train, but the couple did not stop for a visit, over which Steele-Maitland expressed her regret: "[...] we reached Aleppo. Such a pity we couldn't see it. From the train – as it was nearly full moon we could see a fine citadel and it like Damascus is part of the edge of the desert."[137] The obvious touristic potential of the city remained unexploited.

In May 1931, the delegate of the French High Commission in Aleppo forwarded the sample of a new bilingual review (written in French and Arabic) to the High Commission in Beirut. It was the first publication of the local *Archaeological Society*, which the delegate presented as the result of a French-Syrian collaboration aiming to attract the interest of the inhabitants of Aleppo (*"les éléments locaux"*) in the study of history.[138] Kamil al-Ghazzi, one of the founding members of the *Archaeological Society*, who was also on the editorial board of the review and the author of its Arabic preface, was most likely involved in the drafting of the French preface too. This foreword stated the main purpose of the society and its publication as follows:

> On the occasion of its foundation in January 1930, the "Archaeological Society of Aleppo" declared that its objective is the dissemination of information on archaeological matters regarding Syria, the protection of monuments that constitute the national heritage, and the expansion of tourism in Northern Syria. [...]
> This review will be the journal of the "Archaeological Society", but also the one of the intellectual archaeological and touristic movements in Northern Syria; and the voice of a people which is becoming aware of itself, and which – thanks to the support provided by the French mandatory nation – is striving to reach as fast as possible the intellectual level of the European nations.[139]

In this passage, the author demonstrated the significance of archaeological research and tourism development by referring to notions of cultural heritage, nation-building and the civilisational evolution of nations. He defined Aleppo and its *Archaeological Society* as being part of various networks: spatially, he situated Aleppo at the centre of a region termed "Northern Syria", thus implying the existence of a Syrian entity. Moreover, he predicted that Aleppo would become an international centre, manifest in the fact that tourists and archaeologists visited the region. Such ideas were in line with the mandate treaty, which had conferred the responsibility of enabling Syrians to self-govern to the French mandate power. This implied creating the conditions for political independence as a nation-state.[140] The preface framed the activities of the society in a self-confident manner, but in a way that revealed opportunities for cooperation between the *Archaeological Society* and the High Commission, as the author assumed that they shared the goal of educating broader segments of the population.

In the Arabic version, al-Ghazzi also connected archaeological excavations and tourism, but this time he addressed a local Aleppan public, rather than an international audience:

> The aim of this magazine is: [...] To inform about the quality of these antiquities [of Aleppo] and to provide illustrations of their beauties [...]. This booklet is distributed among the scholarly associations and in public places in the famous Western kingdoms and foreign countries, so the lovers of ancient ruins will see [the illustrations] and they will long to see them [the antiquities] and render themselves there and Aleppo will make a huge financial profit from it, which should not be underestimated: Such a profit as make the regions of Egypt and Beirut and Lebanon and Baalbek and Tadmor and other places, about the historical greatness of which tourists know thanks to photographs and publications on the pages of newspapers and scientific magazines specialised associations have founded for the specific aim of making tourists come to the region from all corners and directions. The benefits from it are so huge that the pen cannot describe it.[141]

Both texts advertised the value of tourism, but the reasoning aimed to convince different audiences in divergent terms. The major argument of the Arabic version, which addressed Aleppo's educated middle and upper classes, was economic: the inhabitants of Aleppo, who were at the time facing a harsh economic crisis, might profit economically from the influx of tourists. Well-known tourist destinations such as Egypt, Beirut and Baalbek had proven the economic potential of tourism. In contrast to the French text, this perspective embedded the city of Aleppo in the spatial context of a larger post-Ottoman region. Locating Aleppo within a national Syrian territory did not play a role here. Rather, al-Ghazzi referred to the notion of an imperial *Bilad al-Sham*, ordered in networks of interconnected centres. For the Aleppan inhabitants addressed in the Arabic text, as for the postcard editors we encountered in the first section of this chapter, interurban networks and cities in competition within a larger region had apparently remained relevant categories of spatial order.[142]

The editors of the review thus sought to address both a French and a local public not only by using their respective languages, but also by drawing on concepts and ideas meaningful to them. According to Keith Watenpaugh, this capacity to translate notions into various idioms characterised the representatives of the modern Aleppan middle classes. Among them was al-Ghazzi, an educated city-dweller, editor of a newspaper, and author of a chronicle of Aleppo (published in 1926), which was based not only on Arab, but also on European, sources. Since 1925, he had been an active member of local archaeological societies. (It might be interesting to note here that the first of these associations, *Jamiʿiyyat ʿAdiyat Halab*, was sponsored among others by the family owning the *Hotel Baron* in Aleppo. Evidently, the hotel owners well understood the link between tourism and archaeology.)[143]

Al-Ghazzi's double-sided reasoning also implied that he was searching for allies to win over with promising arguments. Despite his reputation and qualifications, al-Ghazzi's political influence remained limited. As a result of the French-Syrian policy of "honourable cooperation", Syrian notable families had emerged as the dominant actors in local urban and national political spheres in Syria, while the French consciously circumvented the often nationalist urban middle classes. Inspired by General Lyautey's approach in the French protectorate of Morocco, especially after the Syrian Revolt, the French strategy consisted in finding members of the major notable families who were willing to cooperate with the mandate power. The strategy of cooperating with notables was informed by prejudice about the alleged tribal structures of Arab societies. The mandate administration and their Parisian interlocutors hoped to turn the notables into mediators between the French and local society.[144]

Watenpaugh described how the middle classes, aspiring to participate in politics, attempted to establish relations with Aleppan notables. Although they managed to achieve some sort of communication, the notables did not consider these educated middle classes as persons of equal status. Kamil al-Ghazzi, too, experienced that his educational and historical projects did not attract significant support from locally influential notables.[145] For the patterns of political mobilisation of the notables, the creation of a historical consciousness envisioned in the French preface, as well as the creation of identities in order to mobilise larger groups of the population, were not (yet) relevant. The economic prospect of tourism development by contrast was expected to benefit the urban middle classes rather than the sectors in which the notables or their clientele held interests.

Such ideas spoke more to the French mandate power: the local French delegate enthusiastically forwarded the magazine to the High Commission. Yet while the French representatives approved of the project and granted some subsidies, no cooperation or joint efforts in tourism development resulted from al-Ghazzi's initiative. In the French approach to notable interlocutors, persons such as Kamil al-Ghazzi and their initiatives did not matter.

Neither the French administration nor the notables thus backed his efforts to popularise archaeology and develop tourism more systematically. Not even in 1934, when a major programme of economic development was drafted by High Commissioner Damien de Martel in an attempt to convince Syrian leaders and the population of the beneficial effects of French rule, they attributed significance to his project.[146] Although the programme included measures of strengthening tourism, these measures were inspired by the French repertoire of tourism development applied in several other parts of the empire. Cooperation with local associations such as al-Ghazzi's *Archaeological Society* did not feature in the suggestions. Kamil al-Ghazzi's remark that Egyptian and Lebanese authorities and associations invested immense efforts in tourism development, therefore can be read not only as a reference to a successful model, but also as an expression of mild frustration.

In Syria, the urban middle classes who cared about these showcases of an educated modernity, of notions such as national identity, historical legitimacy and heritage, were not the ones shaping and influencing politics. Like their Egyptian, Palestinian and Lebanese counterparts they published articles in journals, wrote books, conducted excavations and research, and founded circles such as the *Société Archéologique*, but they were not at the core of decision-making. While they attempted to cooperate with politically influential notables, they remained in subordinate positions and depended on the (limited) goodwill and interest of their interlocutors.[147] The Syrian notables, however, did not have an interest in the creation of an integrated nation-state, and therefore were not inclined to adopt strategies aiming at the construction of a national consciousness beyond anti-French leanings. They maintained their system of mobilising supporters, which was based on socio-spatial adherence, rather than on categories such as class and convictions, political strategies and worldviews.[148]

Whereas al-Ghazzi's local urban initiative aimed at the protection of ancient heritage, the development of a historical consciousness, and tourist accommodation, a coherent national strategy for developing tourism did not emerge as a result of the political structure in Mandate Syria. His local initiative remained geographically confined, expressing a local or regional – rather than national – sense of belonging. As the notables shaping national politics did not expect to profit from nationalist principles, discussions about the economic, educational and political potential of tourism did not reach the national level.

The approach of the *Société Archéologique* thus resembled the measures taken by the tourist development associations in the neighbouring countries: advertisements and practical advice in maps and guidebooks combined with efforts to improve accommodation infrastructures. Unfortunately, it appears that the initiatives of the *Société Archéologique d'Alep* did not meet with any response. Neither the High Commission nor the Syrian Government put the proposals into effect, and the few tourists who visited the city did not take into account what al-Ghazzi and his circle had to tell about the city, its history and culture.

As a result, local advocates of tourism development seemed to have difficulties in establishing contact with visitors and shaping their perceptions of Aleppo.

Humphrey Bowman visited the city in 1928, during his third journey to the French Mandate. The Syrian Revolt had marked his previous journey to Beirut and Damascus in 1926, when Bowman's indignation about French practices of rule had permeated his diary entries. Two years later, he described his tourist activities more extensively than during the previous visits. Despite his familiarity with the region, Orientalist expectations and frames shaped his vision of Aleppo. He indulged in the Oriental beauty of the *souks*, manifest in the ornaments of the *khans*, the abundance of goods and the quality of handcrafted goods. Aleppo was, according to him, "unspoilt by Frankish [referring to an Arabic term for Europeans, jd] taste".[149] Not even the markets of Damascus, he stated happily, could rival Aleppo in this respect.[150]

Kamil al-Ghazzi, *Wattar Frères* and others had offered a variety of touristic perspectives on Aleppo: as an archaeological site and monumental witness to ancient empires; or as a city with vibrant street life and modern infrastructures. Yet, without the institutional support of well-organised associations or the mandate power, their interpretations did not reach Bowman, who rather denied to Aleppo the urban modernity that *Wattar Frères* had visualised.

It seems that in Syria, middle-class actors had difficulties in systematically shaping tourism policies, and thereby the narratives spread within the context of tourism. While the High Commission had approved of the proposal of the *Société Archéologique*, the initiative did not result in any form of cooperation or systematic tourism development, and the members did not seem to be involved in the actual communication with tourists. Orientalist visions, as on Thévenet's postcards, continued to circulate. Local perspectives hardly reached potential visitors. Moreover, the lack of recognition from official authorities implied that Syrian historians, archaeologists and intellectuals had greater difficulties in addressing tourists than Egyptian, Lebanese and Yishuvi actors in tourism had. Even in Palestine, the monopoly of access in certain monuments of Jerusalem provided Arab middle-class actors with better access to international visitors.

Palmyra: The desert on display

Palmyra (Tadmor) had the charms of a seemingly abandoned desert town. Very few postcards reached France from the Eastern Syrian desert; if they did so, they often showed the colonnaded street rising over the desert sands (fig. 20).[151] The seeming neglect and disinterest of the local population in historical sites was a common trope in guidebooks and its visual manifestations on postcards often disguised the efforts photographers made to exclude the population and other visitors from their photographs.[152] The trope of the Oriental void serves as a starting point for this final section. As far as tourists were concerned, the suggestion of emptiness on

Figure 20: The ruins of Palmyra (Photographer: Jean Torossian. Reproduced by kind permission of The Fouad Debbas Collection / Sursock Museum, Beirut)

the postcard does not seem entirely misleading. Although guidebooks suggested numerous travel routes which included a stop at this most impressive site of Roman ruins, the hotel in Palmyra was barely profitable. In 1933, High Commissioner Ponsot opined that its unfortunate owner, Pierre d'Andurain, had to be considered one of the nobles who "spends their heritage on equally ruinous as tempting undertakings."[153] Despite the fame of the site, tourists were not yet abundant.[154]

Since the early days of tourism to Palmyra, visitors had considered the *Hotel Zénobie* a symbol of the tourist movement to the region. Norah Rowan-Hamilton, who visited the town in the late 1920s before Pierre d'Andurain invested in the hotel, read its history as a symbol of the decline of the ancient civilisation. The story she told was one of missed opportunities: according to her, a French tour organiser had planned the hotel (though, as it appears, it was actually the Lebanese transport company *Kettaneh*), which was intended to contribute to the emergence of the town as a tourist destination. However, due to the Syrian Revolt, tourists never came and the hotel building was not completed. At the time of Norah Rowan-Hamilton's visit, she concluded, it was in a condition even worse than "Zenobia's ruined city".[155] In her account, the romantic charm of decay and the gloomy atmosphere, as well as the longing for a bygone past, characterised the condition of both the town and of tourism there.

The inhabitants of Palmyra though were part of the attraction that fascinated Rowan-Hamilton during her visit. Referring to tropes of Oriental decline when describing Islamic elements in the town, such as the "squalid Arab village and mosque", in other instances she sketched out the picture of a harmonious ensemble of ancient heritage and contemporary inhabitants. The "fine Greek traits" of the local women, for example, evidently established a lineage from their presumed ancient ancestors to the present-day inhabitants.[156]

Obviously, Rowan-Hamilton's account followed the tradition of the famous explorers touring the Arab East in the nineteenth and early twentieth century.[157]

The 'discovery' of a place fallen into oblivion long ago was a widespread motif, just as the sympathetic approach to the people living in the desert was, who were framed as 'noble savages'. The trope, as identified by Wiedemann, was based on the idea that the isolation of the desert had prevented these people from mixing with supposedly 'inferior' races – an idea apparent in Rowan-Hamilton's remark on the "Greek" faces of the inhabitants.[158] Consequently, the inhabitants of Palmyra served as a picturesque illustration of the scenery, and as a contrast highlighting the enlightened approach of the visitors themselves.[159] Similar to George H. Wiliams' photographs in Egypt, she aimed to re-discover the Syria she was familiar with from other travel reports, the observations of which she confirmed.

As a consequence, the inhabitants of Palmyra, as a living attraction, were supposed to remain within a certain, clearly confined Other space. The desert, presumably populated by archaic tribal societies and difficult to access, was an ideal variety of such a space, which rendered tangible the distinction that members of self-proclaimed modern societies identified between themselves and the allegedly archaic communities inhabiting the Other space. Wang argued that crossing the civilisational boundary and accessing this space for a limited time allowed tourists to assert their own modernity.[160] Elements testifying to changing ways of life in Eastern Syria surprised Rowan-Hamilton, and she reacted to them with a similar mixture of astonishment and irony that we encountered in Williams' descriptions of modern, urban Egypt. She noted with great astonishment, for example, that the sister of one of the local sheikhs had married a "gentleman from Paris".[161] Apparently, the physical and social mobility of the sheikh's sister seemed to contradict her expectations concerning the lives that they were leading. The news about the marriage between the woman from Palmyra and a French man thus must have disconcerted Rowan-Hamilton. While tourists took their own ability to cross civilisational boundaries for granted, they did not expect it to happen on the Other side.

As tourism development was a recent phenomenon in Palmyra, Rowan-Hamilton had to face this irritation. At other tourist sites, the separation of the monumental from the ethnographic sites had already taken place. In Egypt, in Luxor, for example, but also in the Lebanese ruins of Baalbek, the monuments had already been "cleared" from local inhabitants living among and with the ancient ruins. In Palmyra, this process was still underway and the French mandate power did not proceed as smoothly as they expected. From their point of view, the population of Palmyra presented a major obstacle to the touristic accessibility of the site. Policies of heritage preservation justified governmental intervention in local settlement practices. From January 1930 onwards, the ground was to be turned into a so-called "National Archaeological Park", and an entrance fee was levied.[162] Under Damien de Martel, Palmyra was finally classified as a "historical monument", which implied further restrictions with regard to the development of the town: across the entire city and in all places in view of the "archaeological park", housings had to be constructed in "the regional style". Furthermore, the *Service des Antiquités* had to confirm the plans for any new representative building, such as municipality buildings, hotels, mosques and so on, which were to be constructed in Palmyra.[163]

Under the premise of preservation, tourism justified the restructuring of an entire city. For the inhabitants of Palmyra who had lived in or next to the excavation sites, the idea to present Palmyra to a larger audience, meant first and foremost that they had to leave their homes. As the High Commission argued, living amidst the ruins had to be prevented with regard to future settlements, and the existing houses needed to be torn down. The archaeologically significant part of the town was to be evacuated "in the public interest" in October 1929.[164] While the exact number of persons who had to leave their homes is hard to establish, a document indicates that around 70 families were forced to move: in December the authorities issued a decree stating that 68 new houses, a health care centre and a school had to be constructed.[165]

Yet several subsequent decrees insisting on the definite "clearance" of the site suggest that a number of inhabitants refused to leave their homes. The High Commission first set incentives, promising tax exemptions (July 1930) and parcels of land (1934), and finally announced expropriation: houses had to be destroyed in order to "isolate the temple of Baal" (1939).[166] The repeated decrees aiming to expel the inhabitants of Palmyra from the ruins demonstrated the resistance that the authorities faced when putting their plan into action.

Palmyra was not the only such case. In its report to the League of Nations in the year 1927, the mandate administration announced a similar proceeding about the "clearance" of the Crusader fortress Krak des Chevaliers (Qalˁat al-Husn) in the ˁAlawite state.[167] The inhabitants, on the one hand considered a source of interest due to their assumed picturesque qualities, were, on the other hand, separated from their homes amidst the ruins, which were exposed and cleared of seemingly anachronistic elements. Thus, the local population was transformed into an attraction – but an attraction set apart from the ancient ruins. This separation occurred both in narratives and in the spatial order.[168] Similar to travel photographs capturing either the population or ancient monuments, stories of glorious monuments and of mysterious local customs were separated spatially.

In Palmyra, such a separation occurred largely against the will of the inhabitants. Whereas in Aleppo, parts of the population embraced tourism development and considered it a beneficial option for the development of their hometown, such proponents of tourism did not exist in the relatively isolated and much smaller Palmyra. Here, tourism seems to have been the business of foreigners: the hotel had been a project of the Lebanese transport entrepreneur *Kettaneh*, taken over by a Frenchman, and catered to the interests of international travel and transport companies.[169] Moreover, creating tourist mobilities to a strategically important outpost in Eastern Syria was in the interest of the French military.[170] The town, marginal during Ottoman times, grew only during and after the interwar period, as a result of two main economic factors: its position on the oil pipeline from Kirkuk, and the growth of tourism.[171] The population was less heterogeneous, the level of education lower, and the economy less transnationally connected than in Aleppo. The middle classes, which Watenpaugh has described as the main agents of change in Aleppo, barely existed in Palmyra.

The local population therefore neither built up the business of tourism nor profited from its revenues. Rather, tourism emerged as a foreign undertaking, contradicting the vital interests of the inhabitants. As a colonial project, tourism was supported by the colonial state and its administration, rendered possible by urban entrepreneurs from across *Bilad al-Sham*, and carried out by privileged citizens from the imperial metropoles. It integrated the Syrian desert outpost into a larger Syrian entity and created a defining landmark for the French administration, but the processes of territorial consolidation were carried out at the expense of its local inhabitants. Paradoxically, to show the ongoing relevance and greatness of the Roman imperial heritage, the Roman ruins had to be cleared of the persons who lived on its very ground.

Conclusion

The Syrian tourism space as a whole was largely shaped by the French mandate administration, which saw tourism as a means to justify French governance and to tighten its rule across the Syrian territory. Whereas local urban initiatives aimed to foster tourism in their respective towns, a more general strategy of tourism development on a national level did not emerge. Rather, contact between the various levels of policy-making remained limited.

At a local level, representatives of the middle classes in the major Syrian cities had an interest in tourism as part of a larger project of 'modernising' the country. Economic interests often went hand in hand with the idea that tourism was an occasion to present their views on Syria and counter imperialist claims. They formed clubs and associations to promote these interests, and they attempted to propagate the vision of a modern, independent Syria, both among the local population and an outside audience. Printed publications, such as reviews and postcards, spread these visions for the country, strongly influenced by late-Ottoman ideas of an urban civic modernity.

Yet these members of the middle classes played a subordinate role in promoting tourism, because they lacked access to national institutions and the networks necessary for its realisation. Similar to Arab Palestinians, the urban middle-class actors in Syria were largely sidelined by mandate policies, as the French mandatory opted for cooperation with the Syrian notables. Attempts to build relations with either of the sides and thereby secure influence were not successful. Their initiatives remained grounded locally and their media, with the exception of postcards, had a limited range. Their ambitions though resembled the aims of Egyptian advocates of tourism development: The *Archaeological Society of Aleppo* argued that transnational mobilities provided economic opportunities and hoped that tourism would strengthen the significance of the town on a national level. With regard to an international audience, picture postcards of local editors visualised the late-Ottoman modernity of the urban centres and thus their vanguard role in transforming local society while refuting denigrating Orientalist depictions. Due to their lacking influence in processes of decision-making and debate, however, the activities of the Syrian middle classes neither lead to the drafting of a coherent national narrative nor to the creation of a unified tourist and national space.

At a Syrian national level, notables continued to define politics with the general aim of gaining independence from France. New forms of political mobilisation were incorporated into the older forms of adherence, as the example of the scout camp demonstrated. Forms of popular nationalism, conceived of as anti-French agitation, continued to be shaped by personal bonds and alliances, and tourism – or *estivage* – was conceived of as a practice facilitating the exchange with notables from across *Bilad al-Sham*. Notions of drafting a national narrative or creating territorial cohesion by means of tourism were hardly relevant in the national political discourse. For the notables who promoted tourism to Syria, it was a means of supporting a particular clientele or region (such as Bludan and Zabadani), rather than related to an economic or ideological vision for a Syrian nation-state. This markedly differed from neighbouring Lebanon, although in both parts of the Mandate, the *estivage* centres aimed to attract a similar clientele. Only in the late 1930s did the initiative of Fakhri al-Barudi suggest that at least individual notables sought to profit from tourism development as a way of shaping images of Syria on an international level.

The French High Commission, by contrast, launched several initiatives to strengthen the tourism sector in Syria; however, their attempts ignored local needs and were not sustainable. From the perspective of the High Commission, two major arguments rendered tourism development a promising investment: First, the French mandatory perceived tourism as an asset in justifying French governance to the Mandates Commission in Geneva. Potential economic growth was one of the expected benefits of tourism development; the testimonies of travellers to the beneficial effects of the French presence were another one. Therefore, unlike the British mandatory in Palestine, the French High Commission set certain incentives to promote the tourism business, such as the introduction of hotel credits, and paid particular attention to accompanying and guiding tourists across the territory.

Since the French mandatory steered away from granting a high degree of autonomy to local Syrian bodies in general and the urban middle classes in particular, they resorted to a limited number of only partially effective incentives.

Second, tourism development served as a resource in establishing mandate rule and creating a coherent space under French surveillance. The comparison of Aleppo and Palmyra indicated that tourism development was tied to French security interests, prioritised over intentions of fostering economic growth. In a city like Aleppo, where economic development was desperately needed and local interest groups sought the support of the French High Commission, tourism never reached a high priority among French officials. In distant or 'unruly' areas such as Palmyra, the Jabal al-Druze or the ʿAlawite state, tourism was harnessed, as it was intended to serve as a civilian continuation of military conquest and help maintain control.

This pattern resulted from the French strategy of 'pacification', intended to prevent further revolts. Collaborating with members of the major notable families while circumventing the urban, nationalist middle classes were a major pillar of this strategy. The French fear of anti-French national movements resulted in a general ignorance of middle-class visions of a 'modern' country, including notions such as 'national heritage' or 'national history'. Tourism, as a means of territorial control, was imposed as part of the 'pacification' strategies by the French military. As a colonial political tool, it remained devoid of the symbolic, nationalist overtones it had in Egypt (or Lebanon).

While not all tourists approved of the French way of ruling Syria, the French mandate administration remained their major interlocutor, presenting the country, providing access, and offering assistance in planning and undertaking the journey. As a result, tourists either accepted the concept of a *Syrie française* of the French High Commission or departed from the idea of an Oriental (and Orientalist) Syria. The Syrian Revolt had an ambiguous effect in this regard: on the one hand, Syria emerged as a distinct geographical entity in its own right, characterised by its geopolitical relevance. On the other hand, French and British narratives about the revolt reinforced the "colonial gospel": the stereotype of a desert country inhabited by ferocious tribes.[172] The travellers of my sample continued to frame Damascus, Aleppo and Palmyra as situated in a diffuse Eastern continuum. Alternative narratives pertaining mainly to a local or national context were – with the exception of picture postcards – less likely to be consumed by tourists, for whom the idea of a backward Orient merged with the great powers' geopolitical ambitions in the 'Middle East'.

Notes

1   JD PRIVATE COLLECTION, Commission du Tourisme des Etats sous Mandat, *Holidays Off the Beaten Track*, p. 3.

2   The editor, the *Commission de Tourisme du Haut-Commissariat de France* had been founded in 1923 by the French High Commission. Monmarché, *Syrie – Palestine – Iraq – Transjordanie*, p. LXXXII.

3   Pedersen, *The Guardians*, pp. 29–40. Gelvin, *The Modern Middle East*, p. 200.

4   For the British Empire: Migdal, *Strong Societies and Weak States*, pp. 108–116.

5   Gelvin, *The Modern Middle East*, pp. 198–204. Khoury, *Syria and the French Mandate*, pp. 27–70, esp. pp. 55–57. Burke III, 'French Native Policy'. Migdal, *Strong Societies and Weak States*, pp. 108–110.

6   Khoury, *Syria and the French Mandate*, pp. 58–60.

7   Khoury, *Syria and the French Mandate*, pp. 264–265, 340, 346–350, 362–363. Shambrook, *French Imperialism in Syria*, pp. 67–69, 152, 157–159, 247–248.

8   Khoury, *Syria and the French Mandate*, pp. 72–81.

9   I refer to the term of "notables" in the sense Gelvin understands it, as a social elite that has emerged as part of and that was shaped by the Ottoman system of governance: Gelvin, 'The "Politics of Notables"', pp. 26–27. Cf. also Khoury, *Syria and the French Mandate*, pp. 248–251. Cleveland and Bunton, *A History of the Modern Middle East*, pp. 218–224.

10 On tourism in French North Africa cf. Zytnicki and Kazdaghli, *Le Tourisme dans l'Empire français*.

11 Cf. BNF TOLBIAC, Du Taillis, *Le nouveau Maroc*. BNF TOLBIAC, Du Taillis, *Motor-Touring in Morocco*. BNF TOLBIAC, Goud, *Le tourisme au Maroc*. BNF TOLBIAC, Fauré, *Guide pratique de Marrakech*. BNF TOLBIAC, ESSI de Rabat-Salé et sa région, *Guide touristique historique*. BNF TOLBIAC, Syndicat d'Initiative et de Tourisme d'Ifrane, *Ifrane, Maroc*. BNF TOLBIAC, Syndicat d'initiatives et de tourisme „ESSI" Maroc, *Comment venir et voyager au Maroc*.

12 On the term 'middle classes' in this context cf. Watenpaugh, *Being Modern*, pp. 17–26. The implications of 'national heritage' with regard to power relations have been highlighted by Dieter Langewiesche: Langewiesche, 'Was heißt "Erfindung der Nation"?', pp. 616–617.

13 FDC, Thévenet, *Panorama de Hama (Syrie)*. FDC, Wattar Frères, *Hama (Syrie): Panorama générale*.

14 Marie-Hélène Degroise (1947–2012), formerly curator at the Centre des archives d'outre-mer in Aix-en-Provence, published an extremely useful blog, gathering information on an impressive range of photographers and postcard editors in the colonies and overseas territories: Marie-Hélène Degroise, Photographes d'Asie (1840–1944), <http://photographesenoutremerasie.blogspot.de/>, last accessed 05/04/2017.

15 DeRoo, 'Colonial Collecting', p. 85. Schor, 'Cartes postales', p. 12. Jaworski, 'Alte Postkarten als kulturhistorische Quellen'. Gunning, 'The Whole World Within Reach', pp. 27–29. Under the mandate, Lebanon and Syria became members of the Postal Union, which led to major improvements in postal services. The number of post offices in Syria and Lebanon increased from 68 in 1919 to 411 in 1939: Saliba, *Beyrouth*, p. 13. On the postal system cf. Bassil, *French Postal History in Tripoli*. Bassil, *Seventy Years of Postal History*.

16 Mizrahi provides detailed insights into the social background as well as previous experiences of the officers in the *Service des Renseignements*: Mizrahi, *Genèse de l'État mandataire*, pp. 186–202.

17 Moors, 'From "Women's Lib." to "Palestinian Women"', p. 23.

18 Alloula, *The Colonial Harem*, p. 5, footnote 5.

19 Cf. Çelik and Eldem, 'Introduction', p. 12. Sheehi, 'A Social History of Early Arab Photography', pp. 177–179, 190–193. Micklewright, *A Victorian Traveler*, pp. 78–79.

20 Cf. DeRoo, 'Colonial Collecting', p. 85. Micklewright, *A Victorian Traveler*, p. 186 et pass.

21 I am deeply indebted to Wolf-Dieter Lemke for having invited me to see his impressive collection, and sharing knowledge about postcards as well as collectors' practices. Dominik Oesterle, in 2016 intern at the OIB Beirut, has informed me about the collection and Stefan Seeger and Dina Banna from the OIB Library kindly assisted me in establishing contact with Wolf-Dieter Lemke.

22 327 of the postcards were sent by soldiers, 39 by tourists or soldiers on excursions. The sample was compiled between February and July 2017.

23 Between 1928 and 1932, the annual number of postcards sent from the British Mandate surpassed 900,000 (with the exception of 1931). In 1933, senders posted almost 1.4 million postcards in Palestine, and in the peak year of 1936, 3.3 million postcards were sent from Palestine: Jarman, 'Report on the Administration of Palestine 1933', IV, p. 519. Jarman, 'Report on the Administration of Palestine 1936', VI, p. 291.

24 The fact that in my sample most recipients were female corresponds to findings of previous studies, which pointed out that collecting postcards was a bourgeois feminine activity. Schor, 'Cartes postales', p. 12. DeRoo, 'Colonial Collecting', pp. 90–91.

25 There are two works in particular on local postcard producers in *Bilad al-Sham*: el-Hage, *Des photographes à Damas*. Debbas, *Des photographes à Beyrouth*. For Beirut, Debbas has established an overview of motifs: Debbas, *Beyrouth: Notre mémoire*. The following introductions provide overviews of certain postcard series: Tarazi, *Vitrine de l'Orient*. Gavin, *The Image of the East*. Toubia, *Sarrafian Liban*.

26 AKPOOL, Terazi & Churbadji, *Damas. Maison la Municipalité*.

27 On 'types': Alloula, *The Colonial Harem*. Gregory, 'Emperors of the Gaze', pp. 209–212. DeRoo, 'Colonial Collecting', p. 89.

28 On Soubhi & Mounir Aita: Degroise, 'Photographes d'Asie'.

29 AKPOOL, Soubhi S. & Mounir Aïta, *Damas – Le Sérail*.

30 Simon Jackson has demonstrated the significance of the notion of 'development' in French colonial rule: Jackson, 'Mandatory Development'.

31 Palestinian photographic studios, however, barely seemed to be involved in postcard production. el-Hage, 'The Armenian Pioneers'. Lemke, 'Ottoman Photography'. Nassar, 'Early Local Photography in Jerusalem'. Semmerling, *Israeli and Palestinian Postcards*, pp. 6–10. Moors, 'Presenting People'.

32 Debbas, *Beyrouth: Notre mémoire*, p. 10.

33 DeRoo, 'Colonial Collecting', pp. 85–87. Naomi Schor objected that exotic motifs such as the 'type' photographs were not limited to Oriental or colonial contexts. As photographers produced similar views of metropolitan landscapes, she attributed such styles to a general nostalgia for the past in a time of rapid modernisation and an allegedly disappearing 'traditional' way of life: Schor, 'Cartes postales'. In a similar way, John MacKenzie has explained Orientalism more generally: MacKenzie, *Orientalism*, xvi-xviii.

34 AKPOOL, Thévenet, *Deir-es-Zor: Type*. Cf. also: FDC, Thévenet, *En Syrie: Porteuse d'Eau*. FDC, Thévenet, *En Syrie. Types Anazé*. FDC, Thévenet, *Alep: Type du Pays.*, AKPOOL, Thévenet, *Turcomanes*. AKPOOL, Thévenet, *Alep: Type du pays*.

35 AKPOOL, Wattar Frères, *L'Emir Moudchem*.

36 AUB, *Séance Publique de la Chambre des Députés, 28/03/1927*, pp. 503–504.

37 Lemke, 'Ottoman Photography', p. 248.

38 Debbas, *Beyrouth: Notre mémoire*, p. 12. Tarazi, *Vitrine de l'Orient*.

39 Shaw, 'Ottoman Photography', p. 181. Lemke, 'Ottoman Photography', pp. 242, 245–247.

40 Beaugé and Çizgen, *Images d'empire*, pp. 230–240. Wolf-Dieter Lemke has highlighted the model character of these Ottoman buildings, which had similar shapes in various provinces of the empire: Lemke, 'Ottoman Photography', pp. 242–244.

41 The Serail: AKPOOL, Soubhi S. & Mounir Aïta, *Damas – Le Sérail*. FDC, Wattar Frères, *Hama (Syrie): Panorama générale*. FDC, Torossian, *Palmyre*. BARTKO REHER, Edition Gulef, *Damas – Cour de la Grande Mosquée*.

42 Debbas, *Des photographes à Beyrouth*, pp. 51–52.; Debbas, *Beyrouth: Notre mémoire*, p. 12. el-Hage, *Des photographes à Damas*, p. 37.

43 Makdisi, 'Ottoman Orientalism', p. 795.

44 Cf. for landscape painting: Scheid, 'Divinely Imprinting Prints'.

45 Schayegh, *The Middle East and the Making of the Modern World*, pp. 143–145.

46 FDC, Férid, *Colonel Andréa*.

47 Marie-Hélène Degroise remarked that Ferid edited French aerial photographs into postcards: Degroise, 'Photographes d'Asie'.

48 Neep, *Occupying Syria under the French Mandate*, pp. 47, 93.

49 FDC, Férid, *Alentours du Camp du Muséifré*.

50 It is likely that these functions were relevant to the French imperial administrations not only in the Levant, but also in other contexts, in particular North Africa, where the *Compagnie Générale Transatlantique* and its charismatic president John Dal Piaz closely cooperated with the French governor Louis-Hubert Lyautey. On Dal Piaz and tourism development in the Sahara: Berthonnet, 'Le rôle des militaires français'.

51 Sarraut, *La Mise en valeur*, pp. 83–90. On the idea of *"mise en valeur"* in the context of the French civilising mission since the late nineteenth century, cf. Conklin, *A Mission to Civilize*, pp. 37–65.

52 BNF TOLBIAC, *Dix ans de mandat, 1931*, pp. 62–63. On French economic policies in Syria cf. Khoury, *Syria and the French Mandate*, pp. 85–93.

53 ANL, Ministère des Affaires étrangères de la République Française, *Rapport sur la Situation de la Syrie et du Liban, 1923–24*, pp. 36–37.

54 CADN, *Frédéric Martin, à M. le Haut Commissaire, 03/05/1923*.

55 CADN, Compagnie des Messageries Maritimes, *Voyage en France, 1923*.

56 CADN, *Note par le conseiller pour l'instruction publique, 02/06/1923*.

57 ANL, Ministère des Affaires étrangères de la République Française, *Rapport à la Société des Nations, 1926*, p. 149. Pedersen, *The Guardians*, pp. 142–168.

58 ANL, Ministère des Affaires étrangères de la République Française, *Rapport à la Société des Nations, 1928*, pp. 89–90. On the idea of the French civilising mission in the context of mandate rule: Osterhammel, '"The Great Work of Uplifting Mankind"', esp. pp. 372–373.

59 BNF TOLBIAC, *Dix ans de mandat, 1931*, pp. 48–49.

60 Already the Sahara crossing in the 1920s had been celebrated as a step towards opening the desert for ordinary tourists. André Citroën himself had immediately pointed out the possibility of tourism: Berthonnet, 'Le rôle des militaires français', p. 86.

61 CADN, *Lettre de Georges Marie Haardt, 17/10/1930*, p. 3.

62 CADN, *Lettre de Georges Marie Haardt, 17/10/1930*, p. 4. On Berthelot's impact on French policies in Lebanon and Syria: Dakhli, 'L'expertise en terrain colonial', p. 24.

63 Jansen, 'Die Erfindung des Mittelmeerraums', pp. 195–199. Lorcin, 'Rome and France in Africa'. Furlough, 'Une leçon des choses', p. 455.

64 For the ʿAlawite region see BNF TOLBIAC, Maestracci, *La Syrie contemporaine*, pp. 188–191. Khoury, *Syria and the French Mandate*, pp. 57–60. Traboulsi, *A History of Modern Lebanon*, p. 80.

65 Mascle, *Le Djebel Druze*. In 1930 the Capuchins built a Church in Suwayda, headed by Joseph Mascle: <http://capucinsorient.org/couvent-christ-roi-soueida>, last accessed 11/08/2019.

66 CADN, Mascle, *Le Djebel Druze: Aperçu archéologique*, p. 1.

67 CADN, Mascle, *Le Djebel Druze: Aperçu archéologique*, pp. 1, 16.

68 Neep, *Occupying Syria under the French Mandate*, pp. 42–45.

69 E.g. BNF TOLBIAC, Goud, *Le tourisme au Maroc*, pp. 1–2. BNF TOLBIAC, Michel, *Guide officiel de Safi*, pp. 11–12. BNF TOLBIAC, ESSI de Rabat-Salé et sa région, *Guide touristique historique*, p. 56. BNF TOLBIAC, Syndicat d'Initiative et de Tourisme d'Ifrane, *Ifrane, Maroc*, p. 3.

70 JD PRIVATE COLLECTION, Chardon, *Méditerranée Orientale – Egypte*.

71 Berthonnet, 'Le rôle des militaires français', p. 95.

72 Neep, *Occupying Syria under the French Mandate*, pp. 103–105.

73 Furlough, 'Une leçon des choses', pp. 443–445.

74 CADN, *Lettre du délégué p.i. du Haut-Commissaire, 22/11/1927*.

75 In Algeria and Indochina for example: Zytnicki, *L' Algérie, terre de tourisme*, pp. 120–122. Demay, *Tourism and Colonization in Indochina*, pp. 96, 157–158. BNF GALLICA, *De l'or pour la Syrie, 24/10/1931*.

76 The revolt had been crushed by June 1927: Khoury, *Syria and the French Mandate*, p. 204. Khaouam Frères were running hotels in various cities and in the major *estivage* towns around Lebanon and Syria, such as Homs, Hama, Damascus, Beirut, and Baalbek, attracting notables and upper middle-class visitors. They were proud to announce that their hotels lived up to the standards of the renowned grand hotels in Europe and Egypt. LEIDEN UNIVERSITY LIBRARY, Société de Villégiature au Mont Liban, *Guide de la Société de Villégiature, 1925*, p. 25.

77 CADN, *Lettre du délégué p.i. du Haut-Commissaire, 22/11/1927*. On Ahmad Nami's ministry cf. Khoury, *Syria and the French Mandate*, pp. 198–200. He resigned in February 1928: Khoury, *Syria and the French Mandate*, p. 327.

78 Monmarché, *Syrie – Palestine – Iraq – Transjordanie*, p. 298. Thomas Cook & Son, Lumby and Garstang, *Cook's Traveller's Handbook to Palestine, Syria & Iraq, 1934*, p. 305. Thomas Cook & Son et al., *The Traveller's Handbook for Palestine and Syria, 1929*, p. 6. BNF GALLICA, *Orient Palace Hôtel, 08/1934*.

79 Khoury, *Syria and the French Mandate*, pp. 77–79. Thompson, *Colonial Citizens*, pp. 64–66.

80 Shambrook, *French imperialism in Syria*, p. 152.

81 ANL, Ministère des Affaires étrangères de la République Française, *Rapport sur la situation de la Syrie et du Liban, 1923–24*, pp. 36–37. ANL, Ministère des Affaires étrangères de la République Française, *Rapport à la Société des Nations, 1926*, pp. 149–150. JD PRIVATE COLLECTION, Commission du Tourisme des Etats sous Mandat, *Holidays Off the Beaten Track*.

82 The Government, though, did not enjoy great support. The nationalists perceived it as collaborationist and boycotted the parliament sessions in protest. Shambrook, *French Imperialism in Syria*, pp. 116–124.

83 AUB, *Décret No. 1284, 10/06/1933*. The members changed in 1936, however, the measure seems to have been due to administrative reasons rather than to new responsibilities of the commission: AUB, *Décret No. 726, 15/09/1936*.

84 E.g. AUB, *Décret No. 28, 12/04/1937*.

85 Shambrook, *French imperialism in Syria*, pp. 204–211.

86 AUB, *Décret No. 17, 25/04/1936*. Althouth the SET focused in particular on tourism development in Lebanon, it is likely that the SET also published the *Guide d'Estivage en Syrie*, purchased by the Ministry of the Interior in September 1936. In the same year, the SET published a lavishly illustrated brochure on Lebanon: BO, Ministère de l'économie nationale du gouvernement libanais and Société d'encouragement au tourisme au Liban, *Liban, Tourisme – Estivage – Sports d'hiver*. AUB, *Décret No. 735, 19/09/1936*.

87 AUB, *Décret No. 50, 05/08/1936*, p. 210. AUB, *Décret No. 50, 05/08/1936*, p. 212. AUB, *Décret No. 658, 25/08/1936*. AUB, *Annexes au Décret No. 735, 19/09/1936*. AUB, *Décret No. 645, 25/05/1938*.

88 On Vacca (*1891) cf. Gabrieli, 's.v. Vacca de Bosis'.

89   Vacca, 'L'Ufficio nazionale arabo', p. 683.

90   Vacca, 'L'Ufficio nazionale arabo', pp. 683–684.

91   Vacca, 'L'Ufficio nazionale arabo', p. 684.

92   CADN, *Lettre du Président du Conseil, 27/02/1938*.

93   CADN, *Lettre du M. le délégué du Haut-Commissaire, 28/05/1936*, p. 3.

94   CADN, *Lettre de Khalil Kseib, 13/05/1938*. CADN, *Lettre du Conseiller d'Ambassade, 29/07/1938*.

95   CADN, *Lettre du Secrétariat général, 11/03/1927*.

96   For example, M. Brame, 18/03/1923; Claude Anet, 27/03/1923: CADC, *Français en Syrie, Dossier général, 1922–1927*. Some of these visitors, grateful for the assistance, sent reports, brief memories or impressions to the Foreign Office after their return.

97   Potentially dangerous visitors, to many of whom visa were rejected, comprised applicants suspected of supporting Arab nationalist movements (Max von Oppenheim, Georg Kampffmeyer) or being involved in espionage activities, mainly citizens of the wartime enemies Germany and Austria. CADC, *Passeports. Demandes particulières. 1922–1929*.

98   CADC, *Voyages de personnalités – Tourisme 1924–1929*.

99   Both Bowman and Rowan-Hamilton, for example, who visited Syria at the time of the revolt, remained highly critical with regard to French politics in Syria.

100   CADC, *M. Guerlet, Chargé d'Affaires, 15/10/1926*.

101   CADC, Vallotton-Warnery, *"Impressions de Syrie"*. Neep, *Occupying Syria under the French Mandate*, pp. 37, 75–76.

102   CADC, *Voyages de personnalités – Tourisme 1924–1929*. CADC, *Le Ministre des Affaires étrangères, 01/03/1929*. CADC, *Le ministre de France en Roumanie, 8/11/1928*. CADC, *Légation de Roumanie en France, 17/12/1928*. CADC, *M. R. Dollot, Consul de France, 29/05/1929*. CADC, *L'Ambassadeur de France en Pologne, 22/02/1929*.

103   CADC, *Ministère des Affaires étrangères: Note pour la sous-direction*.

104   AKPOOL, Edition Gulef, *Damas – La rue droite*.

105   Daniel Neep provided further details on such photographs documenting the violence of the mandate power. He argued that the excessive violence shown in the photographs was considered to reflect the presumed extreme, irrational violence of colonised societies: Neep, *Occupying Syria under the French Mandate*, pp. 55, 58. While this is a convincing explanation for the photographs of hanged rebels at the gibbet, in my opinion it does not sufficiently explain the uses and intentions of photographs documenting the destruction of buildings and monuments, in particular given the international outrage provoked by the bombardments of Damascus. Similar photographs circulated of the destroyed Alexandria following British bombardment in 1882: Micklewright, *A Victorian Traveler*, pp. 111–118.

106   Cf. BNF TOLBIAC, Roux-Servine, *Sur les routes du Levant*, pp. 62–65.

107   For a concise account of the bombardment of Damascus, as well as the press coverage and the international scandal cf. Pedersen, *The Guardians*, pp. 142–168. Ryan, *Picturing Empire*, Ch. 3 provides an excellent discussion of the use of photography in the context of Victorian military practice.

108   RCSA, Williams, *Travel Diary: Palestine and Syria*, pp. 23–29.

109   RCSA, Williams, *Travel Diary: Palestine and Syria*, p. 23.

110   Pedersen, *The Guardians*, pp. 147–148.

111   BL, Blue Star Line, *Guide to the Mediterranean*, pp. 43–44.

112   Rowan-Hamilton, *Both Sides of the Jordan*, pp. 247–248.

113   Rowan-Hamilton, *Both Sides of the Jordan*, pp. 247–261.

114   MECA, Bowman, *Diary, Vol. Jan 1921–March 1924*, 26-28/09/1922.

115   MECA, Bowman, *Diary, Vol. August 1924–October 1926*, 31/05/1926.

116 NLS, Steele-Maitland, *Diary of Visit to Egypt*, 21/01/1932.

117 Le Gras, *Le Lac Méditerranéen*, pp. 124–130.

118 Le Gras, *Le Lac Méditerranéen*, pp. 125, 128, 132.

119 Le Gras, *Le Lac Méditerranéen*, p. 121.

120 Le Gras, *Le Lac Méditerranéen*, pp. 124, 126–127, 130, 132.

121 ANF, Vacaresco, *La Croisière Bleue, in: Croisières*, pp. 72–75. On Hélène Vacaresco: NA, *French cultural expansion in Egypt, 1929*.

122 Veccia Vaglieri, 'Visita di Amin el-Huseini'. Around a week after Amin al-Husayni had left, his cousin Jamal, the leader of the Palestine Arab Party, also departed for an *estivage* stay in Lebanon and Syria.

123 In history, Bludan is associated notably with the Congress held in the town in 1937, when major pan-Arabists expressed their support for the Palestinian Revolt: Vacca, 'Inaugurazione del II Congresso panarabo'. Regarding the Lebanese summer resorts, the *Palestine Post* regularly reported on the departures of important families from Palestine for their *estivage*: Stanton, 'Locating Palestine's Summer Residence', p. 50.

124 E.g. Zytnicki, *L'Algérie, terre de tourisme*, pp. 120–122. Demay, *Tourism and Colonization in Indochina*, pp. 96, 157–158. BNF GALLICA, *De l'or pour la Syrie, 24/10/1931*.

125 Khoury, *Syria and the French Mandate*, pp. 457–471, 485–493.

126 Among them were tax exemptions for hotel owners, but also minor improvements in the presentation of towns, comfort and security in the region of Bludan and Zabadani: AUB, *Funduq Bludan, 03/04/1936*. AUB, *Décret Législatif No. 50, 05/08/1936*. AUB, *Décret Législatif, 29/06/1937*.

127 AUB, *Funduq Bludan, 03/04/1936*. AUB, *La-Tashjiˤ al-Istiyaf fi Suriya*.

128 AUB, *Iftitah Mukhayyam al-Kashaf fi Bludan*. AUB, *Annexes au décret législatif No. 66, 28/09/1936*, p. 331. AUB, *Décret No. 645, 07/1938*.

129 AUB, *Iftitah Mukhayyam al-Kashaf fi Bludan*. Keith Watenpaugh and Jennifer Dueck have argued for the significance of the Scouts and their movements as political actors in Syria: Watenpaugh, *Being Modern*, pp. 291–298. Dueck, 'A Muslim Jamboree'. Dueck, *Claims of Culture*, pp. 219–220.

130 Gelvin, *Divided Loyalties*, pp. 18–22. Also in 1938, a camp of "young explorers" from Syria, Lebanon and Egypt gathered in the village. Vacca, 'Campeggio panarabo di Esploratori'.

131 CADN, *Le Directeur de la Sûreté générale, 09/06/1932*.

132 Watenpaugh, *Being Modern*, pp. 17–20.

133 For Watenpaugh, the interest in history and archaeology is a central aspect in his characterisation of the middle class: Watenpaugh, *Being Modern*, pp. 302–303.

134 CADN, *Nouvelles touristiques, 05/1931*, pp. 19–20.

135 The early pro-Ottoman resistance movements against the French demonstrated persisting Ottoman loyalties. Watenpaugh, *Being Modern*, pp. 160–173. Provence, 'Ottoman Modernity, Colonialism, and Insurgency'. Schayegh, *The Middle East and the Making of the Modern World*, pp. 141–145. On attempts of establishing the national paradigm in Aleppo cf. Watenpaugh, 'Cleansing the Cosmopolitan City'.

136 CADN, *Nouvelles touristiques, 05/1931*, pp. 19–20.

137 NLS, Steele-Maitland, *Diary of Visit to Egypt*, 22/01/1932.

138 CADN, *Lettre du délégué-adjoint du Haut-Commissariat, 05/05/1931*.

139 CADN, La Direction, *Avant-Propos, 05/1931*.

140 Khoury, *Syria and the French Mandate*, pp. 44–45. Pedersen, 'Eine heilige Aufgabe der Zivilisation'. Schayegh, *The Middle East and the Making of the Modern World*, pp. 134–137.

141 CADN, al-Ghazzi, *al-Muqaddima [Preface]*.

142   Philip Khoury argued that the notables remained in an intermediary position between the (French) empire and (local) populations. Their position changed only after independence: Khoury, *Syria and the French Mandate*, p. 23.

143   Al-Ghazzi also played a role in the reconstitution of the Chamber of Commerce in Aleppo. Watenpaugh, *Being Modern*, pp. 42, 126–127, 198–209.

144   Khoury, *Syria and the French Mandate*, p. 248. Dakhli, 'L'expertise en terrain colonial', p. 20. Burke III, 'French Native Policy'.

145   Watenpaugh, *Being Modern*, pp. 208, 222–229. Watenpaugh, 'Middle-Class Modernity'. Schayegh, *The Middle East and the Making of the Modern World*, pp. 207–211. Charles S. Maier has argued that in many societies the old elites collaborated with the emerging intellectuals and middle classes, yet, without equally sharing power. Maier, 'Leviathan 2.0', pp. 99, 176–185.

146   Shambrook, *French Imperialism in Syria*, p. 152.

147   Similarly, the economic policies of the French High Commission did not seem to have systematically considered local perspectives: In 1933, Norman Burns stated that the High Commission in Beirut regularly exchanged with the economic department of the Lebanese Government, yet, he does not mention a Syrian equivalent. In Syria, the source of advice for the High Commission's economic policies, according to Burns, were the Chambers of Commerce: Burns, *The Tariff of Syria*, p. 18.

148   Schayegh, *The Middle East and the Making of the Modern World*, pp. 200–205. Maier, 'Consigning the Twentieth Century to History'.

149   MECA, Bowman, *Diary, Vol. October 1926–May 1929*, 01/06/1928.

150   MECA, Bowman, *Diary, Vol. October 1926–May 1929*, 04/06/1928.

151   FDC, Torossian, *Palmyre*.

152   For Palmyra, guidebook authors stated that the temples had been "destroyed and invaded" by the houses of the villagers, or that the site had been "occupied by an Arab village". JD PRIVATE COLLECTION, Chardon, *Méditerranée Orientale – Egypte*, p. 158. Thomas Cook & Son, Lumby and Garstang, *Cook's Traveller's Handbook to Palestine, Syria & Iraq, 1934*, p. 333. Similar phrases were used for ancient sites in Egypt: They had been supposedly "wantonly destroyed", or "plundered" by the local inhabitants: Thomas Cook & Son and Elston, *The Traveller's Handbook for Egypt and the Sûdân, 1929*, pp. 133, 186, 469. On this trope in an Egyptian context and how it justified colonial intervention: Colla, *Conflicted Antiquities*, pp. 40–41, 64.

153   CADC, *Le Haut-Commissaire, 04/04/1933*. The "ruinous and tempting undertakings" Ponsot alluded to most likely included d'Andurain's adventurous wife who could not entirely clear herself from the reputation of having instigated the death of three male relatives, her husband among them. Moreover, rumours circulated that she was a spy: d'Andurain, *Marga d'Andurain, 1893–1948*, pp. 5, 10–11.

154   CADC, *Le Haut-Commissaire, 04/04/1933*.

155   Rowan-Hamilton, *Both Sides of the Jordan*, pp. 223–224.

156   Rowan-Hamilton, *Both Sides of the Jordan*, p. 228.

157   James Canton has analysed British travelogues from ca. 1882–2003 more systematically, interesting in this context is esp. his section on the accounts of female travellers: Canton, *From Cairo to Baghdad*, pp. 109–127.

158   Wiedemann, 'Heroen der Wüste', pp. 64–65. Cf. also Canton, *From Cairo to Baghdad*, pp. 140–141. Cf. also for the French imaginaries, connected in particular to the North African Sahara: Roux, *Le Désert de sable*.

159   In Rowan-Hamilton's case, it was her companion 'R', who revealed the implications of the Corinthian capitals he identified: Rowan-Hamilton, *Both Sides of the Jordan*, p. 228.

160   Wang, *Tourism and Modernity*, pp. 19–20.

161 Rowan-Hamilton, *Both Sides of the Jordan*, p. 238.
162 AUB, *Arrêté No. 1622, 04/12/1929*.
163 AUB, *Décret No. 2122, 15/02/1934*. AUB, *Décret No. 2156, 26/02/1934*.
164 AUB, *Arrêté No. 1479, 14/10/1929*.
165 AUB, *Arrêté No. 1637, 12/12/1929*.
166 AUB, *Arrêté No. 2302, 26/07/1930*. AUB, *Décret législatif No. 54, 30/07/1934*. AUB, *Loi no. 97, 11/01/1939*.
167 ANL, Ministère des Affaires étrangères de la République Française, *Rapport à la Société des Nations, 1927*, pp. 230–231.
168 Mitchell, *Colonising Egypt*, pp. 28–33.
169 BNF GALLICA, *L'Intransigeant, 03/06/1933*.
170 A garrison was stationed here: Monmarché, *Syrie – Palestine – Iraq – Transjordanie*, p. 326.
171 Bosworth, 'Tadmur'.
172 Burke III and Yaghoubian, 'Middle Eastern Societies', p. 8.

References

Sources

Akpool Postcard retailer
EDITION GULEF, *Damas – La rue droite*, Akpool.
SOUBHI S. & MOUNIR AÏTA, *Damas – Le Sérail*, postcard, Akpool.
TERAZI & CHURBADJI, *Damas. Maison la Municipalité*, c. 1922, Akpool.
THÉVENET, CLOVIS, *Alep: Type du pays*, Akpool.
THÉVENET, CLOVIS, *Deir-es-Zor: Type*, Akpool.
THÉVENET, CLOVIS, *Turcomanes*, Akpool.
WATTAR FRÈRES, *L'Emir Moudchem avec sa suite*, Akpool.

Archives nationales (France), Pierrefitte-sur-Seine
VACARESCO, HÉLÈNE, *La Croisière Bleue, in: Croisières du Yacht Alphée*, ed. Yvonne Cotnaréanu, Fonds Lucien Romier, Voyages, ANF/408 AP/2.

Archives nationales du Liban, Beirut
MINISTÈRE DES AFFAIRES ÉTRANGÈRES DE LA RÉPUBLIQUE FRANÇAISE, *Rapport à la Société des Nations sur la Situation de la Syrie et du Liban, Année 1926*, Paris: Imprimerie Nationale, 1927, ANL.
MINISTÈRE DES AFFAIRES ÉTRANGÈRES DE LA RÉPUBLIQUE FRANÇAISE, *Rapport à la Société des Nations sur la Situation de la Syrie et du Liban, Année 1927*, Paris: Imprimerie Nationale, 1928, ANL.
MINISTÈRE DES AFFAIRES ÉTRANGÈRES DE LA RÉPUBLIQUE FRANÇAISE, *Rapport à la Société des Nations sur la Situation de la Syrie et du Liban, Année 1928*, Paris: Imprimerie Nationale, 1929, ANL.
MINISTÈRE DES AFFAIRES ÉTRANGÈRES DE LA RÉPUBLIQUE FRANÇAISE, *Rapport sur la Situation de la Syrie et du Liban, Juillet 1923–Juillet 1924*, Paris: Imprimerie Nationale, 1924, ANL.

Bartko Reher, Postcard retailer
EDITION GULEF, *Damas – Cour de la Grande Mosquée*, Bartko Reher.

Bibliothèque nationale de France, Paris

Damas: "Orient Palace Hôtel", Correspondance d'Orient, 08/1934, p. 75, BNF Gallica.

De l'or pour la Syrie. Crise...d'or?, Les Échos (de Damas), 24/10/1931, p. 2, BNF Gallica.

Dix ans de mandat: L'œuvre française en Syrie et au Liban, 1931, BNF Tolbiac/8-O2A-1366.

DU TAILLIS, JEAN, *Le nouveau Maroc*, Ouvrage illustré de dix-neuf photographies de l'auteur et suivi d'un voyage au Riff, Paris: Société d'Éditions géographiques, maritimes et coloniales, 1923, BNF Tolbiac/8-O3J-442.

DU TAILLIS, JEAN, *Motor-Touring in Morocco*, Dunlop-Guide. Paris: Edition des guides du tourisme automobile, Paris: Edition des Guides du tourisme automobile, 1925, BNF Tolbiac/8-O3J-466.

ESSI DE RABAT-SALÉ ET SA RÉGION, *Guide touristique historique édité par le syndicat d'initiatives et de tourisme*, 1931, BNF Tolbiac/GE-FF-14853.

FAURÉ, LOUIS, *Guide pratique de Marrakech*, c. 1930s, BNF Tolbiac/8-O3J-649.

GOUD, HENRY, *Le tourisme au Maroc*, 4ème edition, 1935, BNF Tolbiac/8-O3J-686.

MAESTRACCI, NOËL, *La Syrie contemporaine. Tout ce qu'il faut savoir sur les territoires placés sous mandat français*, Paris: Charles-Lavauzelle, 1930, BNF Tolbiac/8-O2E-657.

MICHEL, G. R., *Guide officiel édité par le Syndicat d'Initiative et de Tourisme ESSI de Safi*, before 1933, BNF Tolbiac/8-O3J-682.

ROUX-SERVINE, LOUIS, *Sur les routes du Levant: Petit manuel du voyage expérimental*, Paris: J. Barreau, 1934, BNF Tolbiac/8-G-13749.

SYNDICAT D'INITIATIVE ET DE TOURISME D'IFRANE, *Ifrane, Maroc*, Éditions Inter-Presse, Imp. Réunies Casablanca, between 1929 and 1935, BNF Tolbiac/8-O3J-745.

SYNDICAT D'INITIATIVES ET DE TOURISME „ESSI" MAROC, *Comment venir et voyager au Maroc*, Alger, 1930, BNF Tolbiac/2004-241500.

Une femme française executée en Arabie, L'Intransigeant, 03/06/1933, p. 3, BNF Gallica.

Bibliothèque Orientale, Université Saint-Joseph, Beirut

MINISTÈRE DE L'ÉCONOMIE NATIONALE DU GOUVERNEMENT LIBANAIS, and SOCIÉTÉ D'ENCOURAGEMENT AU TOURISME AU LIBAN, *Liban. Tourisme – Estivage – Sports d'hiver*, 1938, BO.

British Library, London

BLUE STAR LINE, *Guide to the Mediterranean by S.S. "Arandora Star"*, Season 1929, London, BL.

Centre des Archives diplomatiques de La Courneuve

Français en Syrie – Dossier général 01/07/1922–31/01/1927, Direction des Affaires Politiques et Commerciales, Série E Syrie – Liban, CADC/50 CPCOM/389.

L'Ambassadeur de France en Pologne à Son Excellence M. le Ministre des Affaires Étrangères à Paris, 22/02/1929, Passeports. Demandes particulières. 1922–1929, Communication politique et commerciale, Série E. Syrie – Liban, CADC/50 CPCOM/311.

Le Haut-Commissaire de la République Française en Syrie à son Excellence M. le Ministre des Affaires étrangères, 04/04/1933, Communication politique et commerciale, Série E. Syrie – Liban, Français en Syrie, 1930–1934, CADC/50 CPCOM 605.

Le Ministre de France en Roumanie au Ministre des Affaires étrangères, 08/11/1928, Passeports. Demandes particulières. 1922–1929, Communication politique et commerciale, Série E. Syrie – Liban, CADC/50 CPCOM/311.

Le Ministre des Affaires étrangères/Services des Œuvres françaises à l'étranger au Haut Commissaire de la République Française à Beyrouth, 01/03/1929, Passeports. Demandes particulières. 1922–1929, Communication politique et commerciale, Série E. Syrie – Liban, CADC/50 CPCOM/311.

Légation de Roumanie en France au Ministre des Affaires étrangères, Paris, 17/12/1928, Passeports. Demandes particulières. 1922–1929; Communication politique et commerciale, Série E. Syrie – Liban, CADC/50 CPCOM/311.

M. Guerlet, Chargé d'Affaires de la République Française à Berne, à S.E. M. Briand, Ministre des Affaires étrangères, Paris, 15/10/1926, Français en Syrie – Dossier général 01/07/1922–31/01/1927, Direction des Affaires Politiques et Commerciales, Série E. Syrie – Liban, CADC/50 CPCOM/389.

M. R. Dollot, Consul de France à Trieste à S.E. M. le Ministre des Affaires étrangères, 29/05/1929, Passeports. Demandes particulières. 1922–1929, Communication politique et commerciale, Série E. Syrie – Liban, CADC/50 CPCOM/311.

Note pour la sous-direction d'Afrique et du Levant, n.d., Ministère des Affaires étrangères: Note pour la sous-direction d'Afrique et du Levant, n.d., Passeports. Demandes particulières. 1922–1929, Communication politique et commerciale, Série E. Syrie – Liban, CADC/50 CPCOM/311.

Passeports. Demandes particulières. 1922–1929, Communication politique et commerciale, Série E. Syrie – Liban, CADC/50 CPCOM/311.

VALLOTTON-WARNERY, HENRI, *"Impressions de Syrie"*, Journal de Genève, 15/10/1926, Français en Syrie – Dossier général, 01/07/1922-31/01/1927, Direction des Affaires Politiques et Commerciales, Série E. Syrie – Liban, CADC/50 CPCOM/389.

Voyages de personnalités – Tourisme, 1924–1929, Ministère des Affaires étrangères, Série E. Syrie – Liban, Affaires diverses, CADC/50 CPCOM/418.

Centre des Archives diplomatiques de Nantes

AL-GHAZZI, KAMIL, *al-Muqaddima [Preface]*, in: Majallat al-ʿAdiyat, tusdaru Shahriyan bi-ʿAnayat Jamʿiyya ʿAdiyat Halab, al-ʿAdad al-Awal Shahr Ayar 1931 [Société Archéologique d'Alep (ed.), Revue Archéologique, Mai 1931], Fonds Beyrouth, Cabinet politique, CADN/1 SL/1/V/626.

COMPAGNIE DES MESSAGERIES MARITIMES, *Voyage en France organisé par la Compagnie des Messageries Maritimes pour les étudiants et pour la jeunesse égyptienne syrienne, Saison d'été 1923*, Fonds Beyrouth, Haut Commissariat. Services de l'Instruction publique, CADN/1SL/600/20.

Frédéric Martin, Agent des Messageries Maritimes à M. le Haut Commissaire, Beyrouth, 03/05/1923, Fonds Beyrouth, Haut Commissariat. Service de l'instruction publique, CADN/1SL/600/20.

LA DIRECTION, *Avant-Propos*, in: Société Archéologique d'Alep (ed.), Revue Archéologique, Mai 1931, Fonds Beyrouth, Cabinet politique, CADN/1 SL/1/V/626.

Le Directeur de la Sûreté générale, Inspecteur général des polices à M. le Conseiller du Haut Commissariat aux Relations extérieures/Bureau diplomatique, 09/06/1932, Bureau diplomatique, Règlementation du travail, CADN/1SL/1/V/1406.

Lettre de Georges Marie Haardt au Haut Commissaire Ponsot, 17/10/1930, Fonds Beyrouth, Cabinet Politique, CADN/1 SL/1/V/626.

Lettre de Khalil Kseib au délégué du Haut Commissaire, 13/05/1938, Fonds Beyrouth, Cabinet Politique, CADN/1SL/1/V/626.

Lettre du Conseiller d'Ambassade chargé du Consulat Général à Jérusalem au Haut Commissaire de la République Française, 29/07/1938, Fonds Beyrouth, Cabinet politique, CADN/1 SL/1/V/626.

Lettre du Délégué p.i. du Haut-Commissaire auprès des Etats de Syrie et du Djebel Druze à M. Le Haut Commissaire de la République Française, signé Desloges, 22/11/1927, Fonds Beyrouth, Cabinet Politique, CADN/1 SL/1/V/1573.

Lettre du Délégué-adjoint du Haut-Commissariat pour le Vilayet d'Alep à M. le Haut Commissaire de la République Française, 05/05/931, Fonds Beyrouth, Cabinet politique, CADN/1SL/1/V/626.

Lettre du M. le Délégué du Haut Commissaire auprès de la République syrienne, Damas, à M. Meyrier, Délégué général du Haut Commissaire, Beyrouth, signé G. Fain, 28/05/1936, annexe, Fonds Beyrouth, Cabinet politique, CADN/1 SL/1/V/626.

Lettre du Président du Conseil (Jamil) relative à l'exercice de la profession du guide en Syrie, 27/02/1938, Fonds Beyrouth, Cabinet politique, CADN/1 SL/1/V/626.

Lettre du Secrétariat général, cabinet, à M. Grandguillot, Directeur de l'Office économique et commercial, Alexandrie, signé: P. de Reffye, 11/03/1927, Fonds Beyrouth, Cabinet Politique, CADN/1 SL/1/V/1573.

MASCLE, JOSEPH, *Le Djebel Druze: Aperçu archéologique*, Soueida, 15/12/1930, Fonds Beyrouth, Cabinet Politique, CADN/1 SL/1/V/626.

Note par le Conseiller pour l'instruction publique du Haut Commissaire pour le Secrétaire Général du 02/06/1923, Fonds Beyrouth, Haut Commissariat. Services de l'instruction publique, CADN/1 SL/600/20.

Nouvelles touristiques, Revue Archéologique, 05/1931, Société Archéologique d'Alep (ed.), Revue Archéologique, 05/1931, Fonds Beyrouth, Cabinet politique, CADN/1 SL/1/V/626.

Fouad Debbas Collection, Sursock Museum, Beirut

FÉRID, L., *Alentours du Camp du Muséifré (Septembre 1925)*, FDC.

FÉRID, L., *Colonel Andréa dans la Colonne du Soueïda*, FDC.

THÉVENET, CLOVIS, *Alep: Type du Pays*, FDC.

THÉVENET, CLOVIS, *En Syrie. Types Anazé*, FDC.

THÉVENET, CLOVIS, *En Syrie: Porteuse d'Eau*, FDC.

THÉVENET, CLOVIS, *Panorama de Hama (Syrie)*, FDC.

TOROSSIAN, JEAN, *Palmyre*, FDC.

WATTAR FRÈRES, *Hama (Syrie): Panorama générale*, FDC.

Jafet Library, American University of Beirut

Annexes au décret législatif No. 66, 28/09/1936, Journal officiel de la République Syrienne du 26/11/1936, pp. 319–331, AUB.

Annexes au décret No. 735, 19/09/1936, Journal officiel de l'Etat de Syrie, 26/11/1936, p. 331, AUB.

Arrêté No. 1479, 14/10/1929, Al-Acima: Bulletin officiel bimensuel des actes administratifs, 31/10/1929, p. 8, AUB.

Arrêté No. 1622, 04/12/1929, Al-Acima: Bulletin officiel bimensuel des actes administratifs, 15/12/1929, pp. 10–11, AUB.

Arrêté No. 1637, 12/12/1929, Al-Acima: Bulletin officiel bimensuel des actes administratifs, 15/12/1929, p. 11, AUB.

Arrêté No. 2302, 26/07/1930, Al-Acima: Bulletin officiel bimensuel des actes administratifs, 31/07/1929, p. 123, AUB.

Chambre des Députés, Séance publique du 28/03/1927, Journal officiel du gouvernement égyptien No. 41, 09/05/1927, Supplément, AUB.

Décret législatif No. 50, 05/08/1936, autorisant des douzièmes provisoires pour les mois de Mai, Juin et Juillet 1936, Journal officiel de la Syrie du 27/08/1936, pp. 201–203, AUB.

Décret législatif No. 54, 30/07/1934, Journal officiel de l'Etat de Syrie, 15/09/1934, pp. 182–183, AUB.

Décret législatif, 29/06/1937, ouvrant un crédit de 120 L.S. (Office de Tourisme et de l'Estivage), Journal officiel de la Syrie No. 25, 08/07/1937, p. 113, AUB.

Décret No. 1284, 10/06/1933, constituant la Commission du tourisme, Journal officiel de l'Etat de Syrie, 30/06/1933, p. 166, AUB.

Décret No. 17, 25/04/1936, Journal Officiel de l'Etat de Syrie, 16/07/1936, pp. 163–164, AUB.

Décret No. 2122, 15/02/1934, Journal officiel de l'Etat de Syrie, 28/02/1934, p. 39, AUB.

Décret No. 2156, 26/02/1934, Journal officiel de l'Etat de Syrie, 15/03/1934, p. 52, AUB.

Décret No. 28, 12/04/1937, Journal officiel de l'Etat de Syrie, 22/04/1937, pp. 71–74, AUB.

Décret No. 50, 05/08/1936, Journal officiel de l'Etat de Syrie, 27/08/1936, AUB.

Décret No. 645, 25/05/1938, Journal Officiel de l'Etat de Syrie, 21/07/1938, p. 847, AUB.

Décret No. 658, 25/08/1936, Journal Officiel de l'Etat de Syrie, 03/09/1936, p. 219, AUB.

Décret No. 726, 15/09/1936, Journal officiel de l'Etat de Syrie, 24/09/1936, pp. 243–244, AUB.

Décret No. 735, 19/09/1936, Journal Officiel de l'Etat de Syrie, 24/09/1936, AUB.

Funduq Bludan, [The Bludan Hotel], Alif Ba, 03/04/1936, p. 4, AUB.

Iftitah Mukhayyam al-Kashaf fi Bludan, [The Opening of the Scouts Camp in Bludan], Alif Ba, 22/08/1936, p. 4, AUB.

La-Tashjiˤ al-Istiyaf fi Suriya, [For the Encouragement of Estivage in Syria], Alif Ba, 31/07/1936, p. 7, AUB.

Loi No. 97, 11/01/1939, Journal officiel de l'Etat de Syrie, 06/04/1939, pp. 47–48, AUB.

jd private collection

CHARDON, J., *Méditerranée Orientale – Egypte: Guide du passager pour les escales des croisières et pour les excursions en Syrie et en Égypte*, Paris: Hachette, 1935, jd private collection.

COMMISSION DU TOURISME DES ETATS SOUS MANDAT, *Syria and Lebanon: Holidays Off the Beaten Track*, Beirut 1930, jd private collection.

Middle East Centre Archive, University of Oxford

BOWMAN, HUMPHREY, *Diary, Vol. August 1924–October 1926*, Humphrey E. Bowman Collection, MECA/ GB165-0034.

BOWMAN, HUMPHREY, *Diary, Vol. Jan 1921–March 1924*, Humphrey E. Bowman Collection, MECA/GB 165-0034.

BOWMAN, HUMPHREY, *Diary, Vol. October 1926–May 1929*, Humphrey E. Bowman Collection, MECA/ GB165-0034.

National Archives, Kew

French cultural expansion in Egypt, 1929, Foreign Office, NA/FO 141/624/7.

National Library of Scotland, Edinburgh

STEELE-MAITLAND, MARY, *Diary of Visit to Egypt, Palestine and Turkey, 24/12/1931–29/01/1932*, NLS/ ACC 5553.

Royal Commonwealth Society Archives, Cambridge University Library

WILLIAMS, GEORGE H., *Travel Diary: Palestine and Syria, 1925*, RCSA/RCMS 53/3/2.

Other published sources

BURNS, NORMAN, *The Tariff of Syria: 1919–1932* (Beirut: American Press, 1933).

JARMAN, ROBERT L., 'Colonial Office, Report by His Majesty's Government in the United Kingdom of Great Britain and Northern Ireland to the Council of the League of Nations on the Administration of Palestine and Trans-Jordan for the Year 1933, issued by the Colonial Office, London 1934', in Robert L. Jarman (ed.), *Palestine and Transjordan Administration Reports 1918–1948*, 16 vols. (Archive Editions, 1995), IV, pp. 285–621.

JARMAN, ROBERT L., 'Colonial Office, Report by His Majesty's Government in the United Kingdom of Great Britain and Northern Ireland to the Council on the League of Nations on the Administration of Palestine and Trans-Jordan for the Year 1936, issued by the Colonial Office, London 1937', in Robert L. Jarman (ed.), *Palestine and Transjordan Administration Reports 1918–1948*, 16 vols. (Archive Editions, 1995), VI, p. 1–431.

LE GRAS, CHARLES, *Le Lac Méditerranéen: Pèlerinage à Jérusalem et croisière en Méditerranée orientale, du 5 Avril au 19 Mai 1927* (Avignon: Aubanel Frères, 1927).

MASCLE, JOSEPH, *Le Djebel Druze: Avec carte, itinéraire du touriste et nombreuses illustrations* (Beirut, 1936).

MONMARCHÉ, MARCEL, *Syrie – Palestine – Iraq – Transjordanie* (Les Guides bleus, Paris: Hachette, 1932).

ROWAN-HAMILTON, NORAH, *Both Sides of the Jordan: A Woman's Adventures in the Near East* (London: Herbert Jenkins, 1928).

SARRAUT, ALBERT, *La Mise en valeur des colonies françaises* (Paris: Payot, 1923).

SOCIÉTÉ DE VILLÉGIATURE AU MONT LIBAN, *Guide de la Société de Villégiature au Mont Liban 1925*, Cairo, 1925, Leiden University Library.

THOMAS COOK & SON, and ELSTON, ROY, *The Traveller's Handbook for Egypt and the Sûdân* (London: Simpkin Marshall, 1929).

THOMAS COOK & SON, ELSTON, ROY, LUKE, HARRY CHARLES et al., *The Traveller's Handbook for Palestine and Syria* (2nd edn., London: Simpkin Marshall, 1929).

THOMAS COOK & SON, LUMBY, CHRISTOPHER, and GARSTANG, JOHN, *Cook's Traveller's Handbook to Palestine, Syria & Iraq* (6th edn., London: Simpkin Marshall, 1934).

VACCA, VIRGINIA, 'Inaugurazione del II Congresso panarabo di Bludan in Siria', *Oriente Moderno*, 17/10 (1937), pp. 497–500.

VACCA, VIRGINIA, 'Campeggio panarabo di esploratori presso Damasco', *Oriente Moderno*, 18/8 (1938), pp. 422–423.

VACCA, VIRGINIA, 'L'Ufficio nazionale arabo di stampa e propaganda di Damasco e le sue pubblicazioni', *Oriente Moderno*, 18/12 (1938), pp. 683–689.

VECCIA VAGLIERI, LAURA, 'Visita di Amin el-Huseini alla Siria e sue ragioni politiche', *Oriente Moderno*, 15/10 (1935), pp. 511–512.

Secondary Literature

ALLOULA, MALEK, *The Colonial Harem* (Minneapolis: University of Minnesota Press, 1986).

BASSIL, SEMAAN, *Seventy Years of Postal History at the French Post Office in Beirut: Soixante-dix ans d'histoire postale du bureau de poste français à Beyrouth* (Beirut: Lebanese British Friends of the National Museum, 2009).

BASSIL, SEMAAN, *An Introduction to French Postal History in Tripoli (1852–1914) in the Age of Steamships, French Influence in the Levant, and the Decline of the Ottoman Empire* (Beirut: The Lebanese British Friends of the National Museum, 2013).

BEAUGÉ, GILBERT, and ÇIZGEN, ENGIN (eds.), *Images d'empire: Aux origines de la photographie en Turquie, D'après la Collection Pierre de Gigord* (Istanbul: Institut d'Etudes Françaises, 1993).

BERTHONNET, ARNAUD, 'Le rôle des militaires français dans la mise en valeur d'un tourisme au Sahara de la fin du XIXe siècle aux années 1930', in Colette Zytnicki and Habib Kazdaghli (eds.), *Le Tourisme dans l'Empire français. Un outil de la domination coloniale? Politiques, pratiques et imaginaires (XIXe–XXe siècles)* (Saint-Denis: Publications de la Société française d'histoire d'outre-mer, 2009), pp. 79–96.

BOSWORTH, C. E., 'Tadmur', in Peri J. Bearman, Thierry Bianquis, C. E. Bosworth et al. (eds.), *Encyclopaedia of Islam* (2nd edn., Leiden: Brill) <http://dx.doi.org/10.1163/1573-3912_islam_SIM_7285>, accessed 7 May 2019.

BURKE III, EDMUND, 'A Comparative View of French Native Policy in Morocco and Syria, 1912–1925', *Middle Eastern Studies*, 9/2 (1973), pp. 175–186.

BURKE III, EDMUND, and YAGHOUBIAN, DAVID N., 'Middle Eastern Societies and Ordinary People's Lives: Rethinking Middle Eastern History', in Edmund Burke III and David N. Yaghoubian (eds.), *Struggle and Survival in the Modern Middle East* (2nd edn., Berkeley: University of California Press, 2006), pp. 1–32.

CANTON, JAMES, *From Cairo to Baghdad: British Travellers in Arabia* (London, New York: I.B. Tauris, 2011).

ÇELIK, ZEYNEP, and ELDEM, EDHEM, 'Introduction', in Zeynep Çelik and Edhem Eldem (eds.), *Camera Ottomana. Photography and Modernity in the Ottoman Empire, 1840–1914* (Istanbul: Koç University Press, 2015).

CLEVELAND, WILLIAM L., and BUNTON, MARTIN P., *A History of the Modern Middle East* (4th edn., Boulder: Westview Press, 2009).

COLLA, ELLIOTT, *Conflicted Antiquities: Egyptology, Egyptomania, Egyptian Modernity* (Durham: Duke University Press, 2008).

CONKLIN, ALICE L., *A Mission to Civilize: The Republican Idea of Empire in France and West Africa, 1895–1930* (Stanford: Stanford University Press, 1997).

DAKHLI, LEYLA, 'L'expertise en terrain colonial: Les orientalistes et le mandat français en Syrie et au Liban', *Matériaux pour l'histoire de notre temps*, 99/3 (2010), pp. 20–27.

D'ANDURAIN, JULIE, *Marga d'Andurain, 1893–1948: Une passion pour l'Orient. Le Mari Passeport, nouvelle édition annotée* (Langres: Hémisphères, 2019).

DEBBAS, FOUAD, *Beyrouth: Notre mémoire: Promenade guidée à travers une collection d'images de 1880 à 1930* (3rd edn., Saint-Ouen: Éditions Henri Berger, 1986).

DEBBAS, FOUAD, *Des photographes à Beyrouth, 1840–1918* (Paris. Marval, 2001).

DEGROISE, MARIE-HÉLÈNE, 'Photographes d'Asie (1840–1944)' <http://photographesenoutremerasie.blogspot.de/>, accessed 10 Sep 2021.

DEMAY, ALINE, *Tourism and Colonization in Indochina, 1898–1939* (Newcastle upon Tyne: Cambridge Scholars Publishing, 2014).

DEROO, REBECCA J., 'Colonial Collecting: French Women and Algerian *cartes postales*', in David Prochaska and Jordana Mendelson (eds.), *Postcards. Ephemeral Histories of Modernity* (University Park: Pennsylvania State University Press, 2010), pp. 85–92.

DUECK, JENNIFER M., 'A Muslim Jamboree: Scouting and Youth Culture in Lebanon under the French Mandate', *French Historical Studies*, 30/3 (2007), pp. 485–516.

DUECK, JENNIFER M., *The Claims of Culture at Empire's End: Syria and Lebanon under French Rule* (Oxford, New York: Oxford University Press, 2010).

EL-HAGE, BADR, *Des photographes à Damas, 1840–1918* (Paris: Marval, 2000).

EL-HAGE, BADR, 'The Armenian Pioneers of Middle Eastern Photography', *Jerusalem Quarterly*, 31 (2007), pp. 22–26.

FURLOUGH, ELLEN, 'Une leçon des choses: Tourism, Empire, and the Nation in Interwar France', *French Historical Studies*, 25/3 (2002), pp. 441–475.

GABRIELI, FRANCESCO, 's.v. Vacca de Bosis, Virginia: Arab-Islamic Biographical Archive', 1993 <https://wbis-1degruyter-1com-1hctpu1xm20c9.han.ub.uni-kassel.de/biographic-document/X154-974-4>, accessed 8 Nov 2019.

GAVIN, CARNEY E. S., *The Image of the East: Nineteenth-Century Near Eastern Photographs by Bonfils: From the Collections of the Harvard Semitic Museum* (Chicago: University of Chicago Press, 1982).

GELVIN, JAMES L., *Divided Loyalties: Nationalism and Mass Politics in Syria at the Close of Empire* (Berkeley: University of California Press, 1999).

GELVIN, JAMES L., 'The "Politics of Notables" Forty Years After', *Middle East Studies Association Bulletin*, 40/1 (2006), pp. 19–29.

GELVIN, JAMES L., *The Modern Middle East: A History* (4th edn., New York, Oxford: Oxford University Press, 2016).

GREGORY, DEREK, 'Emperors of the Gaze: Photographic Practices and Productions of Space in Egypt, 1839–1914', in Joan M. Schwartz and James R. Ryan (eds.), *Picturing Place. Photography and the Geographical Imagination* (London: I.B. Tauris, 2009), pp. 195–225.

GUNNING, TOM, "The Whole World Within Reach": Travel Images Without Borders', in Jeffrey Ruoff (ed.), *Virtual Voyages. Cinema and Travel* (Durham: Duke University Press, 2006), pp. 25–41.

JACKSON, SIMON, 'Mandatory Development: The Political Economy of the French Mandate in Syria and Lebanon, 1915–1939', Dissertation (New York, New York University, 2009).

JANSEN, JAN C., 'Die Erfindung des Mittelmeerraums im kolonialen Kontext: Die Inszenierung eines "lateinischen Afrika" beim "Centenaire de l'Algérie française" 1930', in Frithjof Benjamin Schenk (ed.), *Der Süden. Neue Perspektiven auf eine europäische Geschichtsregion* (Frankfurt a.M., New York: Campus, 2007), pp. 175–205.

JAWORSKI, RUDOLF, 'Alte Postkarten als kulturhistorische Quellen', *Geschichte in Wissenschaft und Unterricht*, 51/2 (2000), pp. 88–102.

KHOURY, PHILIP S., *Syria and the French Mandate: The Politics of Arab Nationalism, 1920–1945* (London: I.B. Tauris, 1987).

LANGEWIESCHE, DIETER, 'Was heißt "Erfindung der Nation"? Nationalgeschichte als Artefakt: Oder Geschichtsdeutung als Machtkampf', *Historische Zeitschrift*, 277/3 (2003), pp. 593–617.

LEMKE, WOLF-DIETER, 'Ottoman Photography: Recording and Contributing to Modernity', in Jens Hanssen, Thomas Philipp, and Stefan Weber (eds.), *The Empire in the City. Arab Provincial Capitals in the Late Ottoman Empire* (Würzburg: Ergon, 2002), pp. 237–249.

LORCIN, PATRICIA M. E., 'Rome and France in Africa: Recovering Colonial Algeria's Latin Past', *French Historical Studies*, 25/2 (2002), pp. 295–329.

MACKENZIE, JOHN M., *Orientalism: History, Theory, and the Arts* (Manchester, New York: Manchester University Press, 1995).

MAIER, CHARLES S., 'Consigning the Twentieth Century to History: Alternative Narratives for the Modern Era', *American Historical Review*, 105/3 (2000), pp. 807–831.

MAIER, CHARLES S., 'Leviathan 2.0: Die Erfindung moderner Staatlichkeit', in Emily S. Rosenberg (ed.), *Weltmärkte und Weltkriege. 1870–1945*, Akira Iriye and Jürgen Osterhammel (Geschichte der Welt, München: C.H. Beck, 2012), pp. 33–286.

MAKDISI, USSAMA, 'Ottoman Orientalism', *The American Historical Review*, 107/3 (2002), pp. 768–796.

MICKLEWRIGHT, NANCY, *A Victorian Traveler in the Middle East: The Photography and Travel Writing of Annie Lady Brassey* (Aldershot: Ashgate, 2003).

MIGDAL, JOEL S., *Strong Societies and Weak States: State-Society Relations and State Capabilities in the Third World* (Princeton: Princeton University Press, 1988).

MITCHELL, TIMOTHY, *Colonising Egypt* (Berkeley: University of California Press, 1988).

MIZRAHI, JEAN-DAVID, *Genèse de l'État mandataire: Service des renseignements et bandes armées en Syrie et au Liban dans les années 1920* (Paris: Sorbonne, 2003).

MOORS, ANNELIES, 'From "Women's Lib." to "Palestinian Women": The Politics of Picture Postcards in Palestine/Israel', in David Crouch and Nina Lübbren (eds.), *Visual Culture and Tourism* (Oxford, New York: Berg, 2006), pp. 23–39.

MOORS, ANNELIES, 'Presenting People: The Politics of Picture Postcards of Palestine/Israel', in David Prochaska and Jordana Mendelson (eds.), *Postcards. Ephemeral Histories of Modernity* (University Park: Pennsylvania State University Press, 2010), pp. 93–105.

NASSAR, ISSAM, 'Early Local Photography in Jerusalem', *History of Photography*, 27/4 (2003), pp. 320–332.

NEEP, DANIEL, *Occupying Syria under the French Mandate: Insurgency, Space and State Formation* (Cambridge, New York: Cambridge University Press, 2012).

OSTERHAMMEL, JÜRGEN, '"The Great Work of Uplifting Mankind": Zivilisierungsmissionen und Moderne', in Boris Barth and Jürgen Osterhammel (eds.), *Zivilisierungsmissionen. Imperiale Weltverbesserung seit dem 18. Jahrhundert* (Konstanz: UVK Verlagsgesellschaft, 2005), pp. 363–425.

PEDERSEN, SUSAN, 'Eine heilige Aufgabe der Zivilisation: Das Mandatssystem des Völkerbunds in neuem Licht', *Jahrbuch des Wissenschaftskollegs zu Berlin*, 2005/2006, pp. 218–234.

PEDERSEN, SUSAN, *The Guardians: The League of Nations and the Crisis of Empire* (Oxford, New York: Oxford University Press, 2015).

PROVENCE, MICHAEL, 'Ottoman Modernity, Colonialism, and Insurgency in the Interwar Arab East', *International Journal of Middle East Studies*, 43/2 (2011), pp. 205–225.

ROUX, MICHEL, *Le Désert de sable: Le Sahara dans l'imaginaire des Français, 1900–1994* (Paris: L'Harmattan, 1996).

RYAN, JAMES R., *Picturing Empire: Photography and the Visualization of the British Empire* (London: Reaktion Books, 1997).

SALIBA, ROBERT, *Beyrouth: Architectures aux sources de la modernité, 1920–1940* (Marseilles: Parenthèses, 2009).

SCHAYEGH, CYRUS, *The Middle East and the Making of the Modern World* (Cambridge, Mass.: Harvard University Press, 2017).

SCHEID, KIRSTEN, 'Divinely Imprinting Prints, or, How Pictures Became Influential Persons in Mandate Lebanon', in Cyrus Schayegh and Andrew Arsan (eds.), *The Routledge Handbook of the History of the Middle East Mandates* (London, New York: Routledge, 2015), pp. 349–369.

SCHOR, NAOMI, 'Cartes postales: Representing Paris 1900', in David Prochaska and Jordana Mendelson (eds.), *Postcards. Ephemeral Histories of Modernity* (University Park: Pennsylvania State University Press, 2010), pp. 1–23.

SEMMERLING, TIM JON, *Israeli and Palestinian Postcards: Presentations of National Self* (Austin: University of Texas Press, 2004).

SHAMBROOK, PETER A., *French Imperialism in Syria, 1927–1936* (Reading: Ithaca Press, 1998).

SHAW, WENDY M. K., 'Ottoman Photography of the Late Nineteenth Century: An 'Innocent' Modernism?', *History of Photography*, 33/1 (2009), pp. 80–93.

SHEEHI, STEPHEN, 'A Social History of Early Arab Photography or a Prolegomenon to an Archaeology of the Lebanese Imago', *International Journal of Middle East Studies*, 39/2 (2007), pp. 177–208.

STANTON, ANDREA L., 'Locating Palestine's Summer Residence: Mandate Tourism and National Identity', *Journal of Palestine Studies*, 47/2 (2018), pp. 44–62.

TARAZI, CAMILLE MICHEL, *Vitrine de l'Orient: Maison Tarazi, fondée à Beyrouth en 1862* (Beirut: Les Éditions de la Revue Phénicienne, 2015).

THOMPSON, ELIZABETH, *Colonial Citizens: Republican Rights, Paternal Privilege, and Gender in French Syria and Lebanon* (New York: Columbia University Press, 2000).

TOUBIA, SAMI, *Sarrafian: Liban 1900–1930* (Mansourieh: Editions Aleph, 2008).

TRABOULSI, FAWWAZ, *A History of Modern Lebanon* (2nd edn., London, New York: Pluto Press, 2012).

WANG, NING, *Tourism and Modernity: A Sociological Analysis* (Amsterdam, New York: Pergamon, 2000).

WATENPAUGH, KEITH D., 'Middle-Class Modernity and the Persistence of the Politics of Notables in Inter-War Syria', *International Journal of Middle East Studies*, 35/2 (2003), 257-286.

WATENPAUGH, KEITH DAVID, 'Cleansing the Cosmopolitan City: Historicism, Journalism and the Arab Nation in the Post-Ottoman Eastern Mediterranean', *Social History*, 30/1 (2005), pp. 1–24.

WATENPAUGH, KEITH DAVID, *Being Modern in the Middle East: Revolution, Nationalism, Colonialism, and the Arab Middle Class* (Princeton: Princeton University Press, 2006).

WIEDEMANN, FELIX, 'Heroen der Wüste: Männlichkeitskult und romantischer Antikolonialismus im europäischen Beduinenbild des 19. und frühen 20. Jahrhunderts', *Ariadne. Forum für Frauen- und Geschlechtergeschichte*, 56 (2009), pp. 62–67.

ZYTNICKI, COLETTE, *L' Algérie, terre de tourisme: Histoire d'un loisir colonial* (Paris: Vendémiaire, 2016).

ZYTNICKI, COLETTE, and KAZDAGHLI, HABIB (eds.), *Le Tourisme dans l'Empire français: Un outil de la domination coloniale? Politiques, pratiques et imaginaires (XIXe–XXe siècles)* (Saint-Denis: Publications de la Société française d'histoire d'outre-mer, 2009).

CHAPTER 5

Lebanon: The tourist nation-state

Abstract

Already in the early days of the mandate, a broad consensus existed in Lebanon about the importance of tourism. While initially tourism was perceived as a major economic potential by Lebanese actors and the mandate power alike, it was increasingly understood as a vital political resource. The chapter shows not only the contradictions tourism development generated for the mandate power, but also the opposition among Lebanese lower classes against tourism development. Over time, however, tourism contributed to both an imaginary and a material consolidation of the national territory, stabilising the Lebanese borders and contributing to the acknowledgement of Lebanon in both a regional and a global context.

Keywords: tourism; Lebanese history; French Mandate; nationalism; middle class; modernity; state formation

Nature, sparing with goods of the earth and of the underground, has provided us instead with plenty of goods under its most blue coat, with its rarest beauties: climate and light, water and light, green landscapes and more light, the historical memory and still the Light! This is the true wealth of our country, the gold mine, which is not hidden in the ground, but spread all over and our eyes look at it but they do not see.[1]

This call for a greater importance of the tourist industry in Lebanon was published in 1919 in the francophone magazine *Revue phénicienne*, which advocated the creation of a separate Lebanese state under French guidance. The quotation indicated that the development of the tourist sector was not only an economical affair, but a matter of national pride. Fouad al-Khoury, the author of the article, presented Lebanon's unique characteristics, thereby dissociating the country from its surrounding areas: the scarcity of agricultural lands and mineral resources was compensated by the beauty of the Lebanese landscapes, its climate, the abundance of water and light, and its history. In the following paragraphs, al-Khoury argued that similar conditions had rendered Switzerland an important tourist destination and a wealthy country. In 1919, a definition of 'Lebanese' properties and a comparison with the small yet successful Switzerland was a political statement. The future territorial order of the post-Ottoman Arab lands was still subject to debate, and a Lebanese nation-state was a vision pursued by some inhabitants only.

From the outset, the Lebanese nation-state was defined as a tourist space. Deputies in the Chamber as well as entrepreneurs used tourism as a resource to position Lebanon both as a leisure resort for the Arab world and as a modern Mediterranean nation on an international level. Such a conception of Lebanon materialised in museums and newspaper articles, in parliamentary debates, in itineraries and advertisements, thereby reaching out to an audience well beyond the intellectual and political circles who had defended and shaped the idea in the long nineteenth century.[2] To the main advocates of tourism to Lebanon, tourism served as a resource to create and establish the idea of a Lebanese nation distinct from Greater Syria or *Bilad al-Sham*, acknowledged not only by an imagined national audience, but also by observers from the imperial and trans-imperial spheres.[3] These spheres, however, were not merely receptive audiences. French imperial actors hoped to disseminate their own conceptions in the context of tourism, and tourists adopted and rejected, shaped and modified the Lebanese idea, ultimately helping to turn it into practised reality. The creation of the Lebanese state was a process intertwined with the creation of Lebanon as a tourist destination.

The opening section of the chapter analyses tourist brochures and guidebooks issued by Lebanese authors and institutions. The brochures, edited with the aim of promoting Lebanon as a tourist destination, contributed to the construction of an idea about a distinct Lebanese nation. The second part focuses on the actors behind such publications. For both French administrators and high-ranking Lebanese politicians, entrepreneurs and intellectuals, tourism was of major concern. The final section shows how selected places on the Lebanese tourist map were integrated into the tourist and the national landscape.

In comparison to the previous case studies, one peculiarity of Lebanon was the high amount of Arab *estiveurs* (*mustaf,* pl. *mustafun*) who had been travelling to the Lebanese mountain area for decades. Generally, the visitors returned to the same locations each summer. In these villages, Beirut's notables and their relatives stayed next to émigré families visiting their hometowns, families from Egypt, Palestine, Syria and Iraq, and European administrative or military staff, who partly adopted estivation practices.[4]

In Lebanon, the *estivage* travel was much more important than the European tourism movement, regarding both the number of people on the move and the financial benefits it generated. Its significance is illustrated by a remark of Frédéric Martin, the Beirut agent of the *Messageries Maritimes,* who stated in 1925 that the (European) tourist season had to be carefully planned, as it should not interfere with the *estiveurs'* movements and thereby impede the more profitable transregional traffic. The otherwise 'dead season', he concluded, should be promoted as the tourist season: winter.[5] The custom of wintering in Mediterranean countries, popular among the European upper class, was introduced in Lebanon, but attracted fewer numbers than the much more well-established Egypt.[6]

Compared to mass tourism after World War II, the number of visitors in the interwar period remained modest. In 1937, when the peak of visitors during the interwar period was reached, Constantin Joannidès, Martin's successor at the *Messageries Maritimes*, counted fewer than 20,000 tourists and *estiveurs*.[7] Still, the French Mandate can be considered the formative period of twentieth-century tourism concerning the establishment of hotel and transport infrastructures, as well as narratives of tourism promotion and nation-building. Despite various crises shaking the world economy and regional politics during the mandate period which hindered the smooth development of tourism, in total, the number of tourists was rising.

In this respect, troubles in the potential destinations had equally detrimental consequences for the tourism sector as did crises in the visitors' countries of origin. The fact that 1925 was a bad *estivage* season,[8] for example, was presumably due to the unstable situation in the most important country of origin of potential *estiveurs*: Egypt (and less to the outbreak of the Syrian Revolt, which occurred at a time when most *estiveurs* had already arrived in their summer resorts). As far as the Euro-American tourists were concerned, two events triggered a sharp decrease in their numbers: the Syrian Revolt in 1925/1926, and the crisis of the world economy in the aftermath of 1929.[9] While tourism recovered quickly after the revolt had been crushed, the economic crisis had a more profound impact as it concerned European, American, and Arab visitors. Since Egypt in particular was harshly affected by the crisis, a large part of the Egyptian bourgeoisie refrained from its summering practices. On the other hand, the growing influx of *estiveurs* from Iraq and Palestine compensated for at least a part of the losses, as both countries profited from a more stable economic situation and improved railway connections.[10] From the mid-1930s, when the effects of the economic crisis eased off in Egypt, the overall number of summer guests increased steadily.[11] A reduction of 2,000 *estiveurs* in 1936 was most likely due to the Palestinian Revolt, but thanks to the Arab *estiveurs* the overall number remained at a high level of more than 8,000. By 1938, the numbers had more than doubled before tourism experienced a collapse after the outbreak of World War II. Joannidès, the agent of the *Messageries Martimes* reported that the Lebanese Government had adopted additional measures to maintain the *estivage* traffic between Lebanon and Iraq, Palestine and Egypt, as the state of war and the defeat of France had seriously damaged the *estivage* business and "very few" *estiveurs* had come to visit Lebanon during summer.[12]

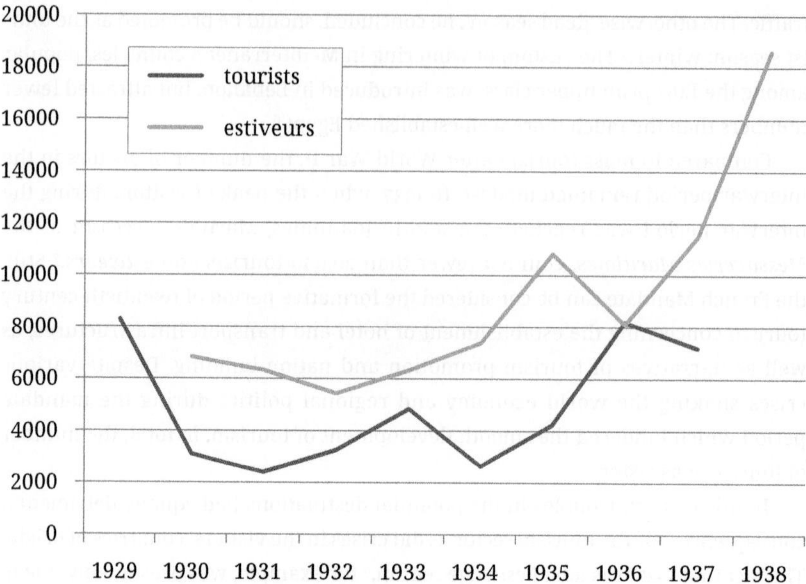

Table 1: Tourists and estiveurs visiting Lebanon, c. 1929–1938[13]

Brochures: Outlining a nation

Despite the continued popularity of well-known European guidebooks such as the *Baedeker*, *Cook's* handbooks or the various publications of the *Guide bleu* series, Lebanese associations offered an increasing number of alternative publications for tourists during the mandate period. Barely differing from European brochures and guides in the kind of information they provided, the small guides, brochures, leaflets or magazines disseminated additional, contrasting, sometimes contradictory, views of Lebanon as a tourist destination. Mostly written in French, English or Arabic, the publications addressed either tourists or *estiveurs*, and sometimes both. The traveller would obtain these brochures for example from local travel agencies, information bureaus for tourists, or local municipalities.[14] Until the mid-1920s, non-governmental groups in particular issued such brochures, notably the *Société de Villégiature au Mont Liban* (VML), an interest group addressing potential *estiveurs*, which held close relations to Lebanese politics and received subsidies from the government.[15]

From around 1928 onwards – that is, after the establishment of the Lebanese Republic – the VML's activities merged with governmental institutions, which became increasingly involved in tourism propaganda. By co-editing and publishing such brochures, the Lebanese Government not only supported an important business, but also aimed to shape the perspectives of visitors. Guidebooks and brochures described, defined and thus established more firmly the recently created State of Greater Lebanon.

Mount Lebanon: Summer resort of the Arab world

In these publications, three basic narratives can be distinguished, each of which viewed Lebanon in a different spatial context. The illustrated cover pages of three brochures can serve to introduce these narratives. The first brochure, *Estivage in Lebanon*, was published in the second half of the 1920s and addressed Arab *estiveurs*.[16] The cover illustration (fig. 21), a drawing printed in black on light brown paper, invited the potential *estiveur* to adopt the perspective of a man whose back was turned to the spectator. His upright posture suggested that he was contemplating the landscape in front of him. Several natural elements framing the drawing symbolised a rich variety of landscapes: a pine tree in a meadow, seaside palm trees, as well as rough cliffs, sparsely covered with bushes, stood close to each other, and thus within easy reach of the visitor. The solid stone bridge in the background was another hint at this easy accessibility, as it connected the different elements. Moreover, it directed the spectator's gaze towards the river, which flowed into the calm sea in the background. The artist, Nicholas Strekalovsky, visualised *estivage* in Lebanon as the romantic contemplation of idyllic landscapes, characterised by lush vegetation and water in abundance. These reminiscences of European Romanticism might be attributed to the painter's Russian origin; however, local literary traditions had popularised similar notions, which served as central tropes in the textual narratives of the brochure.[17] The narrative visualised on this brochure identified Lebanon with the unspoilt nature of its mountains. The Mediterranean coast was discernible in the background, whereas Mount Lebanon as a distinct landscape that provided an escape from crowded, bustling cities appeared as the core of the country.

Greater Syria: Lebanon as part of an integrated *Bilad al-Sham*

In the second example, the authors presented Lebanon as an integral part of the Eastern Mediterranean region. The editors of the VML, who issued the *Tourist's Book for Egypt, Palestine, Syria and the Lebanon* in 1924, confronted the tourist with a complex cover page that combined several elements (fig. 22). A special Art Deco font highlighted the words "The Lebanon" in the title, thereby emphasising the focus of the guidebook, while a sketchy map of the region indicated that the neighbouring countries of Egypt, Palestine and Syria would be covered too. The brochure seems to have been an early example of a brochure not only targeting *estiveurs* but aiming to draw the attention of European tourists on Mediterranean

Figure 21: "*Estivage* in Lebanon" (Illustration: Nicholas Strekalovsky. Reproduced by kind permission of the Umam Foundation, Beirut)

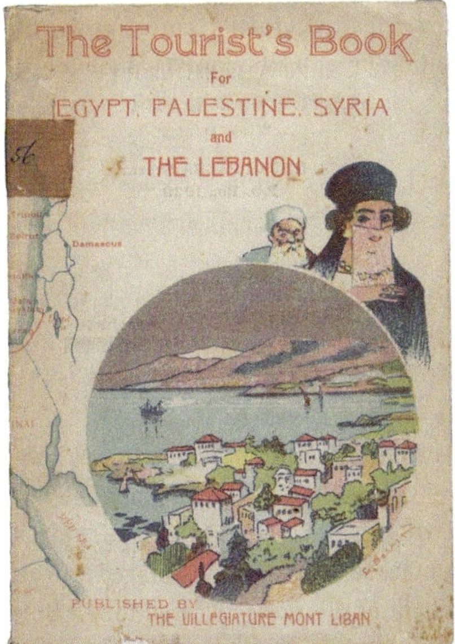

Figure 22: Société de Villégiature au Mont Liban, *The Tourist's Book*, 1924 (Reproduced by kind permission of Harvard University Library)

round trips to the attractions of Lebanon. A circular shape at the bottom of the page presented a view of Beirut. Red-roofed buildings, resembling one another in colour, shape and height, jutting out from trees and green spaces, evoked the impression of a small and well-ordered town. In the background, a snow-capped mountain range framed the picture. The sea was light blue and calm; a few sailing boats added to the impression of a peaceful place. Although the illustration captured traces of human activity, it corresponded with the previous cover page in the absence of persons and in the overarching impression of a picturesque place, close to nature.

The presumed inhabitants appeared outside the circular frame: an elderly man in the background wearing a white turban, next to a woman in the foreground, whose head and face were partly covered by dark blue headgear and a transparent veil. Their clothing identified them as 'Orientals'. On the back cover, the editors presented Egypt to the potential visitor in a similar arrangement. A circular shape revealed a glimpse of Egypt's attractions: namely, the colossi of Memnon, situated in the Necropolis of Thebes. A seemingly tiny man in a white *galabiyya* positioned next to a colossus testified to the enormous height of the statues. The statues, rather than the person, dominated the desert landscape. Outside the circle, a second man in a *galabiyya*, this time holding a camel on the lead, represented another 'type' of the local population. The illustration thus offered the potential visitor the expected clichés of

Oriental inhabitants and Pharaonic attractions. Lebanon, visually represented on the front cover, was situated in the broader continuum of the famous Egyptian tourist space, but it emerged, according to the editors, as a place to visit in its own right.

> The situation of the two countries is unique. [...] these countries [Lebanon and Syria] formed, in olden days, the link between the East and the West, and contributed considerably to the world's civilisation.
> At the present day, they are trying to keep up to the old tradition, but in the reverse direction. The motor car has replaced the caravan, and western produce and passengers from Beirut and Damascus are already being transported in it across the desert to Baghdad.[18]

The traveller who read this introduction in the VML's aforementioned *Tourist book* was informed that Syria and Lebanon were two distinct "countries", but nonetheless constituted one cultural entity, which was characterised by its positioning at a crossroads of civilisations and had been marked by exchange and trade.[19] This topos, repeated in various brochures throughout the 1920s, allowed for flexible positioning of *Bilad al-Sham* between the Mediterranean and the Orient.[20]

The narratives, often shifting between presenting Lebanon as an integral part of Syria or rather as a Mediterranean country reflected contemporary debates surrounding the location, as well as the cultural and civilisational belonging of Lebanon.[21] By the mid-1920s, fractures between Lebanon and Syria were visible, but not yet dominant.

Extracting Lebanon

By the 1930s, the visual representation of Lebanon in tourist brochures had changed. Brochures now pointed to a distinct and unique nation, rather than framing Lebanon as part of *Bilad al-Sham*. From now on, either cedars or the ruins of Baalbek, the two newly defined Lebanese national symbols, served as cover-page illustrations. In 1937, the Ministry of Tourism and the interest group *Société d'encouragement au tourisme au Liban* (Society to Encourage Touring in Lebanon; SET) opted for the ruins.[22] For the cover page, the painter Sylvio Castelot had drawn an assemblage of the iconic elements of the temples of Baalbek, in reality situated across different parts of the site (fig. 23): the six freestanding pillars of the Temple of Jupiter; the lion's head (originally part of the roof decoration); the massive pillar leaning against the wall of the Temple of Bacchus; and *Hajar al-hubla* (the Stone of the Pregnant Woman), an enormous monolith situated on the outskirts of Baalbek. Visitors to the ruins of Baalbek appreciated these elements as subjects of photographs and regularly posed in front of or next to them. Castelot and the editors made sure that their target group would be able to identify the motif as

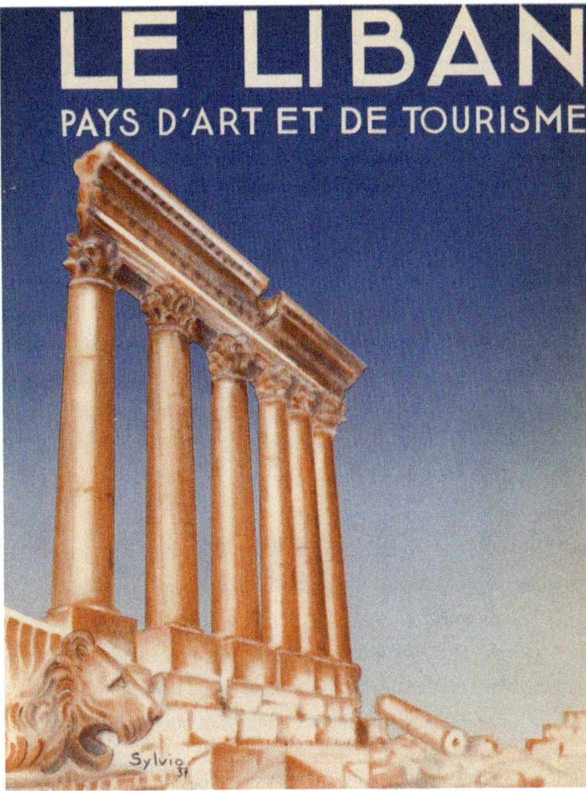

Figure 23: The pillars of Baalbek (Illustration by Sylvio Castelot for the SET. Reproduced by kind permission of the Bibliothèque Nationale de France)

the temples of Baalbek. The impressive archaeological heritage of Baalbek thus symbolised Lebanon, characterised in the magazine title as "the country of art and tourism", as a destination in its own right.[23]

The detachment of a Lebanese entity from a Syrian one was expressed by guidebook authors across three dimensions: geography, history, and race. At the same time, the process comprised two movements: a dissociation from a broader Greater Syria, and an integration into the Mediterranean space. While throughout the late 1920s and early 1930s, brochures and guidebooks qualified 'Syria and Lebanon' as a unity, reiterating the topos that the region was the "cradle of humanity",[24] from the 1930s onwards authors both implicitly and explicitly created a divide between the two parts.

First, Syrian towns and places were gradually omitted from brochures and guides. Whereas the early brochures had advertised an extensive 'tour of the Orient', including visits to sights and places in neighbouring countries such as Palestine, Egypt and Syria, attention was increasingly focused on the Lebanese territory. This specialisation led to a noticeable omission of Syrian places. Ancient cities, but also *estivage* centres such as Bludan, disappeared from the suggested itineraries despite Bludan's geographical proximity to the Lebanese *estivage* region. The exclusion of locations that were now defined as Syrian corresponded to a focus on Lebanese locations in advertisements. The brochures included sights which had so far been considered secondary (at best). Consequently, travellers consulting the brochures became acquainted with a much greater number of Lebanese villages, while simultaneously losing sight of places across the borders.

Second, this spatial integration of the Lebanese territory was related to a new definition of Lebanese history. When the Lebanese *Services Economiques* introduced Lebanon in 1935 to potential tourists and *estiveurs*, the authors enumerated the names of famous cities and *estivage* villages. The list mentioned the Mediterranean coastal towns, the Beqaa Valley and the Mount Lebanon region, before emphasising the cedars of Northern Lebanon in a rhetorical climax:

> At its foot [of the Mount Lebanon range], along the Western slope, the Mediterranean and its towns: Tripoli, Byblos, Berytus, Sidon, Tyre. Extending its eastern slope, there is a plateau: the Bekaa, and a city: Baalbeck.
>
> And the ridges are dotted with villages, among the pine trees, leaning against the slopes of the vineyards among the grapes, nestled against the valleys at the banks of the springs, villages the names of which resemble musical notes: Ehden, Hasroun, Ghazir, Reifoun, Bikfaya, Dhour-Choueir, Beitmery, Broumana, Aley, Sofar, Hammana, Beiteddine, Deir-el-Kamar, Jezzine!... And above the sea, above the towns and villages, over there, in the North, so close to the perpetual snow, the Cedars!
>
> This is Lebanon. This is ancient Phoenicia.[25]

The concluding remarks not only redefined the accumulation of these locations as 'the Lebanon', but also assumed continuity between the Lebanese present and a presumed historical entity of ancient Phoenicia. The orthography the authors chose for the name of the capital further accentuated the Phoenician origins: whereas the reader would expect to read *Beyrouth* in a French text, the authors instead opted for a spelling that pointed out its Phoenician roots, *Béryte*.[26] By narrowing down the historical focus on Lebanon's Phoenician past, the authors created a narrative that explained the integrity of the territory and justified Lebanon's distinction from Syria.[27] References to other ancient cultures were rare in the Lebanese guidebooks of the mid-1930s, although tourists were well aware of – and sometimes more interested in – Roman history or the history of the Crusaders, more prominent in brochures and guidebooks such as the *Guide bleu* or *Cook's Handbook*.[28]

From the perspective of Lebanese nationalists, however, these historical narratives posed two problems if history was to serve as justification for an autonomous Lebanese national entity. The Roman and Crusade historical periods highlighted the shared Syro-Lebanese past, and they were commonly referred to in French imperialist propaganda.[29] If Lebanese cultural and political representatives referred to the medieval or Roman history of the region, they often added references to Phoenician origins or biblical myths and legends, as if to prove in the same breath that the civilisation of Lebanon was even more ancient than the alleged ancestors of the French Empire. Just as important, the Phoenician narrative justified the merging of the coastal districts and the Beqaa with the former Ottoman governorate of Mount Lebanon into the so-called 'Greater Lebanon'.[30] To the liberal Phoenicianists, the ancient reference stood for a shared pre-Christian, pre-Muslim heritage and a geographical determinism aiming at integrating the predominantly Muslim inhabitants of the coast and the Beqaa region into the Lebanese nation.[31] While the Phoenician myth thus integrated 'Greater Lebanon', it separated Lebanon from its Syrian hinterland.

A third narrative assumed distinct racial origins of the Lebanese and Syrian populations. Such notions had been popularised by the francophone poet and intellectual Charles Corm. Corm, a central proponent of Lebanese particularism,[32] and the first agent of *Ford* motorcars in the French Mandate, had a noteworthy interest in tourism.[33] In the aforementioned brochure *"Le Liban: Pays d'art et de tourisme"*, edited by the Ministry of Tourism, Corm published an article in which he established a lineage from the Crusaders to the current inhabitants of the Mount Lebanon region:

> [...] finally the towns of Ehden, Becharré, and Hasroun, which kept the last Crusaders in their impregnable gorges, the blue eyes and blond hair, the healthy blood and the chivalrous soul of which are still to be found among our fellow men from the region of the Cedars [...].[34]

Corm, an ardent defender of cooperation with the French and supporter of the mandate, assumed blood ties between the Lebanese and the European Crusaders, which he claimed to be phenotypically visible, and recognisable in the special character of the Lebanese. This highly exclusivist claim of a racially distinct Lebanese nation differed widely from the liberal Phoenicianism described above.

The assumption of racial homogeneity uniting the Lebanese with their alleged European ancestors was also expressed in the brochure published by Rizkallah Gobril, an entrepreneur offering guided tours and owner of the transport company *Messageries Automobiles Beyrouth/Itinéraires Rizkallah*. It is remarkable in particular for it clearly distinguished the Lebanese population from the Syrian one along the lines of race and religion: Gobril not only identified Damascus as the "political centre of Syria's Muslim part",[35] thereby implying that Lebanon corresponded to the Christian section, but according to him, religious belonging indicated racial differences:[36] Gobril explained to his readers that unlike the Arab populations in neighbouring countries, the Phoenicians, ancestors of the Lebanese, were not of Semitic origin. Rather, they had descended from the tribes of Canaan.[37] While the theory established the presumed connection of the Lebanese with the land they inhabited, it also introduced a dividing line between the Lebanese (Christian and Canaanite-Phoenician) and the Syrian (Muslim and Semitic) populations.

Positions such as those expressed by Gobril and Corm, who supported a strong French influence in Lebanon, became rare in the 1930s.[38] Given the widespread impression among Lebanese politicians and intellectuals that France administered the Mandate contrary to Lebanese interests, many former supporters of a Franco-Lebanese cooperation turned away from the mandatory power and demanded full independence, which often implied the idea of pragmatic, peaceful relations with the neighbouring Arab countries.[39] Accordingly, most of the Lebanese guidebooks and brochures published in this period developed narratives that demonstrated their autonomy as a fully developed, civilised nation – a claim manifest on most cover pages adorned with one of the two Lebanese national symbols.

To represent Lebanon, the editors showcased either the temples of Baalbek or the cedar, as in the brochure *In the country of the cedars* (*Au pays des cèdres*) published by the *Commission du tourisme* in 1931.[40] The significance of this choice is two-fold: first, the commission's decision to identify Lebanon as a tourist desti-nation with its national symbol expressed a self-confident claim to nationhood. Second, the tree symbolised a Phoenicianist, nationalist concept of Lebanon. The cover page revealed that Syria was no longer included in the guidebook, and no longer part of the presented tourist space.[41]

In contrast to the cedar, the ruins of Baalbek were a more ambiguous national symbol. In the cedar, the idea of a distinct, unique and autonomous Lebanese nation crystallised. The six pillars of the Temple of Jupiter, by contrast, were of Roman origin, and throughout the nineteenth century, successive political visions derived legitimacy from the site, as Ussama Makdisi demonstrated: a European Orientalist, an imperial Ottoman, and a Lebanese nationalist interpretation. The imperial origin of the Roman temples also suited French narratives promoting a continued Roman-French tradition of imperial rule in the Mediterranean. Finally, the Roman tradition of the ruins also referred to a shared Syro-Lebanese history.[42]

The aforementioned brochure *Le Liban: Pays d'art et de tourisme*, published by the Ministry of Tourism and the SET, demonstrates the ambiguity of the symbol decorating the cover page. At the same time, this very ambiguity may well have rendered the pillars a particularly popular symbol in tourism advertising, as both French imperialists and representatives of a Lebanese-nationalist reading promoted the site, albeit for different reasons. In the foreword to the brochure, the

French High Commissioner Damien de Martel interpreted the ruins as a symbol of legitimate imperial rule. From his point of view, the ruins stood for Lebanon as part of an integrated Mediterranean region shaped by various empires throughout the centuries, thereby implying the legitimacy of the French presence in the Mediterranean.[43] In the same magazine, the French Orientalist René Dussaud contradicted such an interpretation. He stressed that Baalbek's relevance derived not only from its Roman origin, but also from its testament to the significance of local Syrian and Phoenician traditions.[44] His contribution offered an alternative reading of the ruins, claiming that the autochthonous influence had shaped and marked the Roman buildings. Similar interpretations were spread by Michel Alouf, conservator of the ruins of Baalbek.[45] In order to turn the Roman temple into a truly national symbol, Lebanese nationalists like him had to explicate its meaning.

The various definitions of Lebanon thus reflected current political debates over the situation of the country in a regional and international framework. Early brochures aimed to establish Lebanon on the mental maps of tourists, who were primarily interested in Egypt and the 'Holy Land'. Lebanese guidebook authors posited the reasons why their country merited a visit. In the late 1920s, authors additionally attempted to provide their visitors with a positive vision of the political and civilisational position of the country. They contradicted widespread Orientalist clichés and claimed a position among the 'advanced' nations. Although the brochures communicated various nuances and competing ideas of Lebanese nationalist discourses, references to Greater Syria ceased.

The adoption of a Lebanese framework in tourist brochures was a gradual process; individuals distanced themselves at different moments from alternative conceptions framing Lebanon as part of an Arab Eastern or a French Mediterranean continuum. Still, the emergence of a Lebanese nation in tourist brochures did not just mirror political and social debates during the mandate period. Rather, it seems that Lebanese nationalists in particular consciously used such publications as a means of shaping imaginations of a sovereign Lebanese nation.

High seasons, low seasons: The Franco-Lebanese relationship in terms of tourism

In Lebanon, high-ranking politicians, prominent entrepreneurs and influential intellectuals were involved in tourism to a greater degree than was the case in Egypt, Palestine and Syria, and many of these protagonists were Neo-Phoenicianists arguing in favour of a Lebanese state.[46] The extraordinary commitment to tourism development not only among ardent nationalists, but also among political and industrial leaders such as Albert Naccache, was unique to Lebanon, and it was certainly related to the important economic role the *estivage* sector played for Mount Lebanon after the decline of the silk industry.[47] This commitment translated into powerful tourism advertisement campaigns and a vigorous realisation of measures, which allowed them to turn tourism development into a Lebanese national project and build the bases for a future sovereign state.

In contrast to Syria, the Lebanese actors did not leave the field open for the French mandate administration and French colonial circles promoting tourism to the Mandate. In some cases, French imperial and Lebanese national policies of tourism development reinforced each other, as in the promotion of Baalbek. In other cases, tensions emerged. This section traces the dynamics of the Franco-Lebanese relationship in tourism and identifies three turning points in this relationship. Around 1923, the French mandate administration started fostering tourism development, mainly for economic reasons. Around 1928, national actors gained more political autonomy in the context of the French "politics of cooperation" resulting from the Syrian Revolt, which allowed them to create Lebanese institutions for tourism development.[48] Finally, in the mid-1930s, tensions grew between the French High Commission and Lebanese actors, and the nationalist discourse began to dominate the Lebanese political sphere and its tourism policies alike.

The institutional framework of Franco-Lebanese political interaction changed at various times during the mandate period. In the early years, both the executive and the legislative power remained with the French governor, appointed by the High Commissioner, and French 'advisors'. The first High Commissioner, General Henri Gouraud, appointed the Administrative Commission which consisted of Lebanese representatives and had merely consultative functions.[49] In March 1922, the Representative Council replaced the Administrative Commission. The council consisted of members elected by universal male suffrage, modified by a quota system in order to guarantee the representation of the most important religious communities in Lebanon. It did not include members who were fundamentally opposed to the Franco-Lebanese project – mainly Sunni Muslims – because they had boycotted the election of the council.[50] Although the members of the Representative Council discussed laws and policies, its function remained advisory.

In spring 1926, the mandatory power consented to a constitution for Lebanon, which was adopted in May. Whereas this event should not be mistaken as a starting point for a smooth way to independence, the constitution signified an important political rupture, as it marked a localisation if not of power, then of the political debate.[51] The new-born Lebanese Republic resembled the French model in its institutional structure: two chambers replaced the Representative Council, the Senate and the Chamber of Deputies, and the president was vested with important executive powers. Notwithstanding the republican model, the French mandatory power remained in control over domains such as foreign affairs and the military;

and the president of the new Lebanese Republic was answerable to the High Commissioner.[52] Both the President of the Republic and the Chamber of Deputies could initiate legislative projects, at least in internal Lebanese affairs. In practice, as in Syria, the French High Commissioner had extensive rights to veto legislation as well as the authority to dissolve the constitution, and later developments proved that the High Commissioner was willing to use this influence. Moreover, the French advisors 'assisting' Lebanese government offices retained a decisive position within the Lebanese bureaucracy. Finally, the Lebanese cabinets changed frequently, which slowed down political processes.[53]

The real change in Lebanese politics and boost in governmental activity did not occur before 1928.[54] Leyla Dakhli showed that around this time, the French administration established a new imperial strategy based on cultural cooperation. Subsequently, cultural institutions were no longer administered directly by the High Commission, and cooperation was to replace the tight control. The result, Dakhli concluded, was ambiguous: on the one hand, the new strategy implied a more profound colonisation; on the other hand, local political representatives enjoyed some room for manoeuvre in shaping the colonial state.[55] Whereas in the Syrian part of the Mandate, tourism development was intertwined with policies of colonisation, in Lebanon, nationalist actors were the driving forces behind tourism development – pursuing their own ambitions.

Tourism and the 'mise en valeur' of Greater Lebanon, c. 1920–1928

From the early days of the mandate, the French administration declared tourism development to be one of its primary tasks. High Commissioner Henri Gouraud stated in his speech on 1 September 1922, the first anniversary of the foundation of Greater Lebanon: "The agricultural production of Greater Lebanon, methodically developed, will be able to guarantee its existence, but the true source of wealth, the treasure to be exploited, is the country itself." Thanks to its landscapes and its rich history, he prophesied, Lebanon would become what Switzerland was to Europe: a peaceful mountain retreat for the Middle East: "Therefore, gentlemen, let's open wide the doors of this country."[56]

Official declarations aside, the actual policies of the High Commissioner and the French administration with regard to tourism primarily administered and controlled existing practices rather than systematically developing the tourist industry. Building on continuities from Ottoman times, travellers and *estiveurs* from neighbouring countries were identified as the main target group in the early years of the mandate. Although the High Commission had already expressed the aim of attracting a French public in 1921, cross-Mediterranean tourism was not yet promoted on a larger scale.[57] Rather, the High Commission rejected the Lebanese demand for a reduction in visa and passport fees as well as the creation of a special passport for tourists.[58] In the early 1920s, the main legislative measures concerning tourism regulated gambling in *estivage* villages (apparently to levy taxes more systematically) and aimed to guarantee hygienic conditions in hotels and pensions.[59] The focus on such minor projects regulating the existing business rather than aiming to expand it was unsurprising, given that the recently established French High Commission spent its first years repeatedly restructuring the administration of the territory under mandate. Moreover, limited resources and the French priority of controlling mobilities seem to have prevented several initiatives from being put into effect and as a consequence of the war years, economies in both Europe and the Middle East were recovering slowly, and were regularly hit by economic turbulences, which reduced the number of potential travellers.[60] Accordingly, entrepreneurs such as the shipping company *Messageries Maritimes* had a pessimist outlook on the touristic potential of the region.[61]

Yet, the Beirut agent of the *Messageries Maritimes* had the impression that the rather passive approach of the High Commission to tourism changed in 1923. According to him, the High Commission aimed to foster the business of tourism by setting incentives for the modernisation of hotels and supporting the construction of a large casino hotel. While he expected that it would take several years before the measures would take effect, he approved of the more active approach of the High Commission.[62]

The renewed interest in tourism development was most likely inspired by the suggestions of the Minister of the Colonies, Albert Sarraut, who had published his ideas on the *"mise en valeur des colonies"* ("rendering the colonies profitable") in that same year. Sarraut had been Governor General in French Indochina, where he had tried to introduce large-scale tourism even before World War I. In order to foster tourism development, he established a cooperation with the *Touring Club de France* and reorganised the *Services économiques* in Indochina. After the war, Sarraut served as Minister of the Colonies and propagated the idea of rendering the colonies economically profitable. From his point of view, tourism had both economic and propagandistic benefits, as it would spread knowledge about *la France outre-mer* ('France overseas') and raise the awareness and acceptance of the overseas empire among the French population and an international audience.[63]

Around 1925, similar to Sarrail's measures in Indochina, the High Commission launched its projects of tourism development by means of institution-building.[64] New institutions brought together French and Lebanese actors from various fields. Although pursuing partly diverging interests, they agreed upon the necessity of developing tourism and *estivage* for economic reasons. In March 1925, the Tourist Office (*Office du tourisme et de la villégiature*) took up activities in Lebanon, which was in charge of both advertisements and quality controls. The office would not only publish leaflets, posters and brochures, but also guarantee the quality of tourists' experiences, for example, by providing hotel classifications. Its members reflected this double function: along with the Director of the *Services Economiques*, who presided over the office, were the Director of Public Works and the Director of Hygiene, delegates of the Syro-Lebanese *Touring Club*, as well as a member of the International Chamber of Navigation.[65]

The creation of the *Office du tourisme* seems to have been a reaction of the High Commission to the complaints of *estiveurs:* although 1924 had been a good year for the Lebanese hotel industry, the guests, in particular from Iraq and Egypt, had complained that the comfort and amenities of Lebanese hotels and pensions did not meet international standards.[66] In order to remedy the shortcomings, the French Governor of Lebanon, Vandenberg, and his successor Léon Cayla adopted stricter regulations and controls on the one hand, and set incentives for the creation of new accommodation facilities on the other.[67] A *Comité du tourisme*, which united "high-ranking personalities from the political and the economic spheres" under the patronage of the High Commission, rewarded hotel owners investing in the modernisation of their establishments.[68] A decree assured tax exemptions for the first one hundred villas constructed for *estivage* purposes, and announced similar benefits for the first ten hotels to be built in Beirut.[69] The activities seem to have paid off. By 1925, several modern hotels opened for business: the *Hotel Sursock* in Souk el-Gharb, *Hotel Aouad* in Hasroun, *Hotel Kayrouz* next to the Cedars, and *Hotel Tubet* in ꜤAyn Zhalta.[70] The members of the Representative Council expressed their satisfaction about the work of the *Services Economiques* and its director, Dr Alphonse Ayyub: "We have every good reason to hope that with his contribution, *estivage* will progress rapidly here."[71]

Besides *estivage*, Léon Cayla perceived European cultural tourism as having real potential for increasing Lebanese revenues. The project of constructing a modern and beautifully situated hotel in Beirut was designed especially to attract foreign visitors, who until then knew Beirut mainly as a transit hub:

> We wish [....] to make the numerous tourists stay with us as long as possible. They disembark in Beirut for a stopover or transit at all seasons, and profit from it by visiting the picturesque sites and the magnificent vestiges of the ancient civilisations Lebanon is so rich in. Therefore, the government has invited tenders for the construction of a large hotel [...].[72]

In addition to the "picturesque sites" attracting *estiveurs*, Cayla presented Lebanon's historical sites as economic assets that would attract tourists to Lebanon. Since briefly before Cayla's speech Syrian opponents had risen against the French mandate, the economic potential in European sightseeing tourism may be read as a justification of French imperial policies and a promise addressed to the Lebanese.

In comparison to the French authorities, the members of the Lebanese Representative Council had a more vital interest in tourism. Repeated debates concerning tourism indicate that initiatives aiming to strengthen the business in the Mandate were often launched by the members of the Representative Council and their non-governmental allies in the Lebanese economic and public spheres. Their interest in tourism was three-fold: *estivage* was considered a central pillar of the Lebanese economy, the attraction of sightseeing tourists would contribute to the international prestige of Lebanon and was hence a step towards sovereignty, and both tourism and estivage would contribute to defining a Lebanese state acknowledged in Lebanon, among the neighbouring Arab countries and on an international stage.

Since its constitution in 1922, the members of the Representative Council had launched various initiatives to promote travel, in particular *estivage*, to Lebanon, to contribute to the economic recovery after the war. In late 1922, debates centred on credits to strengthen the two main industries: silk production and *estivage*. The latter had the unanimous support of the deputies: there were debates about the right strategies and measures for strengthening tourism, but the representatives did not disagree on the general necessity of fostering the business.

An important interlocutor in the early to mid-1920s was the Egyptian-based *Société de villégiature au Mont Liban* (VML), who collaborated with the Representative Council, for example, in setting up (and financing) advertising measures.[73] The interest group united exiled Lebanese of nationalist leanings and had promoted *estivage* in the Mount Lebanon region in Cairo even before World War I.[74] It seems to have been the major forum for exchange on tourism affairs between private actors and the Representative Council in the early period and all the more important, as the Egyptian clientele was considered the most important for the Lebanese *estivage* industry.[75] Complaints about insufficient services, as mentioned above, were taken seriously and Lebanese initiatives and politicians aimed to enhance the comfort of *estiveurs*, particularly in the luxurious resorts. In Ehden, for example, a post office opened, and 'estivage offices' were created to take care of travellers in Hasroun and Beit Meri.[76] Some of the famous summer villages were officially declared *centre d'estivage* in that time: Falougha, Chouit, ʿAraya and Hammana in the Matn district, as well as ʿAbey, Beit ed-Dine and Deir al-Qamar in the Chouf region, received this status between 1926 and 1928.[77]

At the same time, strengthening *estivage* became an important argument for councillors attempting to direct funds to certain areas – generally the regions of origin of their families or their political strongholds.[78] The major investments were related to the construction of roads. Tourism in Lebanon relied on the automobile and the extension and improvement of the Lebanese road network, particularly in the mountains, were a condition for the expansion of *estivage*. Plans to improve the road network regularly prompted debates over which connections should be given priority.[79] Sufficient funds were not always available, but it is significant that the councillors did not generally object to the argument – in contrast to Egypt, where deputies occasionally rejected investments in infrastructures of tourism. Whereas they suspected that mainly international corporations would profit from tourism, in Lebanon tourism was a local business and therefore of vital interest to the members of the Representative Council.

The idea to address European tourists on sightseeing tours as another target group alongside *estiveurs* was being promoted by Lebanese actors since the early 1920s and already in this time, Lebanese politicians argued that international tourism was not only of economic relevance, but also a matter of international prestige. In a meeting of the Representative Council, Emir Fu'ad Arslan, a member of a family of Ottoman bureaucrats from Mount Lebanon,[80] brought up the "question of antiquities", referring to the excavations undertaken by the French Egyptologist Pierre Montet in Byblos, who had discovered Phoenician tombs in 1922. Afraid that these and other excavated treasures would be transferred to French museums, Arslan insisted on keeping them inside the country, so they could be presented to the public in the projected archaeological museum. Touristic exploitation had to be combined with the preservation of the identified national heritage: "if not, our independence would remain fiction."[81] Thus, from Arslan's point of view, the exploration and exhibition of the historical heritage were directly related to the acknowledgement of Lebanese sovereignty.

An even more immediate relation between tourism and state-building was assumed by Michel Chiha, who identified tourism and *estivage* not only as the pillars of the Lebanese economy but as a *raison d'être* of the Lebanese nation. In most studies on the history of Lebanon, Chiha is considered as a sort of *eminence grise*, involved in basically all state affairs of Lebanon and yet a barely visible actor, who exercised power by influencing or controlling networks, rather than by direct authority. His actual influence is therefore hard to establish; however, his impact on shaping the economic policies of Lebanon – as an intermediary economy based on trade and services, profiting from its position between East and West – is somewhat uncontested.[82] Chiha identified Lebanon's landscapes and cultural monuments as its major economic resources:

> [...] in a country like ours, where we aim to sell among other products fresh air and charming places, pleasurable stays and feasts for the eyes, the "superfluous" in the words of Voltaire "is an absolute necessity". All countries addressing the traveller and the tourist, inviting them to *estivage* and recreation, are forced to make such arguments and to speak of comfort and luxury without implying prodigality. We offer freshness and beauty, admirable ancient memories, views of Baalbek and Byblos for example, just like other countries offer coal and cotton.[83]

In these lines, the idea of the Lebanese service economy was declared to be a quasi-natural choice. The landscapes and monuments of Lebanon were not only the country's heritage, but also its destiny.

There was a model Chiha and his group of economically liberal nationalists referred to: Switzerland. While they built on the topos of Lebanon as a 'Lebanese Switzerland', coined by nineteenth-century travellers referring to the mountainous landscape, its implications shifted in the brochures and guidebooks during the 1920s. Again, the attraction of Egyptian visitors seems to have played a role in the adoption of this metaphor: In 1919, Albert Naccache described the growing numbers of Egyptians spending their summer holidays in Europe, especially Switzerland. The aim for Lebanon, he concluded, was to reach a Swiss standard and attract this Egyptian clientele.[84] In this sense, deputies referred to the Swiss model in order to justify investments in infrastructures associated with tourism, for example asphalted roads, transport services and the improvement of accommodation facilities.[85] The circle around Chiha applied the metaphor to the economic sphere more generally, inspired by the Swiss reputation of great natural beauty and a service-oriented economy based on banking and tourism.[86] Over time, advocates of tourism development gradually emancipated themselves from the Swiss model: whereas tourism development in Switzerland and Lebanon had been conceived of as a teacher-and-disciple relationship, from the late 1920s on, guidebook authors proclaimed the Lebanese superiority over its Alpine model, emphasising, for example, the healthier Mediterranean climate.[87]

The policies of tourism thus aimed to achieve recognition from the French mandatory and the League of Nations as well as from the neighbouring Arab states. Throughout the period, the press remained an important forum for contact and exchange across *Bilad al-Sham*.[88] Debates between and within the inhabitants of the newly established states took form through articles and letters to the editors.[89] In some Lebanese newspapers, the local news sections regularly reported on numbers and origins of visitors. Thereby, they situated the respective towns in the larger spatial framework of a *Shami* tourist space and presented Lebanon as profoundly embedded in the wider region.[90] Such articles suggested that Lebanon was not isolated and cut off from its neighbours, as opponents to the formation of Greater Lebanon might have feared, but continued to be profoundly embedded in the wider region – by means of communication and tourist movements.[91]

Maintaining these connections in times of crisis was vital for the Lebanese advocates of tourism development. Again, the newspapers had a vital function. During the Syrian Revolt, articles intended to reassure potential visitors abroad about the security in the *estivage* villages. On 17 June 1926, at the beginning of the *estivage* season, the illustrated daily newspaper *al-Maʿrid* published a letter to the editor, written by the Egyptian efendi Ernesto Naʿmatullah. In his letter, published under the headline *Testimony of an Egyptian estiveur*, Naʿmatullah described his initial doubts about spending the summer in a potentially dangerous and unstable Lebanon. However, he explained that when he finally made his way to Bikfaya with his family, he felt perfectly secure during his entire stay and invited his readers to join him.[92] The publication of the letter was an obvious statement addressed to potential *estiveurs* afraid of the violent clashes between French forces and pro-Syrian insurrectionists during the revolt.[93] It is not certain that the letter was authentic. It may have been an invented or commissioned testimony as an attempt to deny reports on the problematic situation of security within the Mandate. After all, the Lebanese *Commission du tourisme et de la villégiature* complained in February 1927 about the negative press coverage in Egypt and Palestine and expressed their intention to react by launching a press campaign to rebut the allegations.[94] Other than the notables in the previous chapter, gathering in the summer resorts as a centre of *Shami* sociability, the journalists and editors thus drafted another idea of Lebanon: as a transnationally connected, yet distinct spatial entity.

Moreover, there seems to have been a second, domestic dimension to the press campaign. In August 1926, *al-Maʿrid* covered the journey of a delegation of journalists to the *estivage* villages of northern Lebanon. A series of articles and black-and-white photographs documented their trip via ʿAley to places such as Hasroun, Ehden, the Qadisha valley, Bsharri and the Cedars.[95] Journalists and editors of the most important Lebanese newspapers, *al-Ahrar, L'Orient, al-Jawaʾib* and *al-Maʿrid*, but also representatives of the famous Egyptian daily *al-Ahram*, participated in the journey and covered the attractions of northern Lebanon. In addition, the delegation comprised high-ranking personalities from Beirut's political sphere, such as Muhammad bik Hamid, the Egyptian consul, the deputy Michel Zaccour, and Dr Ayoub Tabet, the new director of the *Services économiques*. Both the articles and the photographs in the richly illustrated *al-Maʿrid*, edited by the Lebanese nationalist Zaccour, showed refreshing water currents and lush green hills, as well as modern hotels guaranteeing the comfort of travellers.[96] Although it regularly published entertaining content on pages four to six, this report was printed as a serial report on the cover pages of the paper, thus presented as a contribution to the political section of the paper.

Two reasons explain this categorisation. First, the tour obviously sought to restore the reputation of the *estivage* centres during the Syrian Revolt. Just like Ernesto Naʿmatullah's contribution, they demonstrated that security and comfort reigned throughout these locations. Second, the advertisement tour not only strengthened the *estivage* villages located in the more remote areas in the province of Tripoli; it also tied the province more closely to the narratives of the emerging state. While the debate whether Tripoli should remain a part of the Lebanese Republic or be attached to the Syrian part of the Mandate went on even after the adoption of the constitution, the tour presented the region as a core part of Lebanon.[97] The expansion of the tourism business and narratives of a unique mountain range characterising the country strengthened the economic and symbolic ties between the state of Greater Lebanon and the district of Tripoli.

Cooperation, c. 1928–1934

The resistance of local elites to colonial rule in both the French Levant and North Africa, particularly pronounced in the case of the Syrian Revolt and the Moroccan Rif War, necessitated a change in French imperial policies. The so-called "politics of cooperation" combined military control with opportunities for participation. Leyla Dakhli argued that these opportunities granted to local actors existed especially in the field of cultural politics, and identified Philippe Berthelot as a key figure behind this policy, who was at the time general secretary of the Minister of Foreign Affairs in Paris. Berthelot took initial steps in the immediate aftermath of the Syrian Revolt, whereas the major change of direction occurred in 1927/1928: cultural institutions were removed from the High Commissioner's direct control; instead, French and local institutions were to cooperate. As Dakhli has argued, this shift aimed to affect the construction of identities in the region.[98] The new framework had noticeable effects on tourism development, which also profited from an increasing number of visitors and additional stops in Beirut the shipping lines introduced.[99] Lebanese actors, in particular, attributing economic and political significance to the sector, founded new associations and launched measures to foster travel.

First and foremost, however, the policy was a French imperial initiative and the topic of tourism development in the Mandate started to resonate in the colonial networks of the French metropole. On 13 February 1928, Georges Philippar, president of the *Messageries Maritimes*, delivered a speech, accompanied by a set of slides, on *Tourism to Syria* at the *Ecole des Hautes Etudes Sociales.* Philippar, who had climbed the career ladder within the *Messageries Maritimes* since he joined the company in 1912, was well-connected in French colonial circles. He was fond of travel literature and interested in tourism, and had published on tourism to Indochina before.[100] His speech on tourism to the French Mandate of Syria was covered by the newspaper *Cahiers du sud* from Marseille, and the text was subsequently published in *Panorama,* a monthly magazine for "education and colonial vulgarisation".[101] Philippar compared the situation of tourism in Syria and Greece, the ancient sites of which he considered equally attractive from the perspective of visitors. He observed that while Greece had become a well-established and highly frequented tourist destination, Lebanon and Syria were much less popular among European tourists. According to him, four main reasons accounted for this difference: a lack of publicity; the well-organised tourism to neighbouring and competing Palestine; the mediocre state of Syrian hotels; and, most importantly, the preference of the travelling Frenchman for places he knew, particularly if they were related to classical antiquity and French cultural tradition. Consequently, Philippar concluded that a better promotion of Syria was necessary, in particular by visual means.[102] Moreover, he outlined a narrative that incorporated Syria in a French history of conquest and domination (the Crusades), but also in the French literary tradition (from Chateaubriand and Renan to Maurice Barrès). Such a narrative would demonstrate the bonds between France and Syria, he argued, and might attract a greater number of French travellers.[103]

Similar opinions about the necessity and adequate means of strengthening tourism circulated widely in the French pro-colonial press. Even though travel did not dominate the debates and publications in the imperial circles, it was continuously addressed.[104] The lavishly illustrated magazine *Le monde colonial illustré*, for example, regularly published articles on tourism to various parts of the French Empire. The destinations of Syria and Lebanon were covered in the years 1933 and 1935.[105] A closer look at the editors and authors of the publication sheds light on the connection between business interests, colonial circles and tourism. Georges Philippar, for example, was not only a member of its *comité de patronage* (the honorary editorial board); the president of the *Messageries Maritimes* had already been presented as an eminent and important pro-colonialist in 1929.[106] Numerous other high-ranking personalities from pro-colonial economic, political and intellectual spheres served as members of the magazine's *comité de patronage*: Henri Gouraud, former High Commissioner of the Levant; Albert Sarraut, the former Minister of the Colonies who had promoted the concept of *"mise en valeur"*; and Théodore Steeg, who had served as resident general in Morocco. Some of the members performed functions that were directly linked to tourism: in this respect, Georges Philippar was joined by Edmond Chaix, president of the *Touring Club de France*, and an early advocate of travel to Indochina. Moreover, Marcel Olivier figured on the list, at the time president of the shipping company *Compagnie Générale Transatlantique*, the former president of which, John dal Piaz, had been one of the most prolific pioneers in organising tourism to French North Africa.[107]

The *Monde colonial illustré* thus brought together pro-colonial actors from various fields, and it spread their ideas and experiences among a larger audience. The editorial board of the magazine demonstrated the tight connection between politics, tourism and the public, and it showed how tourism development policies and strategies for harnessing tourism circulated in the French Empire. In French Algeria, for example, the colonial authorities introduced measures to foster tourism that strikingly resembled strategies adopted later in Lebanon. Already in 1919, a council named *Congrès général du tourisme d'Afrique du Nord* was established in Algeria, intending to foster tourism in the country and thereby strengthen the economic development. It comprised representatives of different economic branches, local representatives and government officials under the patronage of the *Gouverneur Général d'Algérie*, a position comparable to the High Commissioner in the mandates. Not only the institution, but also the measures applied in Lebanon and Algeria seem familiar: first, the tourist destinations in Algeria obtained the special status of "tourist station", corresponding to the Lebanese *centres d'estivage*, as both enjoyed relief from resort taxes. Second, the *Crédit hotelier* provided financial loans for the renovation or construction of tourist hotels in both countries, which was perceived by Constantin Joannidès as a key instrument contributing to the rise in *estiveurs* in the late 1920s.[108] Third, both in Algeria and in Lebanon tourist organisations promoted new tourist practices such as winter tourism and skiing.[109]

Taking into account that tourism development was perceived as a contribution to monitoring space and was therefore part of the French 'pacification' strategies, such parallels are hardly surprising. As argued in the previous chapter, French colonial administrations had a vital interest in setting incentives for tourist movements. From this angle, the French colonial empire constituted a coherent space of colonial tourism, where tourists served as civilian allies of the French strategies of counterinsurgency. However, to perceive tourism as a colonial enterprise, or even a "tool of colonial domination",[110] covers only a part of the Lebanese reality.

After all, the various committees and commissions consisted mainly of Lebanese actors who pursued their own agendas, some of which differed largely from the colonial vision of both tourism and Lebanon. For example, the *Commission consultative du tourisme et de la villégiature*, established in December 1929, gathered various high-ranking actors from the Lebanese public sphere. In contrast to the *Comité du tourisme* which was controlled by the High Commission, it was associated with the Lebanese Government. Initially, Under-Secretary of State of Economic Affairs Gabriel Menassa presided over the *Commission*, but a few months later the presidency shifted to the Minister of Finance.[111] In Lebanon, the economic relevance of tourism had been uncontested. The French politics of cooperation now offered Lebanese actors room for manoeuvre which they used to institutionalise tourism development. In contrast to the Tourist Development Associations in the neighbouring countries, the *Commission du tourisme* was the direct concern of an important ministry. From now on, Lebanese actors from various fields pursued a relatively consistent policy of stimulating both international and *estivage* tourism, which in the long run turned into a policy of strengthening the Lebanese nation-state.

For the time being, the *Commission du tourisme* examined and proposed measures to support the tourism sector, gave advice on the use of credits, established hotel classifications and conferred the status of *estivage* centres. Its members, supposed to meet once a month, made recommendations concerning the budget of *estivage* municipalities, regulations of gambling, public spectacles and public health, as well as the improvement of the *estivage* towns. The 17 members of the *Commission* represented various interests in tourism. Some were owners or employees of transport companies and travel agencies, for example Constantin Joannidès, agent of the *Messageries Maritimes*; Norman Nairn, who had founded the motor transport company *Nairn Transport* together with his brother Gerald; ʿAbdallah Zehil, agent of the shipping company *Fabre Lines* and author of a tourist guidebook; and Antoine Kettaneh, entrepreneur in the transport business since 1923, who had close ties to *Cook's* travel agency.[112] Writers or intellectuals were also among the members, for example Gebran Tuéni, journalist and founder of the newspapers *al-Ahrar* and *al-Nahar*, who was nominated for Minister of Education in 1931; Georges Vayssié, editor of the francophone journal *La Syrie* and president of the *Touring Club du Liban*, who had already been a member of the earlier *Comité du tourisme* established in 1923;[113] Jacques Tabet, an influential Phoenicianist, former publisher of *La Syrie*, as well as the first treasurer and director of the Lebanese National Museum. As Director of the National Museum, Tabet had crucially contributed to establishing a national narrative that circumvented the role of Islamic civilisations, focusing on the pre-Islamic past and the Christian tradition instead.[114] Another member, the pan-Arabist Khayr al-Din al-Ahdab from Tripoli, was a political opponent of Tabet. He edited the newspaper *al-Ahl al-Jadid*, served as deputy and became prime minister in the 1930s.[115]

Just like the French colonialist networks, national bodies such as the *Commission du tourisme* contributed to the circulation of ideas. Moreover, the institutionalised exchange seemed to strengthen the position of tourism and turn it into a general priority for the commission members. The agent of the *Messageries Maritimes* is a good example in this regard. From the moment Constantin Joannidès joined the *Commission consultative du tourisme et de la villégiature*, he changed the structure of his annual reports to the head office of the *Messageries Maritimes,* dedicating much more space to the topic of tourism. His elaborations on tourism in the report on 1929 fall somewhere between the content of a tourist brochure and a tourism development plan. Joannidès regretted that most tourists limited their trip to Beirut and just a few excursions, for "it would be inexcusable to leave the Levant without having seen Damascus, Aleppo, Palmyra, the Krak des Chevaliers, Baalbek, and the Lebanon" – all the more so as the excursions could be easily combined with a trip to Egypt and some stops in Palestine.[116] Joannidès wrote this report in January 1930, thus shortly after the creation of the *Commission consultative du tourisme.* The concurrence of events seems to have been more than mere coincidence. As the agent of a shipping company, the omission of sights in the Syro-Lebanese hinterland should not have been of great interest to him, yet in this regard, the reports presumably reflected overarching debates about tourism development led in the context of his work for the *Commission du tourisme*.[117] The suggestions Joannidès presented to the head office to advertise the region would not only serve the economic targets of the company, but they were in the interest of tourism development more generally.

The activities of the *Commission du tourisme* came to an end in the spring of 1932, when High Commissioner Henri Ponsot suspended the constitution and declared the dissolution of the Lebanese Chamber of Deputies. During the following period of non-constitutional government, no further projects to foster tourism were launched by the Government, which was then installed by the High Commissioner.[118] Moreover, around the same time, the repercussions of the economic crisis hit the tourism sector. While the number of European and American tourists had plunged from almost 8,300 in 1929 to a mere 2,400 in 1931 (cf. also table 1), the major downturn in *estivage* movements occurred in 1932: while in 1931, more than 6,300 *estiveurs* had visited Lebanon (compared to 6,800 in 1929), only 5,400 summer guests visited Lebanon in 1932. Joannidès considered the 1931 season of tourism and *estivage* rather satisfying given the circumstances,[119] whereas the decreasing number of summer guests in 1932 concerned him to a greater extent. Due to the crisis, Egyptian *estiveurs* in particular refrained from taking their summer holidays, and the growing numbers of *estiveurs* from Iraq and Palestine could not compensate for the losses of the *Messageries Maritimes*: those summer guests arrived by rail or motorcar rather than by the sea.[120] After 1932, the Lebanese *estivage* sector gradually recovered, and, as will be discussed later in this chapter, the immigration of central European, bourgeois Jews to Palestine played a significant role here.[121]

Confrontation, c. 1935–1939

Despite a slow yet steady growth in *estiveurs*, Joannidès started to openly criticise the new High Commissioner, Damien de Martel, whom he accused of ignorance with regard to local society and of neglecting the mandate economy.[122] In his report on 1936, he held the French mismanagement responsible for the difficult economic situation of Lebanon:

> Lebanon could be rich thanks to its *estivage* and its historical sites, but the Authorities would have had to take care of the two main resources that might bring considerable revenues, if the mountain hotels were properly equipped and the means of transport less expensive.[123]

He followed with a list of proposals, identifying restrictions and harsh controls imposed on travellers as the major fault of the French mandate authorities. Accordingly, most of his suggestions aimed at facilitating administrative procedures and making visas more accessible. Moreover, in order to attract a greater number of middle-class tourists, the agent requested more modest accommodation facilities rather than luxurious hotels ("*grands palaces*").[124]

Joannidès' critical stance on the High Commissioner de Martel and French mandate policies reflected the increasingly strained relations between the mandate authorities and their Lebanese subjects during the mid-1930s. By that time, even actors who had initially supported French mandate rule complained about the French mandatory power and called for an independent Lebanon.[125] The political debates echoed in the sphere of tourism development and the national paradigm increasingly shaped respective policies.

The main institution that stood for such a Lebanese nationalist approach to tourism development during the late 1930s was the *Société d'encouragement au tourisme* (SET), which was founded in 1935 and declared to be of general benefit in June 1936.[126] The SET was not an official government institution, yet it received subsidies from the Lebanese state and cooperated with political bodies on a regular basis.[127] Among its first projects, for example, was the issuing of a special Lebanese postal stamp. Part of the revenues from the airmail stamp which advertised travelling to Lebanon, were to be redirected to the SET.[128] Even though a causality cannot be proven, it is an interesting coincidence that the decision to issue the stamp was signed a few days before the Government of the Popular Front in France introduced paid vacations. After all, airmail from the Mandate would be addressed to destinations overseas in general, and to France in particular. Since the stamp would reach a large European audience, its authors might have hoped to compensate for losses of visitors from the surrounding countries: in 1936, the outbreak of the Palestinian Revolt had resulted in a decrease in the number of summer guests by around 2,600 (cf. table 1).[129]

Lebanese political support for the tourism sector also manifested itself institutionally. Khayr al-Din al-Ahdab, formerly a member of the *Commission du tourisme*, established the first Lebanese Ministry of Tourism and *Estivage* in early 1937. The Minister of Tourism, Habib Abu Shahla, closely collaborated with the SET.[130] Although the ministry was abolished in the following cabinet, Constantin Joannidès described a growing activity of the Lebanese Government in the field of tourism. Many of the measures he mentioned resembled strategies adopted in the 1920s and were focused on accommodation and transport. The Lebanese Government aimed to improve roads; it regulated the transportation sector and provided financial incentives for entrepreneurs to invest in apartments and modern hotels.[131]

A new initiative intended to consolidate winter tourism. Already in 1922, Emir Fu'ad Arslan had suggested the advertisement of winter tourism. Inspired by a winter sports society from Beirut's wealthy suburb of Bir Hasan, he had argued in the Representative Council that promotion of winter sports would allow Lebanon to target European winter guests, thereby reaching out beyond the *estivage*

Figure 24: Skiing in Lebanon. Title page of the brochure *Liban. Tourism – Estivage – Winter Sports*, edited by the SET
Credits: Reproduced by kind permission of the Bibliothèque Orientale, Université Saint-Joseph, Beirut

clientele.[132] Reverting to such discussions, the SET and its associates now promoted skiing in Lebanon; the intended target group, however, had now shifted from European winter guests to a regional clientele from neighbouring countries.[133] The idea resonated in the Chamber of Deputies in June 1939, when the deputies adopted legislation aiming to develop not only tourism and *estivage*, but also *hivernage*, winter holidays. Accordingly, the trilingual illustrated magazine *"Liban – Tourisme, estivage, sports d'hiver"*, published by the SET around 1938/39 with texts in French, English and Arabic, dedicated eight entire pages to the Lebanese ski resorts. Moreover, the cover page for Arabic readers showed a picture of a young woman, cross-country skiing against a background of snow-laden cedars and hill slopes (fig. 24).[134] Whereas in 1922 Fu'ad Arslan had mostly European visitors in mind, in the 1930s the advertisements for winter tourism seemed to target the Arab bourgeoisie – and Yishuvi tourists, as the following section will show.

The fact that the woman on the cover illustration was skiing between the snowy cedar trees might have remained a footnote, had there not been another piece of legislation adopted around the same time which aimed to protect Lebanese monuments and sights. The text explicated that the formula "monuments and sights" included natural sights of artistic, urbanist, tourist or public interest, and stated that high age, special beauty or historical interest also justified the protection of trees. This sentence obviously referred to the Lebanese cedar, the official national symbol since 1926. In addition, the law stipulated that a list of all protected places should be drawn up.[135] Thus, the law not only demonstrated a political decision to develop tourism; it also indicated that a Lebanese space was to be defined. The classification of places within a given framework would outline a nation-state and prioritise certain places and 'typical' sights. By listing what rendered its national space specific and specifically interesting, the resulting inventory for tourists would correspond to an inventory of the nation.

These new impulses, which bound together tourism and the national idea, as well as the resulting efforts, were seriously hampered after 1939. In the summer of that year – before the outbreak of World War II – the Government further encouraged *estivage* and, during the first winter season that was statistically recorded, a certain amount of tourist traffic persisted. However, the French defeat in June 1940 brought the business to a standstill.[136]

Topography of a tourist nation-state

If we imagined the Lebanese territory of tourism as a topographical map on which height represents a highly frequented tourism or *estivage* area, Baalbek would appear as a major, solitary peak. Already in the nineteenth century, the ruins of

Baalbek had become an excursion site attracting numerous visitors, many of whom recorded their own presence on the site by carving their names and the year of their visit on the walls of the ancient temples. (Even Michel Alouf, inhabitant and historian of Baalbek and later official conservator of the ruins, did so.) The resulting 'graffiti', written both in Latin and Arabic letters, fascinated later tourists, and the inscriptions became a typical motif captured in photographs.[137] Both by taking photographs of the traces of former visitors and by visually documenting their own presence, the tourists situated themselves within a long series of admirers of ancient beauty.

Whether this attraction was situated on Lebanese or Syrian territory, however, did not matter to most European tourists whose diaries and reports confirmed the complaint of the agent of the *Messageries maritimes* Constantin Joannidès: most travellers visiting the French Mandates restricted their visit to the best-known places like Beirut, Baalbek and Damascus.[138] In this context, 'Lebanon' was a minor stop on their long Oriental journey and they described it as an integral, though special, part of Syria. Even in 1935, the former marine officer Chardon, author of the French *Guide bleu* on the Eastern Mediterranean and above suspicion of any pan-Arab sympathies, judged the administrative division of Syria and Palestine an "artificial separation", and subsumed Lebanon under the heading of "Syria" without further notice.[139]

Such a perceived unity corresponded to the travelling practices of tourists following itineraries from Damascus to Baalbek and then further on to Aleppo, or from Beirut via Byblos to the Krak des Chevaliers (Qalʿat al-Husn).[140] The border between Lebanon and Syria did not represent a barrier and it could be crossed easily by most visitors; in this respect it differed from trans-imperial border crossings between British Mandate Palestine and the French Mandate of Syria.[141]

Lebanese tourist brochures therefore attempted to introduce new "standard" sights to the checklists of visitors: "The standard roundtrip, the one of travellers in a hurry, comprises Beirut, Byblos, the Cedars, Tripoli, the Krak des Chevaliers, Baalbek, and Damascus."[142] This section seeks for the dynamics leading to the establishment of such tourist places. While many of the secondary tourist sites would not have become mandatory stops by the end of the mandate, Lebanese promoters of tourism achieved another goal: the shaping of the Lebanese national space.

Beirut: A tradition of modernity

In the touristic landscape of Lebanon, Beirut was a city of transit, rather than a place to stay. From the point of view of most European visitors Beirut lacked authenticity: it was too Mediterranean to be Oriental; too modern to be Eastern. In a conscious shaping of narratives reminiscent of the Zionist efforts in Palestine, the Beiruti middle classes declared their modernity to be of interest to the tourist.

Moreover, they projected such a self-conception onto the territory of Lebanon – where it encountered resistance, as the following case studies will show.

European tourists passed the port city of Beirut on their way to the ruins of Baalbek or the heartland of Syria; *estiveurs* from Egypt or Palestine arriving by ship or train crossed it before heading to the summer resorts. These groups, as well as businessmen traversing the city, had profited from the city's hospitality sector since the nineteenth century. In 1893, 25 hotels were counted by a traveller, and guests reported that Beirut was the first city in the Near East where pensions met European standards. Already at that time, travellers described Beirut as a transit zone between an allegedly strange and inscrutable Orient and the well-ordered and familiar Occident.[143]

From the point of view of tourists, however, the good reputation of the hotels in Beirut and the comfort they offered did not compensate for the absence of sights. When George H. Williams reached the capital of the French Mandate in March 1925, the first impression he recorded in his diary reflected an opinion widely shared in travel reports: "Beyrouth does not seem an interesting place." Even his local guide gave up: "My guide practically told me that there was nothing to see in the town, and suggested that I should go to Acre."[144] Before Williams left for Palestine ahead of time, he visited the National Museum, recommended by most guidebooks because it exhibited sarcophagi and excavation findings from Byblos and Nahr al-Kalb.

It had become commonplace to state that Beirut was not of much interest as there were "very few relics of antiquity [...] that are of consequence".[145] Moreover, the authors of guidebooks and diaries regretted the absence of *'couleur locale'*. *Cook's Handbook*, for example, warned readers that the bazaar had nothing in common with what they would imagine being Arab markets, that the houses looked European rather than Oriental, and that streets and places were orderly, broad and properly illuminated.[146] Indeed, at the turn of the century, the joint efforts of Beiruti notables and Ottoman imperial elites had led to a change in the urban structure of the new provincial capital. After the war, the old city and the coastline in particular were transformed according to French architectural models.[147] On the one hand, such familiar urban structures disappointed many visitors; on the other, *Cook's Handbook* stated that on the return trip, the traveller would appreciate the quasi-Occidental character of Beirut "after narrow ill-lighted alleys in other parts of Syria, for that very reason."[148]

The frustration of travellers is captured best in the idea of a double transition that was frequently expressed in both visual and textual narratives. Most travellers considered Beirut as a city of transit in a spatio-temporal sense, as a gate leading them towards an authentic, somewhat backward Orient visualised in scenes of entering the harbour and an almost mythical scenery.[149] Yet, they discovered a city in plain transition, expressed in the topos of modernity. In the photograph album of the aforementioned French traveller who had visited the Qadisha Valley, the two photographs representing Beirut exemplified these topoi. She took the first snapshot, showing the port of Beirut, as her ship was about to enter the harbour. The second one captured the *Avenue des Français* (fig. 25).[150] The author seems to have taken both photographs hurriedly, as neither of them seemed carefully structured. The first one, *Le Port de Beyrouth*, showed her arrival at the port in landscape format, with small wooden boats, some of them sailing boats, lying at anchor. The sky dominated around three-quarters of the picture. Whereas the authors of travelogues typically described the situation of Beirut between the Mediterranean Sea and the snow-capped mountains, this image instead stressed the maritime character of the city.

Mediterranean modernity was represented in the second snapshot, which had been taken on land, presumably from a balcony. The promenade of Beirut, the *Avenue des Français*, dominated the photograph. Not only did the promenade and its rows of palm trees resemble its famous cousin in Nice, the *Avenue des Anglais*; the entire street view could just as well have been a scene from the French Riviera. The photographer presented a broad pavement, populated by *flâneurs*; a street on which cars (and not carriages) signalled the arrival of modernity (as did the accidentally captured electricity cables across the picture).[151] Only the faint contours of the mountain range in the background provided a subtle hint at the Lebanese locality.

These two impressions of Beirut as a port city and a bourgeois society of *flâneurs*, dated 20 March 1931, remained the only comment of the traveller on the character of the city. The photographs on the following album page, taken in the Mount Lebanon region, were dated 22nd. Although the tourist seems to have spent at least one more day in Beirut, she did not discover any other places worth being preserved on paper. In contrast, she kept eight photographs of the rough landscape of the Mount Lebanon mountain range in her album. Either she shared Williams' point of view that there was nothing much to see in the capital or she felt that by representing the themes of her own transit and the modernity of the place, she had sufficiently grasped its character. She had crossed boundaries, but instead of accessing the past the boundary turned out to be spatial rather than temporal, signalling a variety of quite familiar modernity.

Both the French High Commission and the Lebanese political representatives of Beirut exhibited and cultivated the modernity of Beirut. Particularly for the latter, similar to their Egyptian counterparts, claiming equality with European countries was a politically vital issue. As the mandate treaties promised self-government once a sufficient level of civilisational maturity had been achieved, members of Parliament regularly put forward arguments that demonstrated the country's civilisation.[152] These debates resonated in brochures and guides. Bodies associated

Figure 25: Album page showing impressions from Beirut (Photographer: anonymous. Reproduced by kind permission of the Bibliothèque Nationale de France)

with the Lebanese Government in particular insisted on the contemporary local culture, contradicting clichés of a timeless Orient.[153] Such arguments were put forward, among others, by the *Commission du tourisme* established by the Lebanese Government, as well as by its successor organisation, the *Société d'encouragement au tourisme* (SET).[154] The authors claimed the country's modernity, refuting Orientalist assumptions about immobile and backward Eastern societies.

The intermediary position between East and West, tradition and modernity, was claimed by authors both for Beirut and for Lebanon as a whole. In the brochure *Syria and Lebanon: Holidays off the beaten track* the authors presented the city as follows:

Spread out in terrace-fashion on the hills of a cape from which it dominates the maritime horizon [...] lies Beirut, one of the gates to Asia, which has made great strides during the last twelve years. It is Mahometan and Christian, this political and intellectual metropolis, which is the port for Damascus and Southern Syria. It plays today the same role as Tyre and Sidon did in the past. It brilliantly renews the ties with the civilisatory traditions of Phenicia.[155]

Referring to topoi from the late nineteenth century, the introduction situated Beirut in a crossroads position position, both from a cultural and an economic perspective; a place combining the advantageous properties of the regions it connected.[156] The passage is reflective of the nationalism of Michel Chiha, the three core ideas of which it summarised. First, the economic potential of Lebanon resounded in the conception of the region as "gate to Asia" and the progress achieved since the last war. Second, the reference to the multireligious population of Beirut demonstrated its integrative potential and the peaceful interreligious cohabitation. Third, as the most important element of the rhetorical climax, the authors accentuated the Phoenician ancestry. The ancient, pre-Islamic and maritime Phoenician culture played an important role in the invented tradition of the nationalists and the economic liberalists that would later be called the *Cénacle Libanais*.[157] Thus, the brochures promoted a vision of the nation as a modern, liberal and multireligious society, the core of which was the trading centre of Beirut.

Unlike in other colonial cities around the Mediterranean, the French mandatory power did not apply its urban politics of conservation and orientalisation in the Levantine capital, a consequence of the history of Beirut rather than the choice of the French.[158] As the city had gained its significance only in the late Ottoman Empire, an ancient or medieval centre as in Aleppo or Marrakech did not exist, and war-time damage had required the reconstruction of the centre.[159] Moreover, by the late Ottoman period, Beirut had already been deliberately shaped as an emblem of modernity by its own inhabitants and by the Ottoman administration.[160] The Ottoman modernity of the city, manifest for example in the *Serail*, the clock tower, or public places such as the *Place des Canons/Place des Martyrs*, was preserved and replenished under the French administration. Additional emblematic buildings communicated the belonging of Beirut to a modern, secular, republican – and French – world, in particular the Parliament and the National Museum.[161] The modernity of the capital, vital for the self-conception of its inhabitants, continued to define its development. For prestigious hotel buildings, the Orientalist style was not an option, even though it was popular and widespread in other parts of the world, not least in Egypt. Instead, the famous and prestigious *Hôtel Saint-Georges*, designed by the French and Lebanese architects Auguste Perret and Antoine Tabet, which opened in 1932, was not only a modern hotel with regard to the amenities it offered, but also in its modernist architecture. Thanks to its prominent location on the coast of Beirut, its modernist façade defined the impressions of travellers who approached the city from the sea.[162]

In the 1930s, the hotel embodied the modernity the inhabitants of Beirut claimed for themselves.[163] In the brochure *Le Liban: Pays d'art et de tourisme*, edited by the SET in collaboration with the Ministry of Tourism in 1937, a large photograph of the *Hôtel Saint-Georges* figured as an illustration of the article on sports (fig. 26). The

Figure 26: Modern Lebanon. The swimming pool of the *Saint Georges Hotel*, Beirut (Photographer: René Zuber. Reproduced by kind permission of the Fonds photographique René Zuber and the Bibliothèque Nationale de France)

black-and-white photograph showed a young 'modern' woman with a bob haircut in a white swimming costume, who posed on a diving board in front of the hotel. The position of the camera seemed to be just above the surface of the water, so the gentle waves in the foreground conveyed the impression that the observer was in the pool, waiting for the woman to jump. On the same page, the photograph of a young male model complemented the first picture. In the posture of an ancient athlete and holding a discus, the muscular young man was presented as a *Lebanese athlete*, thus styled as a representative of his nation.[164]

Similar to iconographies of beach tourism in Egypt, the young and well-trained bodies of a male and a female athlete celebrated the youth of the sitters, following contemporary fashions of interwar modernism. The bright modernist hotel in the background emphasised this effect, while the pose of the man, reminiscent of Olympic athletes, and the vivid visualisation of the seawater rooted the protagonists in a Mediterranean setting. Framing Lebanon as a Mediterranean country, as in Egypt, implied presenting it as a modern society. At the same time, this turn towards a Mediterranean modernity already paved the way for post-World War II tourism development, in which beaches and clubs on the coast played an important role.[165] Different from other Mediterranean countries such as Italy, Spain or

Tunisia, however, Lebanon did not adopt the path towards low-cost Mediterranean mass tourism, but continued to use tourism as a resource to position itself as a destination for the Arab world as well as for an international audience.[166]

That the Beirutis' self-proclaimed modernity stood for the Lebanese society as a whole was a claim put forward in the Chamber of Deputies throughout the mandate. The self-conception as 'gate to the Orient' in an active sense, mediating between two worlds, was cherished and widely referred to in the advertisement brochures. Unlike the Egyptian *Tourist Development Association*, the Lebanese organisations did not explicitly redirect the tourists' Oriental gaze. Claiming Mediterranean modernity, the brochures did not identify any supposed backward Other. However, the local narrative promoted a particular vision of a Lebanese-Beiruti modernity inhabitants in other parts of the country did not necessarily identify with.

The coastal towns: Excavating a nation

Although the ancient coastal towns of Lebanon – namely Tripoli, Byblos, Tyre and Sidon – had important roots in antiquity, and despite their familiarity to travellers from biblical stories and the excavations of Phoenician sites, they remained marginal on tourist maps until the mandate period. Already in the early 1860s, French archaeologist Ernest Renan had published the findings of his expedition to the coastal towns in his work *Mission de Phénicie*.[167] Only after Pierre Montet discovered the Ahiram sarcophagus in 1923, famous for its Phoenician inscription, Byblos became a point of reference in tourist guidebooks. However, although many tourists admired the excavated objects in the new National Museum in Beirut, only a minority actually went to see the ancient cities on the Lebanese coast. Of the travellers whose accounts I analysed, only George H. Williams had visited Sidon; Norah Rowan-Hamilton, on her tour with the *British Archaeological Association* in 1928, visited Tripoli as well as Byblos and its necropolis.[168] The cities and their sights were only gradually established as stations on tourist itineraries.

In contrast to the ruins of Baalbek, seeing the coastal towns was by no means considered mandatory by the travellers; they were perfectly satisfied to head home without any souvenir photograph. Some of the smaller guidebooks even discouraged travellers from exploring the towns. In the *Guide Sam: Le livre d'or de l'Orient*, published in 1930, a subsection on tourism in the chapter on Syria briefly mentioned Sidon and Tyre. According to the authors there was "little to see" though, as the towns had allegedly lost their ancient splendour. Byblos at least was mentioned in a brief list of "other places" that might be worth visiting. Similarly, other guides and tour operators classified Tripoli, Sidon and Tyre as less important: *Thomas Cook & Son* did not have agents in any of the towns, and the *Cook Handbook* of 1934 dedicated only a brief historical overview to them.[169]

A comparison with the Palace of Beit ed-Dine indicates that the coastal cities lacked first and foremost a powerful lobby; neither the French nor the Lebanese Government backed their tourist development in the beginning. Even when efforts to present the Lebanese heritage increased after 1928 and the scope widened beyond the acropolis of Baalbek, the Government gave priority to other projects.[170] Among them was the Palace of Beit ed-Dine in the southern Mount Lebanon area, which the Lebanese Government under Charles Debbas granted the status of historical monument in 1930.[171] In the decree declaring the palace a historical monument, Minister of Education Gebran Tuéni suggested establishing a museum inside the building, to exhibit ancient furnishings and allegedly traditional costumes of Lebanon.[172]

The Palace of Beit ed-Dine, evocative of Christian rule in the Mount Lebanon area and therefore a constitutive *lieu de mémoire* of the Lebanese nation-state, was hence to be associated with the presentation of Lebanese 'national' traditions.[173] In April 1935, thus after the end of the period of non-constitutional government, the deputies decided to grant additional credit for the restoration of the palace as well as the palace of the nearby Maronite village Deir al-Qamar, although the picturesque villages had hardly attracted the attention of European tourists until then – all the more so as the mountain roads "obviously have not been constructed for automobiles".[174]

The choice to invest in Beit ed-Dine and Deir al-Qamar was a political one. First, an important lobby of hotel owners and transport entrepreneurs based in the Mount Lebanon region generated political support for tourism development and the creation of infrastructures in the Chamber to the degree that deputies from other regions started complaining about the funds being directed almost exclusively to the mountain area.[175] Second, it seems that the deputies and the French High Commission supported these demands because they substantiated a certain Lebanese-Christian nationalist narrative.[176] The coastal cities, by contrast, did not have such influential advocates.

Certainly, the Lebanese deputies were aware of both the touristic and nationalist potential of the coastal cities. Already in April 1930, during a debate in the Chamber of Deputies over providing additional credit for excavations in Byblos, Yusuf al-Sawda supported investments in further archaeological excavations, which he considered investments in Lebanon's future.[177] Al-Sawda, originally from the Mount Lebanon village of Bikfaya, had studied at the francophone Jesuit University of Saint Joseph (USJ), and promoted a strictly Phoenician national identity.[178] In the debate, he explicitly connected archaeological excavations with tourism development, considering both vital to the nation's future. Apparently impressed by the consequences of the discovery of Tutankhamun's tomb in Egypt in 1922, al-Sawda emphasised: "We are talking about archaeological discoveries. The future of the country depends on them. Egypt offers a good example [...]. In this way we could exactly like Egypt profit from the visits of millions of tourists and visitors."[179]

Despite such initiatives, governmental efforts to capitalise on the heritage of the coastal cities remained limited. They were hardly promoted as tourist attractions, even though their ancient history was likely to attract European visitors, and the Phoenician heritage qualified them for a central position within Lebanese nationalist narratives – at least they exemplified the Phoenicianist narrative much better than the more frequented Baalbek or the Mount Lebanon hinterland. However, in comparison with the Mount Lebanon area or Beirut, the coastal cities lacked a comparable hospitality sector, so tourism development presumably lacked proponents. The *Guide bleu* listed three hotels in Sidon: *Hôtel de Phénicie*, *Hotel de Moutrân* and *Manzar el Djemil*, but commented that all three of them were "mediocre". In Tyre, the authors recommended only one hotel, *El Ouatan*, though still classifying it as "rather rudimentary". No hints at hotels or pensions were given to potential travellers to Byblos, and apart from Beirut, only the port city of Tripoli seemed to be a place where visitors could stay in acceptable conditions: *Hotel Royal* was considered "rather comfortable" and *Hôtel Plazza* was not only clean but also offered "respectable cooking".[180] The coastal towns were lacking interested entrepreneurs advocating tourism and under-represented in the *Commission du tourisme*, as their sole representative was Khayr al-Din al-Ahdab from Tripoli. The lack of influential supporters might thus account for the marginality of the coastal towns in brochures and guidebooks.[181]

From the mid-1930s onwards, however, the cities began to be advertised more prominently. In 1936, the SET issued a special brochure on Byblos; and in the 1937 brochure *Le Liban: Pays d'art et de tourisme*, the Ministry of Tourism dedicated an article to *Byblos – A Phoenician town* and a second one to the monuments of the Crusaders in Lebanon, among other places in Byblos, Sidon and Tyre.[182] This incorporation of the coastal towns into the brochures and advertising campaigns was more than a mere realisation of earlier strategies or a sign of economic necessity. It was part of a political campaign aiming at the solidification of the Greater Lebanon nation-state.

In 1936, the status of "historical monument" was granted to the Crusader castle Qalʿat al-Bahr in Sidon, and to three mosques in Beirut: the ʿUmari Mosque, the Bab al-Saray Mosque, and the Amir Mundhur Mosque. In Tripoli, the castle, the Taynal Mosque and the Lions' Tower were added to the list of historical monuments.[183] The decrees issued between February and July 1936 were largely attributable to political considerations. During the early months of 1936, a general strike of 50 days and large demonstrations had shaken neighbouring Syria. The following concession of the High Commissioner to negotiate a treaty that would ultimately lead to Syrian independence also generated unrest in Lebanon. Demonstrations were held in particular in the coastal towns of Tripoli and Sidon. From March onwards, especially Lebanese Sunni Muslims raised their voices after a meeting they held in Beirut, the so-called 'Conference of the Coast', and expressed their desire to unite with Syria. The 'unionists' demanded that at least the predominantly Muslim coastal towns, as well as the Beqaa Valley, join the Syrian state. By contrast, the predominantly Christian 'protectionists', such as President Emile Eddé, feared an imminent annexation of Lebanon to Syria in the case of Lebanese independence from the French, whom he perceived as protector of the Lebanese. A third group opted for independence from France and an economic orientation towards Syria, without backing political unity. These various demands and activities led to a tense political atmosphere in Lebanon during the spring and summer of 1936.[184] The fact that in such a context, Emile Eddé declared the Qalᶜat al-Bahr in Sidon, several mosques in Beirut, and various monuments in Tripoli to be part of the Lebanese historical heritage expressed a political claim. The coastal towns had been added to the Lebanese territory against the will of most of their inhabitants in order to guarantee the economic viability of the newly created Lebanese state. Since the Lebanese Government feared the secession of these territories, it was vital that they be culturally integrated into the state.[185]

The signatories of the decrees taken into account, such an assumption becomes all the more plausible: it had been signed by President Emile Eddé and Secretary of State Dr Ayoub Tabet, both of whom were ardent Lebanese Christian nationalists.[186] That the third signatory, Subhi Haydar, was a Shiite minister from the Beqaa, testifies to Traboulsi's remark that by late 1936 the conflict was no longer fought along religious or sectarian lines with regard to Lebanese external relations; rather, it was the internal balance of power which was at stake.[187]

The three members of government had a vital interest in preserving the Lebanese state in its present condition, and cultural politics offered them a way to emphasise the legitimacy of their claims. Similar to the brochures published around 1935/1936 by the SET, the inclusion of the Muslim mosques as part of the Lebanese national heritage was not intended as a message to the French mandatory power, but it addressed parts of the population. The tourism development committees, dominated by the third group advocating political independence from France, mainly presented the Phoenician heritage to the French mandatory power as a means of demonstrating Lebanese civilisation and claiming sovereignty. The pro-French president Emile Eddé and his Government, by contrast, had a different focus. As new dynamics framed the political conflict, they incorporated the coastal towns into the Lebanese tourist landscape. This occurred not only later than in other regions, but also involuntarily. The business related to tourism and *estivage* had been marginal along the coast, but more importantly, the integration was imposed. Neither the inhabitants nor their representatives pushed for the inscription of the towns on tourist maps. The driving force was rather the representatives of the nation-state, who were trying to incorporate – by tourist mobilities and by a narrative – the insurgent towns into its territory.

Mount Lebanon: Classifying movements

If the coastal towns, in particular Sidon and Tyre, were the lowlands of the Lebanese topography of tourism, its highest peaks (next to Baalbek) were to be found in the villages of Mount Lebanon. Besides its economic significance, the mountain range of Mount Lebanon played a pivotal role in the imagination of the Lebanese nation.[188] National myths such as the myth of Adonis associated with Nahr Ibrahim (Adonis River) were positioned here, but also the traumatic experience of the civil war in 1860, which marked the origin of a Lebanese national state in many narratives. Moreover, the region was the home of the national symbol, the cedar, connecting claims of ancient origins with notions of recreation.

At the same time, in the 1920s and 1930s, the mountain villages were, besides Beirut and Baalbek, the national and international centres of Lebanon. In the formation of the Lebanese national space, tourism and *estivage* thus suspended the categories of town and hinterland – at least from a Lebanese perspective. This was also due to the deliberate actions of advocates of international exchange and mobility. Many of these advocates chose the international outlook, often guaranteeing their existence, out of pragmatic reasoning rather than ideological conviction. Although questions about access to, participation in, and profits from the tourist national space were barely debated openly, their choice did not remain undisputed.

The French mandatory in particular perceived the mobilities and connectivity shaping the Mount Lebanon area with suspicion. The growing integration of Lebanon into imperial structures, world markets, a globalised culture and a global public sphere caused inner contradictions and outer conflicts for the French High Commission.[189] Due to the specific spatial configuration of Mount Lebanon, these contradictions came to the fore in a particularly pronounced manner. The example of the Mount Lebanon area suggests that these frictions also divided the colonial society along the lines of class and gender, which I will examine more closely by confronting two sorts of movements towards the *estivage* centres: while the first section traces the movements of 'undesirables', the second section considers the mountain villages as a bourgeois space of leisure.

The new borders, established after the end of the Ottoman Empire, were among the most significant changes the end of World War I brought to the region. Although Cyrus Schayegh demonstrated that the mere drawing of borders did not necessarily imply a reduction of mobility, the issue was subject to negotiation and an important source of conflict between the French mandate authorities, the French institutions in the metropole, different groups of travellers and Lebanese actors.[190] Throughout the mandate period, the French authorities had to constantly redefine, readjust and balance their interest in winning over the mandated populations and international observers by creating an economically self-sufficient mandate on the one hand, and of securing access to the territories on the other.

From a French perspective, the borders needed to be controlled in order to prevent 'undesirables' from reaching the country.[191] Two main categories of 'undesirables' can be distinguished: persons who posed a political threat to the mandate, such as opponents to French rule, and persons who posed a moral threat, in particular solitary female travellers from the lower classes. While attempting to protect the Mandate from such intruders, the French High Commission had to make sure that the French regulation of passports, visas and entry to the Mandate did not hinder the *estivage* business. After all, the mandate government was under pressure from the League of Nations to justify its actions and regularly present reports on the economic situation to the Permanent Mandates Commission.[192]

Public complaints were therefore taken very seriously. In spring 1924, the Egyptian press denounced the difficulties foreigners encountered when they wanted to enter Syria and Lebanon for a stay during the summer months. Given the importance of the *estivage* business, the news caused immediate unrest in Lebanon and multiple complaints were addressed to the mandate authorities, forcing the High Commissioner to react. To avoid protests, he suggested a reduction of visa costs for travellers from Egypt and Palestine, at least for round trips of limited duration, as well as for families.[193] His explanation addressed to the Minister of Foreign Affairs in Paris revealed the contradictions which the French mandate administration faced:

> A general protest is now rising in Lebanon against this demand, and this movement is growing every day, the closer the summer season comes. – Whereas the High Commission has done its best to encourage this movement, and advises the inhabitants to improve the *estivage* stations in order to generate profit, on the other hand, the protesters say, France is putting up barriers in order to keep *estiveurs* from coming to Lebanon.[194]

The French mandatory thus perceived themselves in a dilemma: On the one hand, the Lebanese mountain resorts were an important economic sector of the country and, from the perspective of the Lebanese bourgeoisie, well-integrated into a regional *estivage* space. On the other, these summer practices had a subversive potential from a French point of view. As discussed in the section on Bludan, the mountain villages were perceived as being difficult to survey and French security officers feared that they might turn into gathering points for criminals, rebels and prostitutes.

Since the activities in the resorts could not be monitored systematically, the French tried to control mobilities to the *estivage* centres. This was evident, for example, during the Syrian Revolt in 1925/26, when the High Commissioner was clearly in a dilemma about how to facilitate mobility while monitoring the movements of potential insurgents. Very shortly after the outbreak of the revolt in September 1925, High Commissioner Maurice Sarrail informed the Minister of France in Cairo and the Consul in Jerusalem that until further notice, the facilitated access to the French Mandates for *estiveurs* was suspended.[195] After the revolt had been crushed, the authorities eased the controls and travel restrictions. Moreover, they facilitated the formal requirements for entering the country, a measure which might have contributed to the increase in travellers in the late 1920s.[196]

Other than Arab middle-class tourists and *estiveurs*, political suspects in times of crisis, the staff catering to the summer guests, often lower-class and female, seemed to constitute a permanent threat from the point of view of intelligence officers. Their desire to control, as well as ideas about the gendered foundations of French authority in the Arab world, clashed with self-conceptions of French *libertinage*, Lebanese conceptions of Mount Lebanon as the leisure centre of the Arab East, and the demands of citizens claiming their right to empire-wide mobility. Frictions around these competing notions were not unique to the field of tourism, but the example of the singing and dancing female artists demonstrates that the specific geographical properties of Mount Lebanon were at the basis of different imaginations that proved irreconcilable.

Between at least 1928 and 1938, the mobility of female musical, theatrical and variété artists caused an extensive correspondence between the French High Commission and police inspectors, the Foreign Office in Paris and French consuls around the Mediterranean. The authorities suspected the individuals, who came from various European countries – France, Germany, Austria, Czechoslovakia, Russia and others – of presenting themselves as artists, while actually intending to prostitute themselves on the territory of the Mandate.[197] Although prostitution was not illicit in the Mandate, the mobilities of these women were assumed to potentially cause trouble in the territory under mandate and therefore needed to be carefully monitored.

French authorities feared in particular the moral threat that European, notably French, prostitutes posed to the mandatory. In January 1928, at a time when hotel owners were hoping for a good *estivage* season after the end of the Syrian Revolt, Maugras, the Secretary General of the High Commissioner, alerted the Foreign Office. Maugras inferred that the presence of French women in dancing clubs and similar establishments created a bad impression among the local populations and undermined the "respectability of the French woman".[198] He argued that, ultimately, French prostitutes damaged the reputation of the mandatory power and that, inspired by the British model, the French should deny access to their female citizens aiming to work as artists in cafés, dancing clubs or similar establishments in the Mandates. While Maugras highlighted that such regulations did not imply adopting British prudery, he justified the measure as a matter of respect with regard to local customs.[199] Maugras' idea that in the local Arab society a female prostitute dishonoured the mandatory produced a contradiction identified by Camila Pastor, namely that the mandate administration reproduced a discourse that was considered no longer legitimate in the French metropole.[200] This created disputes – not only with metropolitan citizens, as we will see, but also with the Lebanese, whose customs Maugras claimed to respect.

Some months later, just before the start of the *estivage* season, the Minister of Foreign Affairs informed High Commissioner Ponsot that the suggestions would be put into effect and that visas must be denied to French female dancers, singers and other artists. As it was sometimes difficult to distinguish between artists and women "to whom the lyrical profession is not more than an accessory", he asked the police forces to execute controls with a certain vigour. "Dubious women", he concluded, had to be excluded if their behaviour was disadvantageous to the interests of France.[201]

For the owners of the Lebanese grand hotels, by contrast, concerts and artistic performances were an integral part of *estivage* holidaying and part of their business model.[202] Since the turn of the century, cultural life, particularly in Egypt, had profoundly influenced the entire region and names such as Asmahan, Umm Kulthum and Munira al-Mahdiyya resounded throughout the region and found their audience not least in the *estivage* resorts. Therefore, in May 1932, at the beginning of the *estivage* season, the brothers Nicolas and Elie Gebeili, owners of the *Grand Hôtel d'Aley*, sent a letter to the High Commissioner. Reporters of the paper *al-Maʿrid* had presented the *Grand Hôtel d'Aley* as one of the finest hotels in Lebanon, rivalling European and Egyptian grand hotels. They praised, among others, the splendid soirées its owners organised for their summer guests.[203] The Gebeilis complained that the attempts of the High Commission to more strictly regulate the entries of musicians on mandate territory threatened the basis of their business: after having invested five million francs into the construction of the hotel, they argued, the *estiveurs* increasingly abandoned the Lebanese *estivage* stations which they perceived as "big sanatoria"; that is, boring. According to the Gebeilis, this resulted from the French policy of refusing visas to many of the performing artists on the grounds that a sufficient number of artists already existed in the Mandate.[204]

From the perspective of the General Security, a major problem was that the potential performances of the women would take place in the *estivage* villages and thus at places that were considered difficult to survey. Bouchède, the Director of the General Security, reacted reluctantly. Other than in Beirut, Damascus or Aleppo, he objected, it was impossible to maintain a proper surveillance of artists in the mountain villages. To avoid clandestine prostitution and a "severe risk to public health", he refused to create a precedent with the Gebeili brothers.[205] Special permissions would be granted though to hotels which were able to prove that they would hire "morally impeccable" artists only. Otherwise, he feared that locals would gather in the resorts "to engage in all sorts of vices that were forbidden and strictly controlled in Beirut", and that the summer resorts would lose their bourgeois character.[206] Bouchède, as he saw it, was thus in charge of surveying a confined territory, and in particular mobilities across the Mount Lebanese hinterlands were met with suspicion. His argument that the hotel owners could find a sufficient pool of artists within the French Mandate not only ignored the trajectories characterising the biographies of many artists, but also the fact that for both *estiveurs* and the Lebanese entrepreneurs, an integrated *Bilad al-Sham* in terms of tourism remained an important frame of reference and a centre of exchange, leisure and sociability.

The tension between the French authorities and the owners of hotels and clubs persisted throughout the mandate, as numerous correspondences demonstrate. Regulations seem to have become all the more complex: the potential artists had to prove their artistic impeccability to obtain a visa, but also the respective establishment needed to inform the French officials about the projected event. Even internationally acclaimed stars experienced difficulties in entering the country and performing on stage. When Badiʿa Masabni, a belly dancer originally from Damascus who had achieved fame in Cairo during the 1920s, received an offer to appear on stage in Beirut in 1938 and applied for a visa, the French Consul in Egypt did not have any objections: Masabni was a well-known artist, and she had been performing multiple times with the same troupe of girls.[207] Yet when the news reached the General Security, they immediately intervened: they reasoned that as the General Security had not been notified beforehand, Masabni might enter the country as a tourist without being allowed to perform on stage, yet her troupe would not be able to join her.[208] In the end the case was solved thanks to the support of the French Consulate in Egypt, yet the example shows the tensions resulting from the very different approaches to mobilities among the French High Commission and the Lebanese entrepreneurs in tourism. While both nominally agreed on the necessity of supporting *estivage* practices, its concrete implications – the increasing mobility of women, also of potentially lower-class women, and of travellers to the hinterlands – were met with suspicion and sometimes impeded by the French mandatory.

The debate surrounding the meaning of imperial borders not only divided the mandate authorities and the Lebanese advocates of tourism development, however, but also caused tensions within the French Empire. For the French citizens to whom the authorities had refused entry, the intra-imperial barriers erected by the High Commission seemed at best incomprehensible, if not unacceptable. They expected French citizens to be in a position to profit from the empire, and

Bouchède's restrictive visa policy regarding female travellers was therefore harshly criticised in June 1931 by the secretary general of a syndicate of artists, the *Union artistique de France*. He asserted that while unemployment raged among French artists as a result of the economic crisis, foreigners performed in French overseas territories, whereas the required visas were denied to French citizens.[209] The artist Mathy Chandon, supported by the *Union artistique* and the French deputy Jean Fabry, put forward a similar argument in November 1938, when she requested a visa to perform at the *Parisiana* in Beirut.[210]

Whereas the High Commission repudiated the assumptions, insisting on the necessity to protect the French national prestige, Other mobilities of French citizens across the French Empire thus turned out to be a problem for the authorities.[211] Ideologies of civilisational superiority were increasingly difficult to sustain in a world in which allegedly inferior social classes had started circulating. While the artists and their advocates hoped to profit from the opportunities an imperial labour market offered, the imperial administrations tried to prevent politically suspect as well as morally dubious groups from harnessing these. Accessibility and inaccessibility, restricted and promoted mobilities between the metropole, the mandated territories and their neighbouring countries, had to be constantly renegotiated by a multiplicity of actors, while the empire was caught up in its inner contradictions.

Unaffected by conflicts over the legitimacy of Other mobilities, the Lebanese mountain resorts remained a congregating spot for a global bourgeoisie, who shared practices and codes of conduct regardless of their origins. These guests were not only encouraged by the Lebanese accommodation industry and the Lebanese Government, but also supported by the High Commission. The Arab bourgeoisie who had frequented the Lebanese *estivage* villages since the nineteenth century was now joined by staff members of the imperial administrations and European businesspeople living in the Arab East, as well as by Yishuvi immigrants from neighbouring Palestine.

The descriptions of the hotel amenities and programmes suggest that these visitors enjoyed similar leisure time activities and diversions, regardless of their national backgrounds. In 1938, the SET issued recommendations advising *estivage* villages as to how to attract visitors.[212] Although the authors of the SET claimed a lack of innovative distractions in most villages, in fact the mountain resorts were far less desolate than the SET's plan implied.[213] The *Grand Hôtel d'Aley* of the Gebeili brothers, for example, had an orchestra at its disposal, and offered amenities like a garden, tennis courts, a banqueting hall and a casino to its guests.[214] This standard was met by all the grand hotels in the famous resorts: the *Hôtel Kadri* in Zahlé, the *New Hôtel* and *Hôtel Hajjar* in Souk al-Gharb, the *Grand Hotel Casino* in Sawfar, *Hôtel Abchi* in Ehden, *Grand Hôtel Kassouf* in Dhour el-Choueir, the *Grand Hôtel Beit Méry*, and *Villa Khaouam* in Baalbek. *Hôtel Mon Repos*, next to the Cedars, even had a swimming pool.[215] In general, the *estivage* hotels were better equipped with regard to such facilities than the hotels attracting European tourists, most likely because tourists stayed for shorter periods than the *estiveurs* and would rather spend their days outside sightseeing. This explains at least why the *estivage* hotel *Villa Khaouam* in Baalbek offered a park, tennis courts and an orchestra, whereas the first choice of European tourists, the *Hotel Palmyra*, did not. Similarly, the hotels in Beirut were generally much less equipped with regard to leisure activities.[216]

Whereas Arab bourgeois families from Beirut and the neighbouring countries continued to be the main target group within the *estivage* business, entrepreneurs and interest groups launched various attempts at attracting new groups of potential summer guests. Already in 1927, the VML addressed businessmen working in Egypt, particularly the European and American communities of expatriates. Letters of recommendation written by European or American travellers sought to persuade novices of the *estivage* practice.[217]

Moreover, the idea of developing winter sports was revived in the mid-1930s. The SET promoted skiing as an activity that justified the reputation of Lebanon as the 'Switzerland of the Near East'. As a side effect, winter tourism extended the tourist season. One name was especially connected with the advertisement of winter sports in Lebanon: Philippe Bériel. Bériel, who had served in Tunisia before he joined the High Commission in Beirut in 1922, served as Director of the *Office de Protection de la propriété industrielle et commerciale en Syrie et au Liban* and stayed in this position until at least 1938.[218] From around 1935, he presided over the Section of the Levant of the *Club Alpin Français* (CAF), which numbered approximately 60 members in 1933, and more than 250 in 1937/38, according to him. Throughout the 1930s, Bériel, originally from Lyon, published articles on skiing in various newspapers and magazines, and contacted the French Legation in Egypt to advertise Mount Lebanon as a winter sports destination among the French expatriate community.[219]

His articles not only suggested that throughout the 1930s, skiing turned from a leisure activity of mainly French expatriates to a sport popular among Europeans and Arabs alike, but also that Bériel's perspective on the Mandate changed.[220] Bériel's first article, published in 1932, played with the presumed Orientalist expectations of his readers. Appealing to the curiosity of the reader, he stated that whereas most French associated the country with deserts, palm trees and camels, a mountain range of a height of 3,000 metres defined the Lebanese geography. He suggested in a paternalistic stance that, introduced by the CAF, skiing would allow the local population to better appreciate their own country: "[...] this initiative, meant to [...] demonstrate to the populations, too much used to indolence, the benefit they could, with a little effort, gain from the natural beauties of their country."[221]

Given such a cliché-packed style, it is astounding that in the second article published for the same audience in the very same magazine only three years later,

Bériel's rhetoric changed – and it would remain in his ensuing articles. Refraining from any Orientalist stereotypes, as well as from his condescending tone, he described scenes of Lebanese, Egyptian and Palestinian skiers enjoying winter sports together.[222] His articles, published in the magazine of the *French Alpine Club* as well as in the magazine of the Lebanese Ministry of Tourism, rather focused on giving practical advice and advertised the easy accessibility of the region as well as the modern amenities available in the Lebanese mountain hotels.[223]

Bériel's change of approach in his presentation of winter sports in Lebanon in 1935 might be attributable to the foundation of the SET or, more generally, the renewed Lebanese activity with regard to tourism development. Given Bériel's position and the fact that he contributed to the magazine edited by the Ministry of Tourism, he was certainly in contact with the *Services Economiques*, and was familiar with both its protagonists and their efforts to strengthen tourism. Such networks and the exchange of ideas among Lebanese interest groups seem to have influenced Bériel's presentation of the country. Skiing in Lebanon became – even in his publications targeting a French metropolitan audience – an activity that united an international bourgeoisie.

The efforts of Bériel, the *Club Alpin Français* and the SET met with success. Lebanon became a skiing destination for adherents from the entire region, but in particular Yishuvi immigrants from neighbouring Palestine contributed to the firm establishment of winter tourism. The members of the *Skiing Club of Tel Aviv* organised excursions to Lebanon each winter weekend. The club had been founded, as Bériel put it, "on the burning sands of Tel Aviv" by Jewish immigrants to Palestine. In northern Lebanon, in the region of the Cedars, its members held annual ski races. The use of German terms for racing categories ("Langlaufrennen", "Abfahrtslauf", "Slalomrennen") and the names of the victorious participants, including the chairman of the club, Dr Hermann Badt, indicate the German-speaking origins of the skiers.[224] In the hotels of the ski resorts, they met Egyptians, Lebanese and Syrians, as well as British and French members of the mandate administrations. The CAF and the *Palestine Ski Club*, which counted 200 members in 1938, actively participated in the popularisation of winter sports, and Hebrew was "heard in the dining rooms and echoing over the mountains".[225] Furthermore, since the mid-1930s, the *Palestine Post* regularly reported on the club's activities and its excursions to Lebanon and gave advice on locations and routes to the ski resorts.[226]

The members of the Yishuv were not only fervent adherents of winter sports; they also participated in the *estivage* practices of their Arab neighbours. Although neither the *estivage* brochures nor the Lebanese governmental statistics commented on this fact explicitly, using the neutral and all-encompassing term "*estiveurs* from Palestine", the advertisements of hotels suggested that the Jewish immigrants were an important target group for the *estivage* business. Their significance for the tourist industry was certainly increased further when the Yishuvi sector of the Palestinian economy proved much more stable during the mandate period than in the European or neighbouring Arab countries.[227]

Consequently, attracting Yishuvi *estiveurs* was vital to the Lebanese *estivage* business and contributed to its development in the 1930s, even after the outbreak of the Palestinian Revolt in 1936. In the SET's hotel guides from 1936 and 1937, a number of hoteliers indicated that they served kosher food and that languages such as German, Russian or Hebrew were spoken in their hotels, an obvious acknowledgement to their Jewish clientele.[228] The following table (table 2) is based on these guidebooks and summarises which hotels and pensions offered either kosher food or advertised language skills associated with the Palestinian immigrant community. In some cases, languages such as German and Hebrew were added to a general expression such as "foreign languages"; these other languages probably referred to foreign languages more commonly spoken in the region, such as French, Ottoman Turkish or Greek.

Table 2: Hotels and pensions targeting Yishuvi guests
Hotels (H) and pensions (P) advertising kosher food and/or Hebrew, German or Russian language skills

Place	Name	class	
ˤAley	Grand Hôtel d'Aley	H 1	kosher food, foreign languages [not specified]
	Hôtel Continental	H 3	kosher food, German and Hebrew spoken
	Hôtel Traboulsi	H 3	kosher food, foreign languages and German and Hebrew spoken
	Windsor Hotel	H 2	kosher food, foreign languages and Hebrew spoken
Bécharré	Pension du Liban Nord	P 2	foreign languages and Hebrew
Beitméry	Pension al-Raoudah	P 2	kosher food, French, Turkish, and Hebrew spoken
	Pension Ain-Saadé	P 3	kosher food
Bhamdoun Gare	New Pension	P 1	kosher food
Bhamdoun Village	Hôtel Belle Vue	H 3	kosher food, foreign languages
Deir el-Qamar	Hôtel Henoud	H 3	kosher food, foreign languages, German and Hebrew spoken
Dhour el-Choueir	Grand Hôtel Kassouf	H 1	kosher food, French, English, German spoken

Place	Name	class	
	Hôtel al-Raouda	H 3	kosher food, foreign languages, Hebrew, German spoken
	New Hotel Central	H 3	kosher food, Hebrew spoken
	Pension Khédiviale	P 2	kosher food, foreign languages
	Pension al-Ahram	P 2	foreign languages, Hebrew, German spoken
Jezzine	Pension d'Egypte	P 2	kosher food, foreign languages, German, Hebrew spoken
Ehden	Hôtel Abchi	H 1	kosher food, foreign languages [not specified]
Falougha	Park Hôtel	H 3	kosher food, foreign languages [not specified]
Kleyat	Pension Globus	P 3	kosher food if required, foreign languages
Meyrouba	Hôtel Aïn-Souan	H 3	kosher food, German and Hebrew spoken
Mekkine	Pension Achou	P 2	kosher food, foreign languages [not specified]
Mrouje	Pension Mrouje	P 3	kosher food, foreign languages, German spoken
Nabeh el-Safa	Pension el-Safa	P 3	kosher food, foreign languages [not specified]
Rayfoun	Hôtel Nabeh el-Assal	H 3	kosher food, foreign languages [not specified]
Sawfar	Hôtel d'Egypte	H 3	kosher food, foreign languages, German and Turkish spoken
Souk al-Gharb	Hôtel Hajjar	H 2	kosher food, French, English, German spoken
Zahlé	Hôtel Traboulsi	H 3	kosher food, foreign languages, German and Hebrew spoken
	Hôtel Kadri	H 1	kosher food, Hebrew, German, Russian spoken
	Pension Saadé	P 2	kosher food, Hebrew spoken

Hotels and pensions which added kosher food to their list of commodities in 1937.

Place	Name	Class	
Jezzine	Pension el-Ahram	P 1	kosher food
	Hôtel Kanaan	H 3	kosher food
Meyrouba	Pension des Meyrouba	P 2	kosher food

Due to their special dietary requirements, Jewish visitors to Lebanon were particularly visible in the advertisements; however, they should not be singled out from a large group of *estiveurs* comprising Arab and European nationals. Even though summer guests voiced complaints about the high cost of passports and visas, and about the difficulties they encountered when crossing the borders of the mandates, the number of *estiveurs* was relatively stable during the 1920s and grew from the mid-1930s onwards. An increasing number of new hotels and pensions reflected this growth, and owners invested in the comfort and commodities of the hotels. The new borders fracturing the formerly Ottoman *Bilad al-Sham* did not hinder the mobility of bourgeois *estiveurs* as they were still easily permeable, at least for them. While movements to Lebanon increased despite the newly created borders, mobility was subject to reservations. In times of crisis, borders restricted the movements of bourgeois travellers from neighbouring countries for political reasons; and at all times, potentially suspect travellers of lower classes were easily excluded. For them, the post-Ottoman spatial order materialised even before the highly mobile bourgeoisies experienced its effects.

Baalbek: Appropriations of a place

Tensions related to the question of accessibility not only concerned cross-border mobilities; they also resonated on a local level. Whereas in the case of mobilities, resistance was sparked by a denial of participation, opponents rejected tourist movements and their implications in some localities. Baalbek, the most famous touristic site in Lebanon, was one of these contested places. Its enormous Roman temples had been well known in Europe since the eighteenth century, and a number of nineteenth-century travel writers had popularised the site. German Emperor William II visited the ruins during his journey across the Ottoman Empire in 1898 (organised mainly by *Thomas Cook & Son*), and postcards and photographs show that already at the time, the ruins attracted a number of more ordinary European visitors.[229] During the mandate period, further improved infrastructures increased the accessibility for international tourists, a fact commented upon by various travellers.[230] In 1932, the Scottish tourist Mary Steele-Maitland and her husband arrived by car from Beirut after two and a half hours, her only concern now being the cold:

> We returned to the hotel and got into our car after about 2 ½ hours arrived at Baalbeck. Went to Palmyra Hotel – very cold at Baalbek an icy wind off Lebanon and a slight sprinkling of snow
> We were pleased to see a good stove but I was less pleased to feel the sheets damp and beds very cold. I tried to convince Arthur about this. He was far more interested in the ruins

seen from our window. However we did strip the beds and aired the heavy quilts removing the sheets entirely. With our own Jaeger blankets we were not too badly off.[231]

Her comment not only testifies to the facilitated conditions for accessing Baalbek, but also to new expectations of travellers and tourists with regard to comfort. Such expectations were not restricted to the travelling European aristocracy. All guidebooks, including the ones addressing *estiveurs*, advertised the availability of electricity and running water in the well-equipped hotels and summerhouses.[232] Investments in modern roads, running water, sewage systems and electricity were both a condition for tourism and its consequence. In the centres of *estivage* and tourism, the necessity of attracting visitors often justified spending money on such infrastructures. During a debate in the Lebanese Chamber of Deputies on additional credits for the budget of 1928, the governmental report argued that it was essential to repair the streets, sewage systems and lighting in "this historical town so much frequented by tourists".[233] An objection was raised by Yusuf Salim, scion of a merchant family from Sidon and Member of Parliament since 1925, who reminded the Chamber that it was the municipality, not the state, that was in charge of the aforementioned measures. He argued that tourists rather required further facilitation of access and advocated for the construction of a road which would lead directly to the ruins of Baalbek.[234]

This road, however, did not receive unanimous approval in the town itself. In January 1929, two groups of inhabitants petitioned the French High Commissioner in Beirut. The first group of residents explained that the municipality had planned the construction of a large road linking the ruins of Baalbek to Ra's al-ʿAyn, an idyllic picnic place next to a source much appreciated by *estiveurs*. In order to realise the project, the plaintiffs would be dispossessed of their houses, which had to be pulled down; the same fate had been decided for the Shiite mosque. Moreover, the petitioners accused the municipality of refusing the payment of indemnities and demanded that a French commission estimate the values of their houses, as they did not trust the municipality.[235]

This road project was not the only case in which the interests of residents conflicted with the embellishment of an *estivage* or tourist centre. In spring 1928, a debate over the improvement of Sawfar in the Chamber of Deputies had touched on the social consequences of the *estivage* development. The plan intended to turn the village into an *estivage* centre that would be able to compete with the most modern European resorts and attract "classy foreigners" (*[des] étrangers de marque*).[236] The proposed legislation comprised regulations with regard to construction works, stipulating that the height and façades of new buildings had to be approved by the municipality, and that each of the new houses should be surrounded by green spaces. Whereas its supporters put forward that the entire country would profit from such an embellishment, others, in particular Jamil Talhuq, originally from the *estivage* centre of ⸱Aley known for its luxurious hotels, fervently opposed the project. He argued that if the Chamber passed the law, Sawfar would become "an *estivage* centre for the rich of the earth excluding the prosperous classes and the poor". The lower and even the upper middle classes, he explained, could not afford to build houses corresponding to the regulations, and with the small plots they possessed, they could not afford to leave huge spaces for parks and gardens as required in the project.[237] Whether Talhuq opposed the project due to empathy with the poor or because he feared the competition of Sawfar for his own constituency in ⸱Aley is impossible to discern, but in both Talhuq's intervention and the petition of the inhabitants of Baalbek, the authors outlined a competition between a rich and highly mobile international class of tourists and the immobile local residents. To these residents, the growing numbers of international tourists and *estiveurs* posed a threat, rather than an opportunity.

However, not only the "rich of the earth" and the investors in tourism had an interest in tourism and *estivage*, as Jamil Talhuq suggested. In contrast to the Egyptian case, the Lebanese tourism sector was organised by mostly Lebanese entrepreneurs who had a vital interest in investments and measures destined for the attraction of national and international visitors. From their point of view, tourism was a chance for Lebanon that should not be missed. Such a group of inhabitants of Baalbek addressed another petition to the High Commissioner, expressing their support for the road construction project. One of the petitioners was Michel Alouf, who had already assisted the German archaeologists in the excavation of the ruins and became a member of the *Services des Antiquités* and of the *Institut Français d'Archéologie* under French rule. From 1931 onwards, he was officially in charge of the ruins of Baalbek, where he worked as a tourist guide and served as the local agent of the SET. Among the visitors he had an excellent reputation which he owed in particular to his book on the history of Baalbek. The book had been translated into numerous languages; the French version was published in its fifth edition in 1928.[238] His sons managed the famous *Hotel Palmyra*, which offered a direct view of the ruins and attracted numerous visitors such as the Steele-Maitlands. Moreover, it seems that another member of the family, Nicolas Selim Alouf, opened a cinema (*Empire*) in Baalbek, thus offering modern recreational activities to visitors and inhabitants of the town.[239] Other signatories of the petition were the brothers Khaouam, who owned hotels in various important cities such as Beirut, Damascus, Homs and Hama. In Baalbek they were in charge of *Villa Khaouam*, later *Hôtel de la Source* in Ra's al-⸱Ayn, a hotel frequented by *estiveurs* as opposed to European tourists.[240]

The existence of two camps with contradictory conceptions of the character of Baalbek manifested itself in the dispute over the road project. On one side the Baalbek inhabitants gathered for whom the advent of tourism was a potential.

They were looking forward to the enhancement of the town with electricity and water infrastructures, a good connection to the rest of the country and cultural infrastructures, and they hoped for economic gain. The adherents of the second camp did not expect profits from these investments; on the contrary, some even feared the expropriation of their homes and lands, or rising costs for houses and estates. To them, the change was a challenge or even a threat, rather than a promise.

In their petitions, both sides turned the conflict into a moral problem. Opponents of the road argued in terms of the local community: widows would be dispossessed of their properties, and the municipality apparently planned to tear down a holy place, the local Shiite mosque.[241] Proponents of the project, by contrast, created a universal framework for their argument. First, they rebuffed the argument of their opponents by denouncing their purely selfish (and thereby illegitimate) motivation. Second, they argued that Baalbek's exceptional historical heritage obliged its inhabitants to make the beauty of the site accessible to an outside world. Persons acting out of mere self-interest, they reasoned, should not be able to sabotage the development of Baalbek or the showcasing of the universal heritage of the ruins.[242]

While the supporters of the road obviously pursued their own interests just as the opponents did, their reference to the idea of a universal heritage is significant. We have seen in the case of Palmyra that the accessibility and the alleged protection of universal heritage justified imperial intervention, often in a paternalist stance, as the interests of the local population were considered subordinate to the interests of a global community of visitors. Also in Baalbek, from a French imperial point of view, the ruins testified to the presence of the Roman Empire and thus were interpreted as a monument of imperial grandeur, as we have seen in Damien de Martel's foreword to the SET's tourist brochure (cf. p. 266-267). Policies of heritage preservation demonstrated the fulfilment of the French educational mission in the mandates.[243]

Alouf and the other advocates of Baalbek's tourism development, however, put forward similar arguments of the universal meaning of the site, thereby rejecting both the demands of the opponents of the road project and claims of the mandatory that the protection of heritage required the imperial presence.[244] As for the Lebanese deputies, the acropolis of Baalbek served as a symbol identifying the Lebanese nation-state.[245] The ruins played a significant role anytime the country was presented to an international audience, for example at the colonial exposition in Paris in 1931. While a number of deputies raised doubts about participating in the exposition at all, because they refused to label Lebanon as a colony, a majority voted in favour of the project. Its supporters argued that the exposition was an occasion to "provide proof in France of our political and economic existence".[246] To deliver such proof, the Minister of Education, Gebran Tuéni, decided that among other objects, a model of the temples of Baalbek would be presented to the French audience.[247] Tuéni stressed the significance of the object: "the moral gain is priceless.

[The artist who had shaped the model] has once again proven that the Lebanese can reach the peak of all intellectual and artistic expression".[248] Tuéni expected that greater international awareness about the Lebanese cultural heritage would enhance the prestige of Lebanon and the reputation of the Lebanese people among an international audience.[249] The accessibility of the heritage was therefore vital to the national interest.

The group of petitioners around Michel Alouf could hope for the backing of both the High Commission and the Lebanese Government. The local conflict between a number of house owners and the Shiite community on the one side and the Alouf circle on the other entered into a larger political debate. A global, an imperial and a national frame of reference interfered with the local question as to whether houses should give way to a bypass road. Baalbek appeared on an imagined map of universal heritage sites, which implied the right of an international public to visit it. Indeed, the governing institutions gave this right precedence over that of individual local inhabitants to stay in their homes. Both the imperial power and the national government backed this interest, hoping that the symbol of Baalbek would demonstrate the legitimacy of their claims to the territory. The integration of the place into a new imperial and simultaneously into a new national space after 1920 thus created opportunities for some, who backed and contributed to the process. Unlike the groups opposing the national integration of the coastal towns, the petitioners in Baalbek resisting the process were not motivated by other allegiances. They perceived the conflict to be a local issue of coexistence and responsibility and failed to grasp its national and international dimensions – interests to which their aims were finally subordinated.

Conclusion

Although the new borders introduced on paper after World War I separated a tiny Greater Lebanon from the other parts of *Bilad al-Sham*, movements of tourism and *estivage* to Lebanon increased during the 1920s and 1930s. The closer integration of Lebanon into imperial transportation circuits (both British and French), facilitated the journeys of European tourists and reduced the costs, while the members of the European administrative and military staff in the region created a new reservoir of potential tourists. Other than in Egypt, tourism also enjoyed the strong support of a large part of local society, as the tourism business was mainly organised by Lebanese entrepreneurs and a major economic factor in many regions. In this regard, the Arab *estivage* movements generated higher revenues than European tourism, and administrative hindrances to the trade provoked protests among the population. Difficulties in the *estivage* business, as after the economic depressions affecting Egypt, were partly compensated by tourists from the Yishuv from the 1930s onwards.

From the perspective of the French High Commission, tourism and *estivage* were ambiguous phenomena. As in Syria, the mandate administration backed the movement in principle, notably because of its economic potential, which promised to ease the financial burden of administering the newly acquired territories while it did not menace the French industry. On the other hand, the General Security specifically cautioned against the threats which mobilities posed to the authority of the mandate power. Actual or supposed prostitutes travelling to the Mandates were suspected of spreading not only diseases, but, especially in the case of French women, bore the potential of undermining the reputation of the mandatory. Similar to the mobility of politically suspect individuals in Syria, mobilities of lower-class female travellers required strict control from the point of view of the authorities. Protests against such policies were raised by hotel owners whose business model was partly based on the staging of singers and dancers, but also by French performers who claimed access to the potential professional opportunities of the Empire.

At a national level, in particular those parts of the population who did not profit from the mobility of the regional and transnational bourgeoisies were opposed to further investments in tourism and *estivage*. Some of them, like the residents of Baalbek, feared that the integration of their town into national, transnational and imperial circuits of mobilities and exchange would put them in a disadvantageous position or threaten their way of life and their possessions. Others, for example in the coastal towns, rejected the economic liberalist Beiruti model based on trade and tourism, as they rather imagined a closer economic integration with Syria. However, as both the economic liberalists among the Lebanese nationalists and the French imperialists shared and fostered a Mediterranean outlook, pursuing the same goals for different reasons, it became difficult for pan-Syrian opponents to find allies – both with regard to tourism policies and beyond.

The most committed actors pursuing tourism development in Lebanon belonged to an economically liberal, and politically Lebanese nationalist bourgeoisie. From the outset, they perceived tourism as a resource for implementing their vision of a Lebanese state, even though the focus of their political struggle shifted: whereas in the 1920s they focused on shaping a unique imaginary of Lebanon in contrast to Syria, in the politically heated atmosphere of the mid- and late 1930s the supporters of tourism claimed their independence from the French mandate power and advanced internal coherence, while trying to guard their business model by forging ties of tourism with all surrounding countries.

In contrast to neighbouring Syria and Arab Palestine, the Lebanese middle classes were in a stronger political position. Although the mandate framework applied to Lebanon as well, the French authorities created legislative institutions and government offices that allowed the Lebanese to participate in political decision-making. As in Syria, French advisors existed, yet it appears that in the field of tourism development the Lebanese actors enjoyed a larger room for manoeuvre than their neighbours did.

This allowed deputies and governmental institutions to make use of tourism as a resource for integrating the Lebanese nation-state over time. First, in brochures and information sheets, the mainly Beiruti middle classes envisioned a modern Lebanon, revolving around an allegedly Phoenician economic liberalism, the modernity of Mediterranean Beirut and the landscapes of Mount Lebanon. Such visions shaped the promotion of the country not only at a national level, but also among a regional and an international audience. Similar to Egypt, narratives of modernity were put forward to argue in favour of Lebanese sovereignty. Although the Lebanese middle classes recurred to European vocabularies of nation-building, such as the Phoenician heritage on display in the National Museum, or the assumption of national tradi-tions, they displayed a rather pragmatic use of such narratives. Although widely advertised in tourism propaganda, the Phoenician heritage, for example, remained a somewhat theoretical framework: the core sites of the Lebanese Phoenician past were not developed as tourist destinations before the mid-1930s.

Second, the Lebanese middle classes were perfectly aware of the fact that the national narratives were of relevance not merely in ideological or symbolic debates, but also had serious material repercussions. This potential, of which also Egyptian, Yishuvi and French actors in Syria made use of, was used as a resource for territorially integrating Lebanon in particular during the 1930s.

Tourism development justified the creation of material structures supporting the nation-state: roads connecting the Lebanese 'heartland', Beirut and the Mount Lebanon area, while remote sites such as the temples of Baalbek and the Cedars integrated the periphery. Moreover, it helped to win over a part of the population by offering financial reward and instilling a narrative. Those who benefited materially were more likely to identify with the new state: in this regard, it is significant that the major resistance in Baalbek against mandate rule and in favour of Syrian unity occurred in 1926 and, less violently, in 1929 in the context of the road project. Later on, after increased accessibility had been achieved and tourism began to flourish, the protests dissipated and they did not regain momentum in 1936, although claims for Syrian unity were raised again along the Lebanese coastline. In the centres of protest, particularly Sidon and Tyre, the integration into the tourist space had not yet occurred by 1936, and the overall identification with the nation-state remained unstable.

Thus, in Lebanon, as in Egypt or Yishuvi Palestine, advocates of tourism devel-opment focused on the integration of the national territory in the 1930s. Unlike in these cases, however, it was not domestic tourism that helped create infrastructures and bonds between the centres and the former peripheries, but rather the *estivage* movement, attracting summer guests from the neighbouring countries. Again, the approach of the Lebanese tourism sector seemed pragmatic rather than ideological:

in contrast to Egypt, there was no Orientalising of outside Others – after all, outside Lebanon resided the potential clientele. This included the Jewish immigrants to Palestine, of major importance as the rather positive economic development of the Yishuv during the 1930s helped the Lebanese tourism sector to quickly overcome the effects of the Depression.

In terms of both inside and outside relations, tourism in Lebanon was a vital resource in creating a bourgeois nation-state, as it enabled the processes of claiming, defending and establishing access over territory. By firmly anchoring Lebanon in its regional context, the middle classes also secured its political and economic existence: as the *estivage* centre of both its Yishuvi and its Arab neighbours. While none of the advocates of tourism development anticipated the rupture that would separate Mount Lebanese hotel owners from their Yishuvi customers after 1948, ambitions to use tourism as a resource for maintaining ties with both Europe, the Americas, and the Arab world persisted after World War II.[250]

Notes

[1] al-Khoury, 'L'Industrie hôtelière au Liban', p. 37.

[2] Hakim, *The Origins of the Lebanese National Idea*, pp. 262–264. Hanssen, *Fin de Siècle Beirut*, pp. 55–56. Hourani, *Arabic Thought in the Liberal Age*, pp. 274–279, 285–291 et pass. Kaufman, *Reviving Phoenicia*, p. 110.

[3] Jürgen Osterhammel pointed out that the notion of a 'public sphere' implies a process, as it needs to be created and maintained: Osterhammel, 'Die Weltöffentlichkeit im 20. Jahrhundert', p. 69.

[4] Barakat-Buccianti, 'Beyrouth sous le Mandat français', p. 75.

[5] FRENCH LINES, Compagnie des Messageries Maritimes, *Agence de Beyrouth, Exercice 1925*, Secrétariat, p. 17.

[6] FRENCH LINES, Compagnie des Messageries Maritimes, *Agence de Beyrouth, Exercice 1928*, Secrétariat.

[7] In contrast, Carolyn Gates estimates the number of travellers at 216,000 in 1952 and at more than 544,000 in 1957: Gates, 'The Historical Role of Political Economy', p. 16. For 1937, Gates estimates their number at "less than 30,000". Travellers from the neighbouring countries who did not stay more than ten days in Lebanon, were not counted in the official statistics: FRENCH LINES, Compagnie des Messageries Maritimes, *Agence de Beyrouth, Exercice 1934*, Trafic, p. 5. FRENCH LINES, Compagnie des Messageries Maritimes, *Agence de Beyrouth, Exercice 1935*, Secrétariat, pp. 2–3.

[8] AUB, *Séance du Conseil Représentatif, 27/10/1925*, p. 4.

[9] FRENCH LINES, Compagnie des Messageries Maritimes, *Agence de Beyrouth, Exercice 1927*, Secrétariat. FRENCH LINES, Compagnie des Messageries Maritimes, *Agence de Beyrouth, Exercice 1930*, Trafic, p. 6. FRENCH LINES, Compagnie des Messageries Maritimes, *Agence de Beyrouth, Exercice 1932*, Secrétariat, pp. 3–5.

[10] FRENCH LINES, Compagnie des Messageries Maritimes, *Agence de Beyrouth, Exercice 1932*, Secrétariat, pp. 3–5. Veccia Vaglieri, 'La villeggiatura nel Libano'. Veccia Vaglieri, 'Decadenza della villaggiatura estiva'. N.N., 'Aspetti della crisi', p. 523.

[11] FRENCH LINES, Compagnie des Messageries Maritimes, *Agence de Beyrouth, Exercice 1935*, Secrétariat, pp. 2–3, Trafic, pp. 5–6.

12 FRENCH LINES, Compagnie des Messageries Maritimes, *Agence de Beyrouth, Exercice 1937*, Secrétariat, p. 6; Trafic, pp. 4–5. FRENCH LINES, Compagnie des Messageries Maritimes, *Agence de Beyrouth, Exercice 1938*, Secrétariat, p. 5. FRENCH LINES, Compagnie des Messageries Maritimes, *Agence de Beyrouth, Exercice 1940*, Secrétariat, pp. 2–3; Trafic, p. 5. Veccia Vaglieri, 'Decadenza della villeggiatura nel Libano'.

13 FRENCH LINES, Compagnie des Messageries Maritimes, Agence de Beyrouth, Rapports d'Agence, Exercices 1927–1940.

14 BO, Société de Villégiature au Mont Liban, *Guide de la Société de Villégiature, 1927*, p. 5. AUB, *Séance du Conseil Représentatif, 01/05/1923*, pp. 5–6.

15 AUB, *Séance du Conseil Représentatif, 01/05/1923*, pp. 5–6. LEIDEN UNIVERSITY LIBRARY, Société de Villégiature au Mont Liban, *Guide de la Société de Villégiature, 1925*, title page. I am grateful to Janina Santer, who spotted this brochure and shared it with me. BO, Société de Villégiature au Mont Liban, *Guide de la Société de Villégiature, 1927*, title page. Janina Santer has provided a detailed analysis of the VML's members and activities in her M.A. thesis: Santer, 'Imagining Lebanon', pp. 44–46.

16 UMAM FOUNDATION, *al-Istiyaf fi Lubnan*. The Umam Foundation estimates the date of publication to be 1926. As it was printed during the presidency of Charles Debbas, the date range is 01/09/1926–02/01/1934.

17 The Russian origin of Nicholas Strekalovsky is mentioned in the description of a box of watercolor paintings from the 1930s, archived at the Smithsonian Institution (Collection ID: SIA.FARU7468). Strekalovsky also authored two advertisement posters for *Misr* airlines, dated 1934 and 1935 (cf. https://www.artothek.de/de/bilder-fotos/strekalovsky.html, last accessed on May 30, 2022). The anti-urban romanticism as well as the focus on the mountain landscape are significant elements in the work of Gibran Khalil Gibran (1882–1931), considered a preeminent representative of Lebanese poetry. Jens Hanssen identified this 'mountain romanticism' as a key element of the post-war 'neo-phoenicianist' nationalism: Hanssen, *Fin de Siècle Beirut*, pp. 231–232.

18 HARVARD UNIVERSITY LIBRARY, Société de Villégiature au Mont Liban, *The Tourist's Book, 1924*, pp. 47–48.

19 Cf. HARVARD UNIVERSITY LIBRARY, Société de Villégiature au Mont Liban, *The Tourist's Book, 1924*, pp. 4–6. FDC, Services économiques, *Le tourisme en Syrie, 1928*, pp. 4–5.

20 Similar in: FDC, Services économiques, *Le tourisme en Syrie, 1928*, pp. 4–7.

21 Alexis Wick convincingly argued that Mediterraneanism served its proponents as an argument to de-orientalise themselves: Wick, 'History, Geography, and the Sea', p. 745.

22 BNF RICHELIEU, Maklouf and Société d'encouragement au tourisme au Liban, *Le Liban: Pays d'art*. The SET received subsidies from Syria as well: AUB, *Décret no. 447, 15/06/1936*.

23 For the role of Baalbek in Lebanese-Phoenicianist nationalism see Kaufman, *Reviving Phoenicia*, pp. 148–151.

24 BO, Gobril, *Visitez le Levant*. BO, Zehil, *Petit Guide historique de la Syrie, 1929*. BNF RICHELIEU, Samné and Bureau syrien et libanais d'informations et de tourisme, *Syrie et Liban, 1932*. BO, Commission du Tourisme des Etats sous Mandat, *Le tourisme en Syrie*. FDC, Commission du tourisme et de villégiature du gouvernement libanais, *Au pays des cèdres*. BNF RICHELIEU, *Syrie – Liban, 1935*. BO, Services économiques du gouvernement libanais, *Le Liban: Pays de tourisme*. BO, Société d'encouragement au tourisme au Liban, *Beyrouth*. BNF RICHELIEU, Maklouf and Société d'encouragement au tourisme au Liban, *Le Liban: Pays d'art*.

25 BO, Services économiques du gouvernement libanais, *Le Liban: Pays de tourisme*.

26 The French spelling "Béryte" is derived from the Phoenician "Be'erot", called "Berytos" by Greeks and the Roman conquerors. Kaufman, *Reviving Phoenicia*, p. 240. Elisséeff, 'Bayrūt'.

27 Hanssen, *Fin de Siècle Beirut*, p. 46. Kaufman, *Reviving Phoenicia*, p. 169. Hobsbawm, 'Introduction', pp. 7–8.

28 Monmarché, *Syrie – Palestine – Iraq – Transjordanie*, pp. XLVII, LIV-LVIII. Thomas Cook & Son, Lumby and Garstang, *Cook's Traveller's Handbook to Palestine, Syria & Iraq, 1934*, pp. 23–35.

29 Jansen, *Erobern und Erinnern*, pp. 285–286. Jansen, 'Die Erfindung des Mittelmeerraums'. Furlough, 'Une leçon des choses', p. 455. A similar attempt was undertaken by Mussolini's imperial propaganda machinery, similarly claiming the succession of the Roman Empire and thus openly competing with the French in the Levant, however, remaining more ambiguous with regard to his approach to the local Muslim populations. Dueck, *Claims of Culture*, pp. 119–120.

30 Kaufman, 'Phoenicianism'. Hakim, *The Origins of the Lebanese National Idea*, pp. 219–222.

31 Hartman and Olsaretti, 'The First Boat'.

32 Ippolito, 'Naissance d'une nation'. Firro, *Inventing Lebanon*, p. 24.

33 In the four issues of the *Revue* that appeared in 1919, two articles dealt with tourism and *estivage*. al-Khoury, 'L'Industrie hôtelière au Liban'. Naccache, 'L'industrie de la villégiature'. Moreover, the guide by the *Services Economiques* quoted verses of Corm's nationalist poem *La montagne inspirée*: BO, Services économiques du gouvernement libanais, *Le Liban: Pays de tourisme*, p. 16.

34 Charles Corm, Terre de souvenirs, in: BNF RICHELIEU, Maklouf and Société d'encouragement au tourisme au Liban, *Le Liban: Pays d'art*.

35 BO, Gobril, *Visitez le Levant*, pp. 53–55. Similar arguments were put forward by Jacques Tabet: Firro, *Inventing Lebanon*, pp. 26–28.

36 'Frankish' in Arabic also used to be a common designation for European foreigners. BO, Gobril, *Visitez le Levant*, pp. 18–20.

37 BO, Gobril, *Visitez le Levant*, p. 13.

38 Besides Corm, there was a second exception to this tendency: Georges Samné, a francophone author, who lived in Paris, where he edited the journal *Correspondance d'Orient*. The co-founder and director of the *Bureau syrien et libanais d'informations et du tourisme* had been an long-standing adherent to *"une Syrie française"*, that is, a united Greater Syria governed by the French colonial power, which was hoped to guarantee both security and influence of the Lebanese Maronite population. In the guide he wrote for the *Bureau syrien*, Samné stressed the beneficial role of the French Mandate with regard to economic and political development. BNF RICHELIEU, Samné and Bureau syrien et libanais d'informations et de tourisme, *Syrie et Liban, 1932*. On Georges Samné cf. Firro, *Inventing Lebanon*, pp. 18–19, 24–25. Kaufman, *Reviving Phoenicia*, pp. 79–84.

39 Traboulsi, *A History of Modern Lebanon*, pp. 96–100.

40 FDC, Commission du tourisme et de villégiature du gouvernement libanais, *Au pays des cèdres*. On the cedar as a symbol of Lebanese separatism cf. Kaufman, *Reviving Phoenicia*, p. 236.

41 According to the authors, the cedar had become the national symbol because it represented the both ancient and young Lebanese nation: FDC, Commission du tourisme et de villégiature du gouvernement libanais, *Au pays des cèdres*, p. 19.

42 Makdisi, 'The "Rediscovery" of Baalbek', pp. 140–142, 153–156.

43 Damien de Martel, "L'éternelle Méditerranée", in BNF RICHELIEU, Maklouf and Société d'encourage-ment au tourisme au Liban, *Le Liban: Pays d'art*.

44 René Dussaud, "Baalbeck", in BNF RICHELIEU, Maklouf and Société d'encouragement au tourisme au Liban, *Le Liban: Pays d'art*.

45 Le Gras, *Le Lac Méditerranéen*, p. 138. Makdisi, 'The "Rediscovery" of Baalbek', pp. 154–155.

46 Firro, *Inventing Lebanon*, pp. 97–98. Traboulsi, *A History of Modern Lebanon*, p. 93. Hanssen has demonstrated for Ottoman Beirut that the institutional participation was perceived as a means of socio-political ascendency among the local notables. Hanssen, *Fin de Siècle Beirut*, pp. 161–162.

47   In 1917, Albert Naccache published an analysis on the economic potential of Mount Lebanon. At the time, he hoped for a revival of the silk industry, benefitting from hydropower plants, but perceived the *estivage* business as the second major pillar of its economy: Naccache, 'Aperçu sur la situation économique', pp. 152–153. Cf. also al-Khoury, 'L'Industrie hôtelière au Liban'.

48   Khoury, *Syria and the French Mandate*, pp. 247–248. Dakhli, 'L'expertise en terrain colonial'. On the French bureaucratic infrastructure in contrast to that of the Ottoman rulers cf. Thompson, *Colonial Citizens*, pp. 58–59. el-Solh, *Lebanon and Arabism*, pp. 2–3.

49   Traboulsi, *A History of Modern Lebanon*, p. 88.

50   Salibi, *The Modern History of Lebanon*, p. 165. Firro, *Inventing Lebanon*, pp. 83–84.

51   On the notion of "localization of power" cf. Hanssen, *Fin de Siècle Beirut*, p. 9.

52   Traboulsi, *A History of Modern Lebanon*, pp. 89–90. Salibi, *The Modern History of Lebanon*, pp. 165–168.

53   Salibi, *The Modern History of Lebanon*, pp. 167–168. AUB, *Constitution Libanaise*, p. 1. Titre II, Art. 18.

54   Dakhli, 'L'expertise en terrain colonial', p. 23.

55   Dakhli, 'L'expertise en terrain colonial', pp. 25–27.

56   AUB, Gouraud, *Discours du 1er Septembre 1922*, p. 5.

57   ANL, Gouraud, *Arrêté No. 885 du 05/06/1921*.

58   AUB, de Sercey, *Lettre du Haut Commissaire*, 05/1923, p. 4.

59   AUB, *Décision No. 1667*, 23/05/1923, p. 3. On imperial fears of mobilities and contagion cf. Huber, *Channelling Mobilities*, pp. 241–271. Slight, *The British Empire and the Hajj*, pp. 78–81. Hanssen, *Fin de Siècle Beirut*, pp. 125–127, 130–132. A legislative project on the taxation of gambling led to an extensive debate in the Representative Council, opposing supporters arguing in favour of the *estivage* industry and opponents rejecting any permission of gambling for moral reasons: AUB, *Séance du Conseil Représentatif du 07/12/1922*, p. 4. AUB, *Séance du Conseil Représentatif du 07/12/1923*, p. 4. AUB, *Séance du Conseil Représentatif du 11/12/1923*, pp. 2–3. AUB, *Séance du Conseil Représentatif du 24/04/1923*, pp. 3–4.

60   Traboulsi, *A History of Modern Lebanon*, pp. 96–98. Khoury, *Syria and the French Mandate*, pp. 85–86.

61   FRENCH LINES, Compagnie des Messageries Maritimes, *Agence de Beyrouth, Exercice 1921*. FRENCH LINES, Compagnie des Messageries Maritimes, *Agence de Beyrouth, Exercice 1922*.

62   FRENCH LINES, Compagnie des Messageries Maritimes, *Agence de Beyrouth, Exercice 1923*.

63   Demay, *Tourism and Colonization in Indochina*, pp. 87–91. Furlough, 'Une leçon des choses'.

64   This impression was shared by contemporary observer Frédéric Martin, agent of the *Messageries Maritimes*: FRENCH LINES, Compagnie des Messageries Maritimes, *Agence de Beyrouth, Exercice 1924*, Secrétariat, p. 4.

65   AUB, *Arrêté No. 3041*, 17/03/1925, p. 2.

66   FRENCH LINES, Compagnie des Messageries Maritimes, *Agence de Beyrouth, Exercice 1924*, pp. 3–4.

67   FRENCH LINES, Compagnie des Messageries Maritimes, *Agence de Beyrouth, Exercice 1924*, pp. 3–4. ANL, *Arrêté No 3042*, 17/03/1925.

68   FRENCH LINES, Philippar, *Le tourisme en Syrie*, 13/02/1929, pp. 56–57.

69   ANL, *Arrêté No. 3013*, 04/03/1925.

70   FRENCH LINES, Compagnie des Messageries Maritimes, *Agence de Beyrouth, Exercice 1925*, Secrétariat, pp. 8–9.

71   AUB, *Conseil Représentatif du 04/11/1924, Journal officiel No. 1821*, p. 4. AUB, *Conseil Représentatif du 04/11/1924, Journal officiel No. 1822*, p. 3.

72   AUB, Cayla, *Discours*, 09/1925, p. 5. FRENCH LINES, Compagnie des Messageries Maritimes, *Agence de Beyrouth, Exercice 1924*, p. 4.

73 AUB, *Séance du Conseil Représentatif*, 31/10/1922, pp. 5–7. AUB, *Séance du Conseil Représentatif*, 30/04/1923, pp. 3–4.

74 Santer, 'Imagining Lebanon', pp. 44–46. Janina Santer also identified the directors of the company – among them Haidar Maalouf who had served as secretary of the *Alliance Libanaise* in Cairo. AUB, *Séance du Conseil Représentatif*, 01/05/1923, pp. 5–6.

75 al-Khoury, 'L'Industrie hôtelière au Liban'. Naccache, 'L'industrie de la villégiature'.

76 AUB, *Arrêté No. 2537*, 26/06/1924, p. 3. AUB, *Arrêté No. 2538*, 26/06/1924, p. 3. AUB, *Arrêté No. 2599*, 02/08/1924, p. 2.

77 ANL, *Marsum Raqm 955, Majmuʿat Qawanin*. ANL, *Marsum Raqm 1955, Majmuʿat Qawanin*. ANL, *Marsum Raqm 2203, Majmuʿat Qawanin*. ANL, *Marsum Raqm 2515, Majmuʿat Qawanin*. ANL, *Marsum Raqm 2713, Majmuʿat Qawanin*. ANL, *Marsum Raqm 2912, Majmuʿat Qawanin*. ANL, *Marsum Raqm 3494, Majmuʿat Qawanin*.

78 For example in Dhour el-Choueir: AUB, *Séance du Conseil Représentatif*, 23/03/1926, pp. 19–20. On the necessity of developing the road network: Naccache, 'L'industrie de la villégiature', pp. 210–212. Kristin Monroe pointed out the importance of the automobile (and therefore, the necessity to improve the road network) to tourism traffic in Lebanon already during the mandate period: Monroe, *The Insecure City*, p. 30, footnote 39.

79 E.g. AUB, *Séance du Conseil Représentatif*, 21/11/1925, pp. 3–4.

80 Hanssen, *Fin de Siècle Beirut*, p. 61.

81 AUB, *Séance du Conseil Représentatif*, 18/11/1922, p. 4. AUB, *Séance du Conseil Représentatif*, 18/11/1922, p. 4. AUB, *Conseil Représentatif*, 20/03/1923, p. 4. AUB, *Conseil Représentatif*, 26/03/1923. AUB, *Conseil Représentatif*, 26/03/1923. BNF GALLICA, Montet, *L'Art phénicien au XVIIIe*, p. 1.

82 Shehadi, 'The Idea of Lebanon', pp. 8–11. Hartman and Olsaretti, 'The First Boat'. Salibi, *The Modern History of Lebanon*, p. 167. Traboulsi, *A History of Modern Lebanon*, pp. 93–96.

83 AUB, *Séance du Conseil Représentatif*, 03/11/1925, p. 4.

84 Naccache, 'L'industrie de la villégiature', pp. 211–213.

85 E.g. AUB, *Séance de la Chambre des Députés*, 18/04/1930, pp. 5–6. AUB, *Séance de la Chambre des Députés*, 04/12/1930, p. 44.

86 Traboulsi, *A History of Modern Lebanon*, p. 93. Hartman and Olsaretti, 'The First Boat', p. 56. Shehadi, 'The Idea of Lebanon'.

87 In 1919, Fouad al-Khoury pointed out the enormous benefits tourism had brought to the previously poor country, similar to Switzerland: al-Khoury, 'L'Industrie hôtelière au Liban', p. 39. In contrast, in 1931 the Lebanese Government's *Commission du tourisme* distanced themselves clearly from the comparison: FDC, Commission du tourisme et de villégiature du gouvernement Libanais, *Au pays des cèdres*, pp. 1–3.

88 Ayalon, *The Press in the Arab Middle East*, pp. 74–75.

89 As noted as well by Veccia Vaglieri: Veccia Vaglieri, 'Dichiarazioni politiche, 07/1927', p. 598.

90 Hanssen, *Fin de Siècle Beirut*, p. 214. Cf. for example AUB, *Mada fi Baalbek*. AUB, *Harak al-Istiyaf*. With regard to Palestinian tourists, Andrea Stanton has argued that the self-conception of Lebanon was a relational one: Stanton, 'Locating Palestine's Summer Residence'. This is convincing, and characterises Lebanon's self-positioning with regard to other countries from the region as well. However, I would not consider it a mutual relationality as Palestine was able to draw on a larger reservoir of international tourists (and pilgrims) as well as the domestic Yishuvi tourism sector, as we have seen in Chapter 4, and the groups were attracted by different imaginaries.

91 Traboulsi, *A History of Modern Lebanon*, pp. 80–82.

92 ANL, *Shuhada Mustaf Misri*, 17/06/1926, p. 6.

93 ANL, *al-Jiniral Falih [?] fi Baalbek*, 23/05/1926.

94   CADN, *Lettre du Secrétariat général*, 11/03/1927.

95   ANL, *al-Sihafiyun fi Masayif al-Shamal*. ANL, *Wafd Niqabat al-Sihafa fi Shamal*.

96   In 1934, the newspaper had dedicated a whole issue to Corm's Lebanese nationalist poem *La montagne inspirée*: Kaufman, "'Tell Us Our History'", pp. 18–19.

97   Traboulsi, *A History of Modern Lebanon*, p. 91.

98   Khoury, *Syria and the French Mandate*, pp. 247–248. Dakhli, 'L'expertise en terrain colonial', pp. 20, 25–26. World Biographical Information System Online, 's.v. Berthelot, Philippe'.

99   CADN, *Le tourisme au Liban*, 03/03/1928.

100   He sat for example on the Administrative Council of the *Compagnie Française du Tourisme* founded in September 1919 to foster tourism to France: CADN, Compagnie Française du Tourisme, *Note sur l'établissement d'un Office de Tourisme*. 's.v. Philippar, Georges'.

101   FRENCH LINES, Philippar, *Le tourisme en Syrie*, 13/02/1929. CADN, *Le tourisme en Syrie*, 01/03/1929.

102   FRENCH LINES, Philippar, *Le tourisme en Syrie*, 13/02/1929, pp. 5–6.

103   CADN, *Le tourisme en Syrie*, 01/03/1929. On similar attempts of linking French and Lebanese history cf. Kaufman, *Reviving Phoenicia*, pp. 85–86.

104   Ageron, 'Le "parti" colonial'. Brunschwig, 'Le parti colonial français'.

105   BNF GALLICA, *Syrie-Liban*, 07/1935. BNF GALLICA, *Le tourisme au Liban*, 02/1935. BNF GALLICA, Marchand, *Lamartine, chantre des Cèdres*. BNF GALLICA, Reizler, *Syrie et Liban – Porte de l'Orient, 1933*.

106   BNF GALLICA, *Comité de patronage*. BNF GALLICA, *Le Mois colonial*, 07/1929.

107   Furlough, 'Une leçon des choses', p. 451.

108   Zytnicki, *L' Algérie, terre de tourisme*, pp. 120–121. FRENCH LINES, Compagnie des Messageries Maritimes, *Agence de Beyrouth, Exercice 1928*, Secrétariat.

109   Zytnicki, *L' Algérie, terre de tourisme*, p. 135.

110   Zytnicki, *L' Algérie, terre de tourisme*, p. 9.

111   Menassa's Office for Economic Affairs had been assigned to the Ministry of Finance in October: AUB, *Decret No. 5742, 12/10/1929*, p. 2.

112   BO, Zehil, *Petit Guide historique de la Syrie*, 1929. FRENCH LINES, Compagnie des Messageries Maritimes, *Agence de Beyrouth, Exercice 1923*. FRENCH LINES, Compagnie des Messageries Maritimes, *Agence de Beyrouth, Exercice 1924*, pp. 4–5. FRENCH LINES, Compagnie des Messageries Maritimes, *Agence de Beyrouth, Exercice 1926*, Secrétariat, p. 3. ANL, *Décret No. 8330, 06/08/1931*.

113   BNF GALLICA, *Décision No. 2142 du 19/11/1923*. BNF GALLICA, *La Correspondence d'Orient 1927*.

114   Kaufman, 'Henri Lammens and Syrian Nationalism', p. 119. Kaufman, 'Phoenicianism', p. 182. Firro, *Inventing Lebanon*, pp. 26–28.

115   Firro, *Inventing Lebanon*, pp. 104–105. Salibi, *The Modern History of Lebanon*, pp. 174, 183.

116   FRENCH LINES, Compagnie des Messageries Maritimes, *Agence de Beyrouth, Exercice 1929*, Trafic, pp. 5–6.

117   FRENCH LINES, Compagnie des Messageries Maritimes, *Agence de Beyrouth, Exercice 1932*, Trafic, p. 5.

118   The background was Muhammad al-Jisr's candidacy for the presidential election: cf. Traboulsi, *A History of Modern Lebanon*, pp. 91–92. Firro, *Inventing Lebanon*, pp. 123–124. Salibi, *The Modern History of Lebanon*, pp. 78, 100–101, 174.

119   FRENCH LINES, Compagnie des Messageries Maritimes, *Agence de Beyrouth, Exercice 1929*, Trafic, pp. 5–6. FRENCH LINES, Compagnie des Messageries Maritimes, *Agence de Beyrouth, Exercice 1931*, Secrétariat, pp. 4–6.

120   FRENCH LINES, Compagnie des Messageries Maritimes, *Agence de Beyrouth, Exercice 1932*, Secrétariat, pp. 3–5. On the relevance of attracting in particular Egyptian summer guests as they were the wealthiest among the *estiveurs*: Naccache, 'L'industrie de la villégiature', p. 213.

[121] FRENCH LINES, Compagnie des Messageries Maritimes, *Agence de Beyrouth, Exercice 1934*, Trafic, p. 5.

[122] FRENCH LINES, Compagnie des Messageries Maritimes, *Agence de Beyrouth, Exercice 1934*, Secrétariat, pp. 2–3.

[123] FRENCH LINES, Compagnie des Messageries Maritimes, *Agence de Beyrouth, Exercice 1936*, Secrétariat, p. 4.

[124] FRENCH LINES, Compagnie des Messageries Maritimes, *Agence de Beyrouth, Exercice 1936*, Secrétariat, pp. 4–5; Trafic, p. 4.

[125] Firro, *Inventing Lebanon*, pp. 136–138. Traboulsi, *A History of Modern Lebanon*, pp. 97–98.

[126] BO, Services économiques du gouvernement libanais, *Le Liban: Pays de tourisme*. AUB, *Décret 564/E, 06/06/1936*.

[127] AUB, *Séance de la Chambre des Députés, 29/11/1935*, pp. 80–81. AUB, *Décret 4028/EC, 24/03/1939*.

[128] AUB, *Arrêté 97/LR, 04/06/1936*.

[129] FRENCH LINES, Compagnie des Messageries Maritimes, *Agence de Beyrouth, Exercice 1936*, Trafic, p. 5.

[130] For example, they jointly issued a brochure addressing a francophone public: BNF RICHELIEU, Maklouf and Société d'encouragement au tourisme au Liban, *Le Liban: Pays d'art*.

[131] AUB, *Décret 784/EC, 02/06/1937*, p. 428. AUB, *Décret 2891/EC, 10/08/1938*.

[132] AUB, *Séance du Conseil Représentatif, 18/11/1922*, p. 4.

[133] FRENCH LINES, Compagnie des Messageries Maritimes, *Agence de Beyrouth, Exercice 1938*, Secrétariat, p. 2. BNF RICHELIEU, Maklouf and Société d'encouragement au tourisme au Liban, *Le Liban: Pays d'art*. BO, Ministère de l'économie nationale du gouvernement libanais and Société d'encouragement au tourisme au Liban, *Liban, Tourisme – Estivage – Sports d'hiver*.

[134] AUB, *Loi du 27/06/1939*. BO, Ministère de l'économie nationale du gouvernement libanais and Société d'encouragement au tourisme au Liban, *Liban, Tourisme – Estivage – Sports d'hiver*.

[135] AUB, *Loi du 08/07/1939*.

[136] FRENCH LINES, Compagnie des Messageries Maritimes, *Agence de Beyrouth, Exercice 1940*, Secrétariat, p. 6.

[137] cf. BNF RICHELIEU, TOPO 3263 Liban Généralités et Baalbek.

[138] FRENCH LINES, Compagnie des Messageries Maritimes, *Agence de Beyrouth, Exercice 1929*, Trafic, p. 5–6.

[139] JD PRIVATE COLLECTION, Chardon, *Méditerranée Orientale – Egypte*, p. 136.

[140] NLS, Steele-Maitland, *Diary of Visit to Egypt*. Rowan-Hamilton, *Both Sides of the Jordan*.

[141] Cf. Rowan-Hamilton, *Both Sides of the Jordan*, pp. 153–164.

[142] BNF RICHELIEU, *Syrie – Liban, 1935*.

[143] Hanssen, *Fin de Siècle Beirut*, pp. 195–196. Borgi, 'Hôtels et hôteliers de Beyrouth', pp. 57–59, 74–75, 79–81.

[144] RCSA, Williams, *Travel Diary: Palestine and Syria*, p. 45.

[145] Thomas Cook & Son, Luke and Garstang, *The Traveller's Handbook for Palestine and Syria, 1924*, p. 378.

[146] The descriptions grasped the urban transformations of Beirut in the late nineteenth and early twentieth centuries, driven by the Ottoman administration and local notables: Hanssen, *Fin de Siècle Beirut*, pp. 4–8.

[147] Hanssen, *Fin de Siècle Beirut*, pp. 18–19, 262–263. Thompson, *Colonial Citizens*, pp. 177–178.

[148] Thomas Cook & Son, Luke and Garstang, *The Traveller's Handbook for Palestine and Syria, 1924*, p. 379.

[149] Groebner has observed that the fabrication is inherent in the notion of the "authentic"; distinct from the "original" as it may (and should be) reproduced: Groebner, *Retroland*, pp. 180–183.

[150] BNF RICHELIEU, *Album de photographies, 1931*, images nos. 26, 27. On the enlargement of the port of Beirut cf. Hanssen, *Fin de Siècle Beirut*, pp. 87–91.

[151] The 'corniche', the *Avenue des Français*, was constructed in 1926 and among the places where the French mandate authorities most visibly inscribed their influence in the urban space. Saliba, *Beyrouth*, p. 13. Hanssen, *Fin de Siècle Beirut*, pp. 18–19.

[152] Pedersen, 'Meaning of the Mandates System', pp. 570, 581.

[153] Said, *Orientalism*, pp. 230–234. Fabian, *Time and the Other*.

[154] BNF RICHELIEU, Maklouf and Société d'encouragement au tourisme au Liban, *Le Liban: Pays d'art*. BO, Société d'encouragement au tourisme au Liban, *Villas et logements d'été*. FDC, Commission du tourisme et de villégiature du gouvernement libanais, *Au pays des cèdres*, p. 4. BO, Services économiques du gouvernement libanais, *Le Liban: Pays de tourisme*, Introduction.

[155] JD PRIVATE COLLECTION, Commission du Tourisme des Etats sous Mandat, *Holidays Off the Beaten Track*, p. 13.

[156] Hanssen, *Fin de Siècle Beirut*, pp. 228–230.

[157] Hartman and Olsaretti, 'The First Boat', pp. 43, 45–46. Shehadi, 'The Idea of Lebanon', pp. 8–10. The *Cénacle libanais* came into being in 1946, a predecessor group existed since 1936: Shehadi, 'The Idea of Lebanon', pp. 13–14.

[158] Cf. Holden, 'Constructiong an Archival Cityscape'. Wright, 'Tradition in the Service of Modernity'.

[159] Hanssen, *Fin de Siècle Beirut*, pp. 220–221.

[160] Hanssen, *Fin de Siècle Beirut*, pp. 225–227.

[161] Barakat-Buccianti, 'Beyrouth sous le Mandat français', p. 75. Ghorayeb, 'L'urbanisme de la ville de Beyrouth'.

[162] Barakat-Buccianti, 'Beyrouth sous le Mandat français', p. 76, dated the inauguration of the hotel to 1934, Chenet, 'Un Sanctuaire de Saint-Georges', p. 7, reported that the hotel was opened in 1933. Yet, contemporary reports dated the inauguration to 1932, among them Shukri Ghanem's Correspondance d'Orient: BNF GALLICA, Achard, *Syrie et Liban*, p. 276, and BNF GALLICA, Reizler, *Syrie et Liban – Porte de l'Orient, 1933*, p. 150. Ching, Jarzombek and Prakash, *A Global History of Architecture*, p. 712.

[163] Hanssen, *Fin de Siècle Beirut*.

[164] BNF RICHELIEU, Maklouf and Société d'encouragement au tourisme au Liban, *Le Liban: Pays d'art*, p. 31.

[165] This turn towards the sea after the Lebanese independence has been described by Zeina Maasri, although her conclusion that both the notion of the "Switzerland of the Middle East" and the "Mediterranean Universalism" were attempts at a Westernisation overlooked the importance of attracting an Arab audience, both before and after World War II: Maasri, 'Troubled Geography', pp. 137–138.

[166] Cf. Segreto, Manera and Pohl, *Europe at the Seaside*. Hazbun, *Beaches, Ruins, Resorts*.

[167] Renan, *Misson de Phénicie*. Kaufman, *Reviving Phoenicia*, p. 22.

[168] Rowan-Hamilton, *Both Sides of the Jordan*. RCSA, Williams, *Travel Diary: Palestine and Syria*.

[169] BNF GALLICA, Lévy, *Le Guide Sam*, pp. 693–695. BNF TOLBIAC, Roux-Servine, *Sur les routes du Levant*, pp. 60, 70–71. JD PRIVATE COLLECTION, Chardon, *Méditerranée Orientale – Egypte*, pp. 169–171. Thomas Cook & Son, Lumby and Garstang, *Cook's Traveller's Handbook to Palestine, Syria & Iraq, 1934*, p. 6.

[170] AUB, *Décret No. 1658, 06/03/1933*, pp. 2–4.

[171] ANL, *Décret No. 7424, 10/11/1930*.

[172] ANL, *Décret No. 7424, 10/11/1930*.

173 "A *lieu de mémoire* is any significant entity, whether material or non-material in nature, which by dint of human will or the work of time has become a symbolic element of the memorial heritage of any community". Nora, 'From lieux de mémoire', p. xvii.

174 AUB, *Décret No. 1712, 15/04/1935*. JD PRIVATE COLLECTION, Chardon, *Méditerranée Orientale – Egypte*, p. 142. Monmarché, *Syrie – Palestine – Iraq – Transjordanie*, pp. 30–31.

175 AUB, *Chambre des Députés, Séance publique, 07/12/1927*, p. 13.

176 JD PRIVATE COLLECTION, Chardon, *Méditerranée Orientale – Egypte*, p. 142. Monmarché, *Syrie – Palestine – Iraq – Transjordanie*, pp. 30–31.

177 AUB, *Séance de la Chambre des Députés, 18/04/1930*, p. 15.

178 Veccia Vaglieri, 's.v. es-Sawdā (Yūsuf)'. Abu Khalil, 's.v. Sawdā, Yusuf as-'. Kaufman, *Reviving Phoenicia*, p. 61.

179 AUB, *Séance de la Chambre des Députés, 18/04/1930*, p. 15. The deputies A. al-Khoury and Jamil al-Khazin argued in favour of the preservation and exhibition of monuments of the Crusaders in Byblos.

180 Monmarché, *Syrie – Palestine – Iraq – Transjordanie*, pp. 401, 422, 455.

181 ANL, *Décret No. 6042, 24/12/1929*.

182 BO, Société d'encouragement au tourisme au Liban, *Byblos*. BNF RICHELIEU, Maklouf and Société d'encouragement au tourisme au Liban, *Le Liban: Pays d'art*, pp. 27–29.

183 AUB, *Décret 145/E, 26/02/1936*. AUB, *Décret 612/E, 16/06/1936*. AUB, *Décret 770/E, 22/07/1936*.

184 Traboulsi, *A History of Modern Lebanon*, pp. 98–100. Salibi, *The Modern History of Lebanon*, pp. 179–180. Firro, *Inventing Lebanon*, p. 168.

185 For a brief, systematising overview over debates about the geographical shape of Lebanon cf. Traboulsi, *A History of Modern Lebanon*, pp. 75, 80–85.

186 Salibi, *The Modern History of Lebanon*, pp. 162, 179. Kaufman, *Reviving Phoenicia*, p. 80. Hakim, *The Origins of the Lebanese National Idea*, p. 214.

187 Traboulsi, *A History of Modern Lebanon*, pp. 102–103.

188 It gained particular relevance after World War I, and replaced Beirut in imagination: Hanssen, *Fin de Siècle Beirut*, p. 233, footnote 66. Cf. also Hourani, 'Ideologies of the Mountain'. Maasri, 'Troubled Geography', p. 134.

189 This research paradigm has been defined by Stoler and Cooper, 'Between Metropole and Colony'.

190 Schayegh, 'The Many Worlds of 'Abud Yasin', pp. 278–280.

191 CADN, *Le Lieutenant-Colonel Bucheton, 26/12/1923*. CADN, *Note de Weygand, 12/12/1923*.

192 Pedersen, 'Meaning of the Mandates System', pp. 568–572.

193 CADN, *Lettre Du Haut-Commissaire p.i., 24/05/1924*.

194 CADN, *Le ministre plénipotentiaire, 04/1924*.

195 CADN, *Général Sarrail à M. Le Ministre de France au Caire, 1925*.

196 CADN, *L'inspecteur général des polices, 27/02/1928*.

197 On the blurry distinction between artists and prostitutes in the French Mandate: Pastor, 'Performers or Prostitutes?'. Znaien, 'La prostitution à Beyrouth', paragr. 21. That migrations of prostitutes coincided with the tourist season has been shown for Egypt by Francesca Biancani: Biancani, 'International Migration and Sex Work', pp. 118–119.

198 CADN, *Le Secrétaire général du HC, 28/01/1928*. Cf. also Znaien, 'La prostitution à Beyrouth', paragr. 11, 23. I am indebted to the anonymous reviewer for having pointed out Znaien's article to me.

199 CADN, *Le Secrétaire général du HC, 28/01/1928*.

200 Pastor, 'Suspect Service', pp. 184–186.

201 CADN, *Le Ministre des Affaires étrangères, 02/04/1928*.

202 Cf. the memoir of Jawhariyyeh, *The Storyteller of Jerusalem*, pp. 158–160.

203   ANL, *Fi Sabil al-Masayif.*

204   CADN, *Lettre de Nicolas et Elie Gebeili, 17/05/1932.*

205   CADN, *Le Directeur de la Sûreté générale, 09/06/1932.* On prostitution and disease, and the necessity of controlling circulation: Znaien, 'La prostitution à Beyrouth', paragr. 23. Huber, *Channelling Mobilities*, pp. 299–305. Huber, 'Multiple Mobilities', pp. 319, 325, 330. Hanssen, *Fin de Siècle Beirut*, pp. 124–129. Pastor, 'Suspect Service', pp. 189–190.

206   CADN, *Note des Relations extérieures, 03/06/1932.* CADN, *Le Directeur de la Sûreté générale, 09/06/1932.* CADN, *Département des Relations extérieures à M. Elie Gebaili, 25/06/1932.*

207   The performance was part of a larger tour of the Levant Masabni undertook at a difficult moment of her career: Cormack, *Midnight in Cairo*, p. 291.

208   CADN, *Le consul général de France au Haut-Commissariat, 18/02/1938.* CADN, *Lettre du Directeur de la Sûreté générale, 25/02/1938.* CADN, *Télégramme du HC au Consulat du Caire, 26/02/1938.*

209   CADN, *Le Secrétaire général de l'Union Artistique, 22/06/1931.*

210   CADN, *Lettre de Jean Fabry, député, 09/11/1931.*

211   CADN, *Le directeur de la Sûreté générale, 31/07/1931.* CADN, *Le Délégué général du Haut-Commissaire, 10/08/1931.* Valeska Huber has coined the term of 'Other mobilities': Huber, *Channelling Mobilities*, pp. 306, 318–319.

212   BO, Société d'encouragement au tourisme au Liban, *Plan général pour le développement*, p. 7.

213   BO, Société d'encouragement au tourisme au Liban, *Plan général pour le développement*, p. 35.

214   BO, Société d'encouragement au tourisme au Liban, *Guide hôtelier pour 1936.*

215   BO, Société d'encouragement au tourisme au Liban, *Guide hôtelier pour 1936.*

216   BO, Société d'encouragement au tourisme au Liban, *Guide hôtelier pour 1936.*

217   LEIDEN UNIVERSITY LIBRARY, Société de Villégiature au Mont Liban, *Guide de la Société de Villégiature, 1925.* BO, Société de Villégiature au Mont Liban, *Guide de la Société de Villégiature, 1927*, p. 9.

218   BNF GALLICA, *Tunisie, 16/02/1922.* BNF GALLICA, *Echos & Nouvelles, 05/05/1934.* BNF GALLICA, *Echos & Nouvelles, 20/10/1936.* BNF GALLICA, *Le Ski au Liban, 03/1933.* BNF GALLICA, *Déclaration No. 58, 15/01/1932.* BNF GALLICA, *Office pour la protection, 15/09/1937.*

219   CADN, *Lettre de Philippe Bériel, 10/12/1937.*

220   In 1933, he suggested that skiing in Lebanon was an activity mainly practised by French officials: BNF GALLICA, *Le Ski au Liban, 03/1933.*

221   FFCAM, Bériel, Philippe, *Le Ski au Liban*, p. 364.

222   Bériel, Philippe, "Sports d'hiver", in: BNF RICHELIEU, Maklouf and Société d'encouragement au tourisme au Liban, *Le Liban: Pays d'art*, pp. 32–34, 33.

223   FFCAM, Bériel, Philipp, *Le Ski au Liban*, pp. 114–115.

224   NLI, *Winter Sport in Lebanon, 02/1937.*

225   NLI, Hoffman, *Ski-time at the Cedars, 18/03/1938.*

226   NLI, *Skiing in the Lebanon, 01/1936.* NLI, *Skiers off to the Lebanon, 02/1938.*

227   Krämer, *Geschichte Palästinas*, pp. 278–279. On Arab *estiveurs* from Palestine and their travels as an indicator of Palestinian-Lebanese relations cf. Stanton, 'Locating Palestine's Summer Residence'.

228   BO, Société d'encouragement au tourisme au Liban, *Guide hôtelier pour 1936.* BO, Société d'encouragement au tourisme au Liban and Ministère de l'économie nationale du gouvernement libanais, *Guide hôtelier du Liban 1937.* Andrea Stanton found mention of a travel guidebook to Lebanon, published in Hebrew in 1937: Stanton, 'Locating Palestine's Summer Residence', p. 57.

229   Scheffler, 'The Kaiser in Baalbek', p. 14.

230   SOAS, Smith, *Travel Diaries, Diary 1929, I, 08/11/1929.*

231   NLS, Steele-Maitland, *Diary of Visit to Egypt, 21/01/1932.*

232   UMAM FOUNDATION, *al-Istiyaffi Lubnan*, pp. 2–4.

[233] AUB, *2ème Séance de la Chambre des Députés, 17/10/1928*, pp. 2–3.

[234] Firro, *Inventing Lebanon*, p. 78. AUB, *2ème Séance de la Chambre des Députés, 17/10/1928*, pp. 4–5.

[235] CADN, *Dossier 5: Habitants de Baalbeck, 1929-1930*.

[236] AUB, *1ère Séance de la Chambre des Députés, 20/03/1928*, pp. 2–3. However, from the minutes it is difficult whether the term of "expropriation" referred literally to an expropriation, or whether it was a rhetorical means. In another debate on Sawfar, the deputy Zayn spoke of "expropriation" when he was talking about rules how houses' facades should be constructed. He implied that not the entire terrain could be used freely in this case.

[237] AUB, *1ère Séance de la Chambre des Députés, 20/03/1928*, p. 3.

[238] ANL, *Décret No. 8203, 30/06/1931*. JD PRIVATE COLLECTION, Alouf, *Histoire de Baalbek*. BNF TOLBIAC, Roux-Servine, *Sur les routes du Levant*, p. 46. SOAS, Smith, *Travel Diaries, Diary 1929, I*, 09/11/1929. BO, Société d'encouragement au tourisme au Liban, *Villas et logements d'été*. Le Gras, *Le Lac Méditerranéen*, p. 135.

[239] JD PRIVATE COLLECTION, Alouf, *Histoire de Baalbek*, pp. 9–10. AUB, *Décret No. 1387/EC*.

[240] HARVARD UNIVERSITY LIBRARY, Société de Villégiature au Mont Liban, *The Tourist's Book, 1924*, p. 63. LEIDEN UNIVERSITY LIBRARY, Société de Villégiature au Mont Liban, *Guide de la Société de Villégiature, 1925*, p. 45. UMAM FOUNDATION, *al-Istiyaf fi Lubnan*, p. 22.

[241] CADN, *Requête d'un groupe de propriétaires, 01/01/1929*. CADN, *Requête adressée au Haut-Commissaire, 01/01/1929*.

[242] CADN, *Télégramme d'un groupe d'habitants, 02/01/1929*. CADN, *Sollicitation d'un groupe d'habitants, 11/1929*.

[243] Makdisi, 'The "Rediscovery" of Baalbek'. Osterhammel, '"The Great Work of Uplifting Mankind"', pp. 372–373. Anderson, *Imagined Communities*, pp. 181–182.

[244] On universal heritage in the context of decolonisation: Rehling, 'Universalismen und Partikularismen im Widerstreit', pp. 417–422.

[245] This trope ("In America, they don't know where Lebanon is, but they have heard about Baalbek") is advanced by inhabitants of Baalbek up to the present day, cf. the interviews projected in Vali Mahlouji's exhibition: Vali Mahlouji, Baalbek: Archives of an Eternity, Sursock Museum Beirut, 28/06/2019–22/09/2019.

[246] Quote by Mohamed Djissr. AUB, *11ème Séance de la Chambre des Députés, 12/11/1930*, p. 10.

[247] AUB, *11ème Séance de la Chambre des Députés, 12/11/1930*, p. 9.

[248] AUB, *23e Séance de la Chambre des Députés, 28/12/1931*, p. 11.

[249] Interesting in this regard is the importance of the Baalbek International Festival for the promotion of tourism after independence, which Zeina Maasri interpreted as an actualisation and animation of the ruins: Maasri, 'Troubled Geography', pp. 136–137.

[250] Buccianti, 'Tourisme et villégiature', pp. 26–27.

References

Sources

Archives nationales du Liban, Beirut
al-Jiniral Falih [?] fi Baalbek, [General Vallée[?] in Baalbek], al-Maʿrid, 23/05/1926, p. 4, ANL.

al-Sihafiyun fi Masayif al-Shamal, [The Journalists in the Estivage Centres of the North], al-Maʿrid, 26/08/1926, pp. 1–2, ANL.

Arrêté No. 3013 du 04/03/1925, ANL.

Arrêté No. 3042 du 17/03/1925, ANL.

Décret No. 6042 du 24/12/1929, Recueil des lois et décrets du gouvernement de la République Libanaise, ANL.

Décret No. 7424 du 10/11/1930, Recueil des lois et décrets du gouvernment de la République Libanaise, ANL.

Décret No. 8203 du 30/06/1931, Recueil des lois et décrets de la République Libanaise, ANL.

Décret No. 8330 du 06/08/1931, Recueil des lois et décrets de la République Libanaise, ANL.

Fi Sabil al-Masayif, [On the Way to the Estivage Centres], al-Maʿrid, 26/08/1926, p. 5, ANL.

GOURAUD, HENRI, *Arrêté No. 885 du 05/06/1921*, Le Haut Commissaire de la République Française en Syrie et au Liban, Actes administratifs, ANL.

Marsum Raqm 1955, Majmuʿat Qawanin wa-Marasim Hukumat al-Jumhuriyya al-Lubnaniyya, Mujallid 1, al-Fasl al-Thalith: Marakiz al-Istiyaf, s. 208, [Decree No. 1955, 29/07/1927, Collection of Laws and Decrees of the Lebanese Republic, Volume 1, Chapter Three: Estivage Centres, p. 208], ANL.

Marsum Raqm 2203, Majmuʿat Qawanin wa-Marasim Hukumat al-Jumhuriyya al-Lubnaniyya, Mujallid 1, al-Fasl al-Thalith: Marakiz al-Istiyaf, s. 208, [Decree No. 2203, 27/09/1927, Collection of Laws and Decrees of the Lebanese Republic, Volume 1, Chapter Three: Estivage Centres, p. 208], ANL.

Marsum Raqm 2515, Majmuʿat Qawanin wa-Marasim Hukumat al-Jumhuriyya al-Lubnaniyya, Mujallid 1, al-Fasl al-Thalith: Marakiz al-Istiyaf, s. 208, [Decree No. 2515, 15/12/1927, Collection of Laws and Decrees of the Lebanese Republic, Volume 1, Chapter Three: Estivage Centres, p. 208], ANL.

Marsum Raqm 2713, Majmuʿat Qawanin wa-Marasim Hukumat al-Jumhuriyya al-Lubnaniyya, Mujallid 1, al-Fasl al-Thalith: Marakiz al-Istiyaf, s. 209, [Decree No. 2713, 28/01/1928, Collection of Laws and Decrees of the Lebanese Republic, Volume 1, Chapter Three: Estivage Centres, p. 209], ANL.

Marsum Raqm 2912, Majmuʿat Qawanin wa-Marasim Hukumat al-Jumhuriyya al-Lubnaniyya, Mujallid 1, al-Fasl al-Thalith: Marakiz al-Istiyaf, s. 209, [Decree No. 2912, 08/03/1928, Collection of Laws and Decrees of the Lebanese Republic, Volume 1, Chapter Three: Estivage Centres, p. 209], ANL.

Marsum Raqm 3494, Majmuʿat Qawanin wa-Marasim Hukumat al-Jumhuriyya al-Lubnaniyya, Mujallid 1, al-Fasl al-Thalith: Marakiz al-Istiyaf, s. 209, [Decree No. 3494, 27/06/1928, Collection of Laws and Decrees of the Lebanese Republic, Volume 1, Chapter Three: Estivage Centres, p. 209], ANL.

Marsum Raqm 955, Majmuʿat Qawanin wa-Marasim Hukumat al-Jumhuriyya al-Lubnaniyya, Mujallid 1, al-Fasl al-Thalith: Marakiz al-Istiyaf, s. 207, [Decree No. 955, 27/12/1926, Collection of Laws and Decrees of the Lebanese Republic, Volume 1, Chapter Three: Estivage Centres, p. 207], ANL.

Shahada Mustaf Misri, [Testimony of an Egyptian Estiveur], al-Maʿrid, 17/06/1926, ANL.

Wafd Niqabat al-Sihafa fi Shamal, [The Delegation of the Association of Journalists in the North], al-Maʿrid, 29/08/1926, p. 1, ANL.

Bibliothèque nationale de France, Paris

ACHARD, E., *Syrie et Liban: Les Etats du Levant en 1932*, Correspondance d'Orient 25.426 (1933), pp. 275–278, BNF Gallica.

Album de photographies, Un Voyage en Orient (Egypte, Syrie, Liban) en 1931, BNF Richelieu/VZ 1082.

Comité de patronage du "Monde colonial illustré", Le Monde colonial illustré, 137 (12/1934), p. 194, BNF Gallica.

Décision No. 2142 du 19/11/1923, Bulletin mensuel des actes administratifs du Haut-Commissariat de la République Française en Syrie et au Liban 3.11 (1923), p. 197, BNF Gallica.

Déclaration No. 58, Bulletin mensuel des actes administratifs du Haut-Commissariat de la République Française en Syrie et au Liban, 15/01/1932, p. 5, BNF Gallica.

Echos & Nouvelles, La Petite Tunisie, 20/10/1936, p. 1, BNF Gallica.

Echos & Nouvelles, La Petite Tunisie, 05/05/1934, p. 2, BNF Gallica.

La Correspondence d'Orient 19.349 (1927), p. 14, BNF Gallica.

Le Mois colonial, Le Monde colonial illustré, 07/1929, p. 196, BNF Gallica.

Le Ski au Liban, Le Monde colonial illustré, 115 (03/1933), p. 48, BNF Gallica.

Le Tourisme au Liban: Le charme de Chtaura, Le Monde colonial illustré, 139 (02/1935), p. 25, BNF Gallica.

LÉVY, SAM, *Le Guide Sam. Le livre d'or de l'Orient: Pour l'expansion économique française dans le Levant*, Garches: Le Guide Sam, 1930, BNF Gallica.

MAKLOUF, IBRAHIM, and SOCIÉTÉ D'ENCOURAGEMENT AU TOURISME AU LIBAN, *Le Liban: Pays d'art et de tourisme*, 1937, BNF Richelieu/Vh 123.

MARCHAND, JEAN, *Lamartine, chantre des Cèdres du Liban: Centenaire du voyage de Lamartine en Syrie*, Le Monde colonial illustré, 119 (07/1933), p. 103, BNF Gallica.

MONTET, PIERRE, *L'Art phénicien au XVIIIe siècle avant J.-C. d'après les récentes trouvailles de Byblos*, Monuments et Mémoires, 27 (1924), pp. 1–30, BNF Gallica.

Office pour la protection de la propriété, Bulletin officiel des actes administratifs du Haut-Commissariat, 15/09/1937, p. 554, BNF Gallica.

REIZLER, STANISLAS, *Syrie et Liban – Porte de l'Orient: Le bilan de la collaboration franco-syrienne*, Le Monde colonial illustré, 122 (10/1933), pp. 147–151, BNF Gallica.

ROUX-SERVINE, LOUIS, *Sur les routes du Levant: Petit manuel du voyage expérimental*, Paris: J. Barreau, 1934, BNF Tolbiac/8-G-13749.

SAMNÉ, GEORGES, and BUREAU SYRIEN ET LIBANAIS D'INFORMATIONS ET DE TOURISME, *Syrie et Liban: Guide du touriste*, c. 1932, BNF Richelieu/TOPO 3264 Syrie Généralités.

Syrie – Liban, ed. Société d'encouragement au tourisme au Liban [?], c. 1935, BNF Richelieu/TOPO 3263 Liban Généralités + Baalbek.

Syrie – Liban: Pays de transit et de tourisme, Le Monde colonial illustré, 144 (07/1935), pp. 137–143, BNF Gallica.

Tunisie, Les Annales Coloniales, 16/02/1922, p. 2, BNF Gallica.

Bibliothèque Orientale, Université Saint-Joseph, Beirut

COMMISSION DU TOURISME DES ETATS SOUS MANDAT, *Le tourisme en Syrie et au Liban*, Beirut 1930, BO/68 C4/43.

GOBRIL, RIZKALLAH, *Visitez le Levant par les Messageries Automobiles Beyrouth/ Itinéraires Rizkallah. Liban – Syrie – Palestine – Transjordanie – Irak – Perse*, Imprimerie Cozma 1928–29, [1926], BO/68 C4/36.

MINISTÈRE DE L'ÉCONOMIE NATIONALE DU GOUVERNEMENT LIBANAIS, and SOCIÉTÉ D'ENCOURAGEMENT AU TOURISME AU LIBAN, *Liban. Tourisme – Estivage – Sports d'hiver*, 1938, BO.

SERVICES ÉCONOMIQUES DU GOUVERNEMENT LIBANAIS, *Le Liban: Pays de tourisme et de villégiature*, Paris: La déesse, 1935, BO/68 C4/39.

SOCIÉTÉ DE VILLÉGIATURE AU MONT LIBAN, *Guide de la Société de Villégiature au Mont Liban pour 1927*, Cairo, 1927, BO/68 C4/46.

SOCIÉTÉ D'ENCOURAGEMENT AU TOURISME AU LIBAN, *Beyrouth*, Beirut: Imprimerie Catholique, c. 1935, BO/68 C4/44.

SOCIÉTÉ D'ENCOURAGEMENT AU TOURISME AU LIBAN, *Byblos*, Beirut: Imprimerie catholique, 1936, BO/68 C4/50.

SOCIÉTÉ D'ENCOURAGEMENT AU TOURISME AU LIBAN, *Guide hôtelier pour 1936*, Beirut: Imprimerie Catholique, 1936, BO/68 C4/35.

SOCIÉTÉ D'ENCOURAGEMENT AU TOURISME AU LIBAN, *Plan général pour le développement de la villégiature et du tourisme au Liban, 21/11/1938*, BO/68 C4/52.

SOCIÉTÉ D'ENCOURAGEMENT AU TOURISME AU LIBAN, *Villas et logements d'été au Liban, N.D.*, BO/68 C4/40.

SOCIÉTÉ D'ENCOURAGEMENT AU TOURISME AU LIBAN, and MINISTÈRE DE L'ÉCONOMIE NATIONALE DU GOUVERNEMENT LIBANAIS, *Guide hôtelier du Liban 1937*, Beirut 1937, BO/68 C4/34.

ZEHIL, ABDALLAH, *Petit guide historique de la Syrie, la Palestine et l'Egpyte*, Marseille 1929, BO/68 C4/48.

Centre des Archives diplomatiques de Nantes

COMPAGNIE FRANÇAISE DU TOURISME, *Note sur l'établissement d'un Office de Tourisme pour la France en pays étranger, s.d.*, Ambassade du Caire, Propagande touristique pour la France, 1920–1938, CADN/353 PO/2/230.

Département des Relations extérieures à M. Elie Gebaili, propriétaire du Grand Hôtel Aley, 25/06/1932, Bureau diplomatique, Règlementation du travail, CADN/1SL/1/V/1406.

Dossier 5: Habitants de Baalbeck, 01/01/1929–16/05/1930, Cabinet politique. Dossiers de principe 1920–1946, Liban, CADN/1SL/1/V/934.

Indésirables, 26/12/1923, Le Lieutenant-Colonel Bucheton, Directeur du contrôle de la Sûreté Générale à M. le Général Weygand, Haut-Commissaire, Cabinet politique, Affaires diverses, Dossier: Passeports 1923–1924, CADN/1 SL/1/V/842.

L'Inspecteur général des polices à M. le Directeur du Service des Renseignements du Levant, signé Vardon, 27/02/1928, Cabinet politique, Affaires diverses, Dossier: Passeports 1928, CADN/1 SL/1/V/842.

Le Consul général de France au Caire au Haut-Commissariat, 18/02/1938, Bureau diplomatique, Règlementation du travail, Dossier: Artistes de Music-Hall en résidence dans les territoires sous mandat, CADN/1SL/1/V/1406.

Le Délégué général du Haut-Commissaire, Helleu, au Secrétaire général de l'Union artistique de France, 10/08/1931, Bureau diplomatique, Règlementation du travail, CADN/1SL/1/V/1406.

Le Directeur de la Sûreté générale au Conseiller aux Affaires politiques, Bureau politique, 31/07/1931, Bureau diplomatique, Règlementation du travail, CADN/1SL/1/V/1406.

Le Directeur de la Sûreté générale, Inspecteur général des polices à M. le Conseiller du Haut Commissariat aux Relations extérieures/Bureau diplomatique, 09/06/1932, Bureau diplomatique, Règlementation du travail, CADN/1SL/1/V/1406.

Le Haut Commissaire, Général Sarrail, à M. le Ministre de France, Le Caire et à M. le Consul général de France à Jérusalem, signé Paul Lépissier, n.d., Cabinet politique, Affaires diverses, Dossier: Passeports 1925, CADN/1 SL/1/V/842.

Le Ministre des Affaires étrangères à Ponsot, 02/04/1928, Bureau diplomatique, Règlementation du travail. Dossier: Prestige de la femme française en Syrie, CADN/1 SL/1/V/1406.

Le Ministre plénipotentiaire, Haut-Commissaire p.i. de la République Française en Syrie et au Liban au Ministre des Affaires étrangères, Paris, 04/1924, Cabinet politique. Affaires diverses, Dossier: Passeports 1923–1924, CADN/1 SL/1/V/842.

Le Secrétaire général de l'Union artistique de France au Commissaire général des Etats sous Mandat, M. Bouchède, Chef de la Sûreté de Beyrouth, 22/06/1931, Bureau diplomatique, Règlementation du travail, CADN/1SL/1/V/1406.

Le Secrétaire général du Haut Commissaire au Ministre des Affaires étrangères, 28/01/1928, Bureau diplomatique, Règlementation du travail. Dossier: Prestige de la femme française en Syrie, CADN/1SL/1/V/1406.

Le Tourisme au Liban, L'Orient, 03/03/1928, Cabinet politique. Dossiers de principe 1920–1946. Cabinet Ponsot, 1928–1931. Classeur E: Tourisme, CADN/1SL/1/V/1573.

Le Tourisme en Syrie, Cahiers du Sud, 01/03/1929, CADN/1 SL/1/V/1573.

Lettre de Jean Fabry, député de Paris, à Ponsot, Haut-Commissaire, 09/11/1931, Bureau diplomatique, Règlementation du travail, CADN/1SL/1/V/1406.

Lettre de Nicolas et Elie Gebeili, adressée au Haut-Commissariat, 17/05/1932, Bureau diplomatique, Règlementation du travail, CADN/1SL/1/V/1406.

Lettre de Philippe Bériel à M. le Ministre de France en Egypte, Le Caire, 10/12/1937, Ambassade de France au Caire, CADN/353 PO/2/230.

Lettre du Directeur de la Sûreté générale au Chef du Bureau diplomatique, 25/02/1938, Bureau diplomatique, Règlementation du travail, Dossier: Artistes de Music-Hall en résidence dans les territoires sous mandat, CADN/1 SL/1/V/1406.

Lettre Du Haut-Commissaire p.i. de la République Française en Syrie et au Liban, à M. Gaillard, Ministre de France en Egypte, Le Caire et M. Ballereau, 24/05/1924. Cabinet politique, Affaires diverses, Dossier: Passeports 1923–1924., CADN/1 SL/1/V/842.

Lettre du Secrétariat général (de Reffye) à M. Grandguillot, Directeur de l'Office économique et commercial, Alexandrie, 11/03/1927, Fonds Beyrouth, Cabinet Politique, CADN/1SL/1/V/1573.

Note de Weygand pour M. le Directeur du contrôle de la Sûreté générale, 12/12/1923, Cabinet politique, Affaires diverses, Dossier: Passeports 1923/24, CADN/1 SL/1/V/842.

Note des Relations extérieures à M. le Directeur de la Sûreté Générale, Inspecteur général des polices, signé Chauvel, 03/06/1932, Bureau diplomatique, Règlementation du travail, CADN/1 SL/1/V/1406.

Requête adressée au Haut-Commissaire, 01/01/1929, Cabinet politique, Dossiers de principe 1920–1946, Liban, CADN/1SL/1/V/934.

Requête d'un groupe de propriétaires à Baalbeck à M. le Colonel, Directeur du Service des Renseignements du Levant, 01/01/1929, Cabinet politique, Dossiers de principe 1920–1946, Liban, CADN/1SL/1/V/934.

Sollicitation d'un groupe d'habitants de Baalbek, auprès du Haut-Commissaire, 11/1929, Cabinet politique, Dossiers de principe 1920–1946, Liban, CADN/1SL/1/V/934.

Télégramme du Haut Commissariat au Consulat du Caire, 26/02/1938, Bureau diplomatique, Règlementation du travail, Dossier: Artistes de Music-Hall en résidence dans les territoires sous mandat, CADN/1SL/1/V/1406.

Télégramme d'un groupe d'habitants de Baalbek au Haut-Commissaire, 02/01/1929, Cabinet politique, Dossiers de principe 1920–1946, Liban, CADN/1SL/1/V/934.

Fédération française des clubs alpins et de montagne, Paris

Bériel, Philipp, *Le Ski au Liban*, La Montagne, 267 (03/1935), pp. 113–115, FFCAM.

Bériel, Philippe, *Le Ski au Liban*, La Montagne, 242 (10/1932), pp. 359–364, FFCAM.

Association French Lines, Le Havre

Compagnie des Messageries Maritimes, *Rapports d'Agence. Agence de Beyrouth, Exercice 1921*, French Lines.

Compagnie des Messageries Maritimes, *Rapports d'Agence. Agence de Beyrouth, Exercice 1922*, French Lines.

Compagnie des Messageries Maritimes, *Rapports d'Agence. Agence de Beyrouth, Exercice 1923*, French Lines.

Compagnie des Messageries Maritimes, *Rapports d'Agence. Agence de Beyrouth, Exercice 1924*, French Lines.

COMPAGNIE DES MESSAGERIES MARITIMES, *Rapports d'Agence. Agence de Beyrouth, Exercice 1925*, French Lines.

COMPAGNIE DES MESSAGERIES MARITIMES, *Rapports d'Agence. Agence de Beyrouth, Exercice 1926*, French Lines.

COMPAGNIE DES MESSAGERIES MARITIMES, *Rapports d'Agence. Agence de Beyrouth, Exercice 1927*, French Lines.

COMPAGNIE DES MESSAGERIES MARITIMES, *Rapports d'Agence. Agence de Beyrouth, Exercice 1928*, French Lines.

COMPAGNIE DES MESSAGERIES MARITIMES, *Rapports d'Agence. Agence de Beyrouth, Exercice 1929*, French Lines.

COMPAGNIE DES MESSAGERIES MARITIMES, *Rapports d'Agence. Agence de Beyrouth, Exercice 1930*, French Lines.

COMPAGNIE DES MESSAGERIES MARITIMES, *Rapports d'Agence. Agence de Beyrouth, Exercice 1931*, French Lines.

COMPAGNIE DES MESSAGERIES MARITIMES, *Rapports d'Agence. Agence de Beyrouth, Exercice 1932*, French Lines.

COMPAGNIE DES MESSAGERIES MARITIMES, *Rapports d'Agence. Agence de Beyrouth, Exercice 1934*, French Lines.

COMPAGNIE DES MESSAGERIES MARITIMES, *Rapports d'Agence. Agence de Beyrouth, Exercice 1935*, French Lines.

COMPAGNIE DES MESSAGERIES MARITIMES, *Rapports d'Agence. Agence de Beyrouth, Exercice 1936*, French Lines.

COMPAGNIE DES MESSAGERIES MARITIMES, *Rapports d'Agence. Agence de Beyrouth, Exercice 1937*, French Lines.

COMPAGNIE DES MESSAGERIES MARITIMES, *Rapports d'Agence. Agence de Beyrouth, Exercice 1938*, French Lines.

COMPAGNIE DES MESSAGERIES MARITIMES, *Rapports d'Agence. Agence de Beyrouth, Exercice 1940*, French Lines.

PHILIPPAR, GEORGES, *Le tourisme en Syrie. Conférence prononcée à l'Ecole des Hautes Etudes Sociales, le 13 Février 1929*, Panorama. Revue mensuelle d'éducation et de vulgarisation coloniale, 15e année, French Lines/1997 002 5044.

Fouad Debbas Collection, Sursock Museum, Beirut

COMMISSION DU TOURISME ET DE VILLÉGIATURE DU GOUVERNEMENT LIBANAIS, *Au pays des cèdres*, 1931, FDC.

SERVICES ÉCONOMIQUES, *Le tourisme en Syrie et au Liban*, updated 1928, FDC.

Institut français du Proche-Orient, Beirut

BUCCIANTI, LILIANE, 'Tourisme et villégiature dans la montagne libanaise: Deux exemples précis. Bhamdoun-gare et Bhamdoun-village', Thèse de doctorat (Toulouse, Université de Toulouse, 1975).

Jafet Library, American University of Beirut

Arrêté No. 97/LR du Haut-Commissaire, 04/06/1936, Journal officiel de la République Libanaise No. 3289, 05/06/1936, pp. 2–3, AUB.

Arrêté No. 2537, 26/06/1924, Journal officiel du Grand Liban, 08/07/1924, AUB.

Arrêté No. 2538, 26/06/1924, Journal officiel du Grand Liban, 08/07/1924, AUB.

Arrêté No. 2599, 02/08/1924, Journal officiel du Grand Liban, 15/08/1924, AUB.

Arrêté No. 3041, 17/03/1925, Journal officiel du Grand Liban No. 1856, 31/03/1925, AUB.

CAYLA, LÉON, *Discours: Anniversaire de la proclamation du Grand Liban*, Journal officiel du Grand Liban No. 1901, 08/09/1925, AUB.

Chambre des Députés, Compte rendu de la 9e Séance publique, 07/12/1927, Journal officiel de la République libanaise, AUB.

Chambre des Députés, Séance du 04/12/1930, Journal officiel de la République Libanaise, AUB.

Chambre des Députés, Séance du 18/04/1930, Journal officiel de la République Libanaise, AUB.

Compte rendu de la 23e Séance, 28/12/1931, Journal officiel de la République Libanaise, Débats parlementaires, AUB.

Compte-rendu de la 11ème séance, 12/11/1930, Journal officiel de la République Libanaise, Chambre des Députés, AUB.

Compte-rendu de la 2ème séance, 17/10/1928, Journal officiel de la République Libanaise, Débats parlementaires, AUB.

Compte-rendu de la Séance de la Chambre des Députés, 29/11/1935, Journal officiel de la République Libanaise, AUB.

Conseil Représentatif, Séance du 01/05/1923, Journal officiel du Grand Liban No. 1669, 12/06/1923, AUB.

Conseil Représentatif, Séance du 03/11/1925, Journal officiel du Grand Liban No. 1928, 11/12/1925, AUB.

Conseil Représentatif, Séance du 04/11/1924, Journal officiel du Grand Liban No. 1821, 28/11/1924, AUB.

Conseil Représentatif, Séance du 04/11/1924, Journal officiel du Grand Liban No. 1822, 02/12/1924, AUB.

Conseil Représentatif, Séance du 18/11/1922, Journal officiel du Grand Liban No. 1618, 15/12/1922, Supplément, AUB.

Conseil Représentatif, Séance du 18/11/1922, Journal officiel du Grand Liban No. 1620, 22/12/1922, Supplément, AUB.

Conseil Représentatif, Séance du 20/03/1923 (suite), Journal officiel du Grand Liban, 30/03/1923, pp. 3–4, AUB.

Conseil Représentatif, Séance du 21/11/1925, Journal officiel du Grand Liban No. 1938, 15/01/1926, AUB.

Conseil Représentatif, Séance du 23/03/1926, Journal officiel du Grand Liban No. 1969, 30/04/1926, Supplément, AUB.

Conseil Représentatif, Séance du 26/03/1923, Journal officiel du Grand Liban, 13/04/1923, p. 4, AUB.

Conseil Représentatif, Séance du 26/03/1923, Journal officiel du Grand Liban, 17/04/1923, pp. 3–4, AUB.

Conseil Représentatif, Séance du 27/10/1925, Journal officiel du Grand Liban No. 1926, 04/12/1925, AUB.

Conseil Représentatif, Séance du 30/04/1923, Journal officiel du Grand Liban No. 1669, 12/06/1923, AUB.

Conseil Représentatif, Séance du 31/10/1922, Journal officiel du Grand Liban No. 1608, 10/11/1922, Supplément, AUB.

Constitution Libanaise, Journal officiel de la République Libanaise, 05/1926, pp. 1–4, AUB.

DE SERCEY, *Lettre du Haut Commissaire p.i. de la R.F. en Syrie et au Liban à M. le Gouverneur du Grand Liban, délégué du Haut Commissaire*, Journal officiel du Grand Liban No. 1663, 22/05/1923, AUB.

Décision No. 1667, 23/05/1923, Journal officiel du Grand Liban No. 1666, 01/06/1923, AUB.

Décret 145/E, 26/02/1936, Journal officiel de la République Libanaise, Lois et décrets, No. 3248, 02/03/2936, pp. 3–4, AUB.

Décret 2891/EC, 10/08/1938, Journal officiel de la République Libanaise No. 3623, 22/08/1938, p. 2311, AUB.

Décret 4028/EC, 24/03/1939, Journal officiel de la République Libanaise No. 3665, 30/03/1939, p. 2879, AUB.

Décret 564/E, 06/06/1936, Journal officiel de la République Libanaise No. 3293, 15/06/1936, p. 3, AUB.

Décret 612/E, 16/06/1936, Journal officiel de la République Libanaise, Lois et décrets, No. 3296, 22/06/1936, pp. 3–4, AUB.

Décret 770/E, 22/07/1936, Journal officiel de la République Libanaise, Lois et décrets, No. 3311, 27/07/1936, pp. 2–4, AUB.

Décret 784/EC, 02/06/1937, Journal officiel de la République Libanaise No. 3459, 07/07/1937, AUB.

Décret No. 1387/EC, Journal officiel de la République Libanaise No. 3514, 12/11/1937, AUB.

Décret No. 1658, 06/03/1933, Journal officiel de la République Libanaise No. 2785, 10/03/1933, AUB.

Décret No. 1712, 15/04/1935, Journal officiel de la République Libanaise, Lois et décrets, No. 3113, 22/04/1935, p. 6, AUB.

Décret No. 447, 15/06/1936, Journal officiel de l'Etat de Syrie, 25/06/1936, AUB.

Decret No. 5742, 12/10/1929, Journal officiel de la République Libanaise No. 2770, 14/10/1929, AUB.

GOURAUD, HENRI, *Discours du 1er Septembre 1922,* Journal officiel du Grand Liban No 1589, 05/09/1922, AUB.

Harak al-Istiyaf, [The Estivage Movement], Jubitr 7–8 (1936), AUB.

Loi du 08/07/1939, Journal officiel de la République Libanaise No. 3695, 13/07/1939, pp. 3191–3195, AUB.

Loi du 27/06/1939, Journal officiel de la République Libanaise No. 3692, 03/07/1939, pp. 3191–3195, AUB.

Mada fi Baalbek, [News from Baalbek], Jubitr 3 (1935), p. 31, AUB.

Procès-verbal de la 1ère séance, 20/03/1928, Journal officiel de la République Libanaise, Débats parlementaires, AUB.

Séance du Conseil Représentatif du 07/12/1922, Journal officiel du Grand Liban No.1629 du 23/01/1923, AUB.

Séance du Conseil Représentatif du 07/12/1922, Journal officiel du Grand Liban No. 1630 du 26/01/1923, AUB.

Séance du Conseil Représentatif du 11/12/1922, Journal officiel du Grand Liban No. 1630 du 26/01/1923, Supplément, AUB.

Séance du Conseil Représentatif du 18/11/1922, Supplément du Journal officiel du Grand Liban No. 1618 du 15/12/1922, AUB.

Séance du Conseil Représentatif du 24/04/1923, Journal officiel du Grand Liban No. 1667 du 05/06/1923, AUB.

jd private collection

ALOUF, MICHEL, *Histoire de Baalbek par un de ses habitants,* Beirut: al-Igtihad, 5th edn., 1928, jd private collection.

CHARDON, J., *Méditerranée Orientale – Egypte: Guide du passager pour les escales des croisières et pour les excursions en Syrie et en Égypte,* Paris: Hachette, 1935, jd private collection.

COMMISSION DU TOURISME DES ETATS SOUS MANDAT, *Syria and Lebanon: Holidays Off the Beaten Track,* Beirut 1930, jd private collection.

National Library of Israel, Jerusalem

HOFFMAN, GAIL, *Ski-Time at the Cedars: Where Snow is Snow,* The Palestine Post, 18/03/1938, p. 9. Historical Jewish Press, NLI.

Skiers off to the Lebanon, The Palestine Post, 01/02/1938, p. 8. Historical Jewish Press, NLI.

Skiing in the Lebanon, The Palestine Post, 07/01/1936, p. 2. Historical Jewish Press, NLI.

Winter Sport in Lebanon, The Palestine Post, 25/02/1937, p. 9. Historical Jewish Press, NLI.

National Library of Scotland, Edinburgh

STEELE-MAITLAND, MARY, *Diary of Visit to Egypt, Palestine and Turkey, 24/12/1931 – 29/01/1932,* NLS/ACC 5553.

Royal Commonwealth Society Archives, Cambridge University Library

WILLIAMS, GEORGE H., *Travel Diary: Palestine and Syria, 1925,* RCSA/RCMS 53/3/2.

School of Oriental and African Studies, London
SMITH, EDWIN W., *Travel Diaries. Diary. 1929. Vol. I.*, SOAS/MMS/17/02/05/02/09/02/02.

Umam Foundation, Beirut
al-Istiyaf fi Lubnan [Estivage in Lebanon], c. 1926, Umam Foundation.

Other published sources

AL-KHOURY, FOUAD, 'L'Industrie hôtelière au Liban', *Revue Phénicienne*, 1919, pp. 37–39.

CHENET, G., 'Un sanctuaire de Saint-Georges en pays alaouite: El Khodr de Mqaté', *La géographie. Revue mensuelle publiée par la Société de Géographie*, 64/1 (1935), pp. 1–9.

JAWHARIYYEH, WASIF, *The Storyteller of Jerusalem: The Life and Times of Wasif Jawhariyyeh, 1904–1948*, Salim Tamari and Issam Nassar (Northampton: Olive Branch Press, 2014).

LE GRAS, CHARLES, *Le Lac Méditerranéen: Pèlerinage à Jèrusalem et croisière en Méditerranée orientale, du 5 Avril au 19 Mai 1927* (Avignon: Aubanel Frères, 1927).

MONMARCHÉ, MARCEL, *Syrie – Palestine – Iraq – Transjordanie* (Les Guides bleus, Paris: Hachette, 1932).

N.N., 'Aspetti della crisi economica, politica, e sociale del Vicino Oriente secondo un giornalista arabo', *Oriente Moderno*, 12/11 (1932), pp. 522–524.

NACCACHE, ALBERT, 'L'industrie de la villégiature au Liban', *Revue Phénicienne*, 1/4–6 (1919), pp. 208–213.

NACCACHE, ALBERT, 'Aperçu sur la situation économique du Mont-Liban', in Pierre Naccache, *Un autre Liban. Les écrits d'Albert Naccache* (Paris: Geuthner, 2018), pp. 79–153.

ROWAN-HAMILTON, NORAH, *Both Sides of the Jordan: A Woman's Adventures in the Near East* (London: Herbert Jenkins, 1928).

SOCIÉTÉ DE VILLÉGIATURE AU MONT LIBAN, *Guide de la Société de Villégiature au Mont Liban 1925*, Cairo, 1925, Leiden University Library.

SOCIÉTÉ DE VILLÉGIATURE AU MONT LIBAN, *The Tourist's Book for Egypt, Palestine, Syria and The Lebanon*, Cairo, 1924, Harvard University Library.

THOMAS COOK & SON, LUKE, HARRY CHARLES, and GARSTANG, JOHN, *The Traveller's Handbook for Palestine and Syria* (London: Simpkin Marshall, 1924).

THOMAS COOK & SON, LUMBY, CHRISTOPHER, and GARSTANG, JOHN, *Cook's Traveller's Handbook to Palestine, Syria & Iraq* (6th edn., London: Simpkin Marshall, 1934).

VECCIA VAGLIERI, LAURA, 'Dichiarazioni politiche del Presidente del Consiglio libanese', *Oriente Moderno*, 7/12 (1927), 596-599.

VECCIA VAGLIERI, LAURA, 'Decadenza della villeggiatura estiva nel Libano', *Oriente Moderno*, 9/9 (1929), p. 408.

VECCIA VAGLIERI, LAURA, 'La villeggiatura nel Libano', *Oriente Moderno*, 12/9 (1932), p. 437.

VECCIA VAGLIERI, LAURA, 'Decadenza della villeggiatura nel Libano', *Oriente Moderno*, 16/9 (1936), p. 508.

Secondary Literature

ABU KHALIL, ASʿAD, 's.v. Sawdā, Yusuf as-: Historical Dictionary of Lebanon, Landam Md., 1998.' <https://wbis.degruyter.com/biographic-document/X172-150-7>, accessed 8 Dec 2018.

AGERON, CHARLES-ROBERT, 'Le "parti" colonial', *L'Histoire*, 69 (1984), pp. 72–81.

ANDERSON, BENEDICT R., *Imagined Communities: Reflections on the Origin and Spread of Nationalism* (2nd edn., London, New York: Verso, 2016).

AYALON, AMI, *The Press in the Arab Middle East: A History* (Oxford, New York: Oxford University Press, 1995).

BARAKAT-BUCCIANTI, LILIANE, 'Beyrouth sous le Mandat français ou la "Nice de l'Orient"', in Colette Zytnicki and Habib Kazdaghli (eds.), *Le Tourisme dans l'Empire français. Un outil de la domination coloniale? Politiques, pratiques et imaginaires (XIXe–XXe siècles)* (Saint-Denis: Publications de la Société française d'histoire d'outre-mer, 2009), pp. 73–78.

BIANCANI, FRANCESCA, 'International Migration and Sex Work in Early Twentieth-Century Cairo', in Liat Kozma, Cyrus Schayegh, and Avner Wishnitzer (eds.), *A Global Middle East. Mobility, Materiality and Culture in the Modern Age, 1880–1940* (London, New York: I.B. Tauris, 2015), pp. 109–133.

BORGI, GEORGES, 'Hôtels et hôteliers de Beyrouth au XIXe siècle', *Kalimat al-Balamand*, 2 (1995), pp. 57–85.

BRUNSCHWIG, HENRI, 'Le parti colonial français', *Revue Francaise d'Histoire d'Outre-Mer*, 1959, pp. 49–83.

CHING, FRANCIS D. K., JARZOMBEK, MARK, and PRAKASH, VIKRAMADITYA, *A Global History of Architecture* (Hoboken: Wiley, 2007).

CORMACK, RAPHAEL, *Midnight in Cairo: The Divas of Egypt's Roaring '20s* (New York: W. W. Norton & Company, 2021).

DAKHLI, LEYLA, 'L'expertise en terrain colonial: Les orientalistes et le mandat français en Syrie et au Liban', *Matériaux pour l'histoire de notre temps*, 99/3 (2010), pp. 20–27.

DEMAY, ALINE, *Tourism and Colonization in Indochina, 1898–1939* (Newcastle upon Tyne: Cambridge Scholars Publishing, 2014).

DUECK, JENNIFER M., *The Claims of Culture at Empire's End: Syria and Lebanon under French Rule* (Oxford, New York: Oxford University Press, 2010).

ELISSÉEFF, NIKITA, 'Bayrūt', in Peri J. Bearman, Thierry Bianquis, C. E. Bosworth et al. (eds.), *Encyclopaedia of Islam*, 12 vols. (2nd edn., Leiden: Brill, 1960-2004) <http://dx.doi.org/10.1163/1573-3912_islam_SIM_1330>, accessed 25 Jan 2018.

EL-SOLH, RAGHID, *Lebanon and Arabism: National Identity and State Formation* (London: I.B. Tauris, 2004).

FABIAN, JOHANNES, *Time and the Other: How Anthropology Makes its Object* (New York: Columbia University Press, 2014).

FIRRO, KAIS M., *Inventing Lebanon: Nationalism and the State under the Mandate* (London: I.B. Tauris, 2003).

FURLOUGH, ELLEN, 'Une leçon des choses: Tourism, Empire, and the Nation in Interwar France', *French Historical Studies*, 25/3 (2002), pp. 441–475.

GATES, CAROLYN, 'The Historical Role of Political Economy in the Development of Modern Lebanon', *Papers on Lebanon*, 10 (1989), pp. 3–37.

GHORAYEB, MARLÈNE, 'L'urbanisme de la ville de Beyrouth sous le mandat français', *Revue du monde musulman et de la Méditerranée*, 73/1 (1994), pp. 327–339.

GROEBNER, VALENTIN, *Retroland: Geschichtstourismus und die Sehnsucht nach dem Authentischen* (Frankfurt a. M.: S. Fischer, 2018).

HAKIM, CAROL, *The Origins of the Lebanese National Idea, 1840–1920* (Berkeley: University of California Press, 2013).

HANSSEN, JENS, *Fin de Siècle Beirut: The Making of an Ottoman Provincial Capital* (Oxford, New York: Oxford University Press, 2005).

HARTMAN, MICHELLE, and OLSARETTI, ALESSANDRO, '"The First Boat and the First Oar": Inventions of Lebanon in the Writings of Michel Chiha', *Radical History Review*, 2003, pp. 36–65.

HAZBUN, WALEED, *Beaches, Ruins, Resorts: The Politics of Tourism in the Arab World* (Minneapolis: University of Minnesota Press, 2008).

HOBSBAWM, ERIC, 'Introduction: Inventing Traditions', in Eric Hobsbawm and Terence Ranger (eds.), *The Invention of Tradition* (Cambridge: Cambridge University Press, 1996), pp. 1–14.

HOLDEN, STACY E., 'Constructiong an Archival Cityscape: Local Views of Colonial Urbanism in the French Protectorate of Morocco', *History in Africa: A Journal of Method*, 34 (2007), pp. 121–132.

HOURANI, ALBERT, 'Ideologies of the Mountain and City', in Roger Owen (ed.), *Essays on the Crisis in Lebanon* (London: Ithaca Press, 1976), pp. 33–42.

HOURANI, ALBERT, *Arabic Thought in the Liberal Age: 1798–1939* (2nd edn., Cambridge, New York: Cambridge University Press, 1983).

HUBER, VALESKA, 'Multiple Mobilities: Über den Umgang mit verschiedenen Mobilitätsformen um 1900', *Geschichte und Gesellschaft*, 36/2 (2010), pp. 317–341.

HUBER, VALESKA, *Channelling Mobilities: Migration and Globalisation in the Suez Canal Region and Beyond, 1869–1914* (Cambridge, New York: Cambridge University Press, 2013).

IPPOLITO, CHRISTOPHE, 'Naissance d'une nation: La Revue Phénicienne au Liban en 1919', in Benoît Tadié, Mansanti, Céline and Hélène Aji (eds.), *Revues modernistes, revues engagées* (Rennes: Presses universitaires de Rennes, 2011).

JANSEN, JAN C., 'Die Erfindung des Mittelmeerraums im kolonialen Kontext: Die Inszenierung eines "lateinischen Afrika" beim "Centenaire de l'Algérie française" 1930', in Frithjof Benjamin Schenk (ed.), *Der Süden. Neue Perspektiven auf eine europäische Geschichtsregion* (Frankfurt a.M., New York: Campus, 2007), pp. 175–205.

JANSEN, JAN C., *Erobern und Erinnern: Symbolpolitik, öffentlicher Raum und französischer Kolonialismus in Algerien, 1830–1950* (München: Oldenbourg, 2013).

KAUFMAN, ASHER, 'Phoenicianism: The Formation of an Identity in Lebanon in 1920', *Middle Eastern Studies*, 37/1 (2001), pp. 173–194.

KAUFMAN, ASHER, *Reviving Phoenicia: In Search for Identity in Lebanon* (London: I.B. Tauris, 2004).

KAUFMAN, ASHER, '"Tell Us Our History": Charles Corm, Mount Lebanon and Lebanese Nationalism', *Middle Eastern Studies*, 40/3 (2004), pp. 1–28.

KAUFMAN, ASHER, 'Henri Lammens and Syrian Nationalism', in Adel Beshara (ed.), *The Origins of Syrian Nationhood. Histories, Pioneers and Identity* (London: Routledge, 2011), pp. 108–122.

KHOURY, PHILIP S., *Syria and the French Mandate: The Politics of Arab Nationalism, 1920–1945* (London: I.B. Tauris, 1987).

KRÄMER, GUDRUN, *Geschichte Palästinas: Von der osmanischen Eroberung bis zur Gründung des Staates Israel* (6th edn., München: C.H. Beck, 2015).

MAASRI, ZEINA, 'Troubled Geography: Imagining Lebanon in 1960s Tourist Promotion', in Kjetil Fallan and Grace Lees-Maffei (eds.), *Designing Worlds. National Design Histories in an Age of Globalization* (New York: Berghahn Books, 2016), pp. 125–140.

MAKDISI, USSAMA, 'The "Rediscovery" of Baalbek: A Metaphor for Empire in the Nineteenth Century', in Hélène S. Sader, Thomas Scheffler, and Angelika Neuwirth (eds.), *Baalbek. Image and Monument 1898–1998* (Halle, Stuttgart: Ergon, 1998), pp. 137–156.

MONROE, KRISTIN V., *The Insecure City: Space, Power, and Mobility in Beirut* (New Brunswick: Rutgers University Press, 2016).

NORA, PIERRE, 'From *lieux de mémoire* to Realms of memory: Preface to the English-language edition', in Pierre Nora (ed.), *Realms of Memory. Rethinking the French Past* (New York: Columbia University Press, 1996), pp. xv–xxiv.

OSTERHAMMEL, JÜRGEN, '"The Great Work of Uplifting Mankind": Zivilisierungsmissionen und Moderne', in Boris Barth and Jürgen Osterhammel (eds.), *Zivilisierungsmissionen. Imperiale Weltverbesserung seit dem 18. Jahrhundert* (Konstanz: UVK Verlagsgesellschaft, 2005), pp. 363–425.

OSTERHAMMEL, JÜRGEN, 'Die Weltöffentlichkeit im 20. Jahrhundert', in Jürgen Osterhammel, *Die Flughöhe der Adler. Historische Essays zur globalen Gegenwart* (München: C.H. Beck, 2017), pp. 54–74.

PASTOR, CAMILA, 'Suspect Service: Prostitution and the Public in the Mandate Mediterranean', in Cyrus Schayegh and Andrew Arsan (eds.), *The Routledge Handbook of the History of the Middle East Mandates* (London, New York: Routledge, 2015), pp. 183–197.

PASTOR, CAMILA, 'Performers or Prostitutes? Artistes during the French Mandate over Syria and Lebanon, 1921–1946', *Journal of Middle East Women's Studies*, 13/2 (2017), pp. 287–311.

PEDERSEN, SUSAN, 'The Meaning of the Mandates System: An Argument', *Geschichte und Gesellschaft*, 32/4 (2006), pp. 560–582.

REHLING, ANDREA, 'Universalismen und Partikularismen im Widerstreit: Zur Genese des UNESCO-Welterbes', *Zeithistorische Forschungen*, 8 (2011), pp. 414–436.

RENAN, ERNEST, *Misson de Phénicie* (Paris: Imprimerie Impériale, 1864).

's.v. Philippar, Georges: World Biographical Information System Online' <https://wbis.degruyter.com/biographic-document/F144310>, accessed 31 Jul 2018.

SAID, EDWARD WILLIAM, *Orientalism* (London: Penguin Books, 2003 [1978]).

SALIBA, ROBERT, *Beyrouth: Architectures aux sources de la modernité, 1920–1940* (Marseilles: Parenthèses, 2009).

SALIBI, KAMAL S., *The Modern History of Lebanon* (Westport: Greenwood, 1976).

SANTER, JANINA, 'Imagining Lebanon: Tourism and the Production of Space under French Mandate (1920–1943)', MA Thesis (Beirut, American University of Beirut, 2019).

SCHAYEGH, CYRUS, 'The Many Worlds of 'Abud Yasin: or: What Narcotics Trafficking in the Interwar Middle East Can Tell Us about Territorialization', *The American Historical Review*, 116/2 (2011), pp. 273–306.

SCHEFFLER, THOMAS, 'The Kaiser in Baalbek: Tourism, Archeology, and the Politics of Imagination', in Hélène S. Sader, Thomas Scheffler, and Angelika Neuwirth (eds.), *Baalbek. Image and Monument 1898–1998* (Halle, Stuttgart: Ergon, 1998), pp. 13–49.

SEGRETO, LUCIANO, MANERA, CARLES, and POHL, MANFRED (eds.), *Europe at the Seaside: The Economic History of Mass Tourism in the Mediterranean* (New York: Berghahn Books, 2009).

SHEHADI, NADIM, 'The Idea of Lebanon: Economy and State in the Cénacle Libanais, 1946–1954', *Papers on Lebanon*, 5 (1987), pp. 3–29.

SLIGHT, JOHN, *The British Empire and the Hajj, 1865–1956* (Cambridge, Mass.: Harvard University Press, 2015).

STANTON, ANDREA L., 'Locating Palestine's Summer Residence: Mandate Tourism and National Identity', *Journal of Palestine Studies*, 47/2 (2018), pp. 44–62.

STOLER, ANN LAURA, and COOPER, FREDERICK, 'Between Metropole and Colony: Rethinking a Research Agenda', in Frederick Cooper and Ann Laura Stoler (eds.), *Tensions of Empire. Colonial Cultures in a Bourgeois World* (Berkeley: University of California Press, 1997), pp. 1–56.

THOMPSON, ELIZABETH, *Colonial Citizens: Republican Rights, Paternal Privilege, and Gender in French Syria and Lebanon* (New York: Columbia University Press, 2000).

TRABOULSI, FAWWAZ, *A History of Modern Lebanon* (2nd edn., London, New York: Pluto Press, 2012).

VECCIA VAGLIERI, LAURA, 's.v. es-Sawdā (Yūsuf): Notizie bio-bibliografiche su autori arabi moderni, Annali 1 (N.S.), 1940, 259-298' <https://wbis.degruyter.com/biographic-document/X148-217-7>, accessed 8 Dec 2018.

WICK, ALEXIS, 'History, Geography, and the Sea', *International Journal of Middle East Studies*, 48/4 (2016), pp. 743–745.

WORLD BIOGRAPHICAL INFORMATION SYSTEM ONLINE, 's.v. Berthelot, Philippe' <https://wbis.degruyter.com/biographic-document/F209566>, accessed 31 Jul 2018.

WRIGHT, GWENDOLYN, 'Tradition in the Service of Modernity: Architecture and Urbanism in French Colonial Policy, 1900–1930', in Frederick Cooper and Ann Laura Stoler (eds.), *Tensions of Empire. Colonial Cultures in a Bourgeois World* (Berkeley: University of California Press, 1997), pp. 322–345.

ZNAIEN, NESSIM, 'La prostitution à Beyrouth sous le mandat français (1920–1943)', *Genre & Histoire*, 11 (2012) <http://journals.openedition.org/genrehistoire/1781>.

ZYTNICKI, COLETTE, *L' Algérie, terre de tourisme: Histoire d'un loisir colonial* (Paris: Vendémiaire, 2016).

Tourist transformations

During the 1920s and 1930s, members of the middle classes in the Arab East grasped at international and domestic tourism as political resources with an urgency that differed from later political approaches to tourism. Embedded in different configurations of power, these middle-class actors perceived a transformative potential in tourism, expecting that it would enable them to create viable nation-states. Tourism, they assumed, would allow them to achieve recognition of their states both from international observers and the population, while creating connections in terms of infrastructures and mobilities that would integrate the national territory and facilitate control.

A middle-class project

The most determined social group in tourism development were actors from the middle classes, who had a good grasp of the opportunities tourism provided. They dominated the tourist spaces as travellers, mediators, tour agents, colonial officials or members of local associations and national governments. Certainly, actors from social groups beyond this global bourgeoisie traversed the tourist spaces. Miss M.B. participated in the tourist movement; young Egyptians encountered travellers in their villages and presented themselves proudly for photographs; cameleers and French singers earned a living from the tourism business; and Arab notables gathered in *estivage* resorts. In contrast to the middle-class actors in tourism, however, they did not intend to harness the political potential of tourism. The systematic shaping of tourist spaces was a middle-class project.

Tourism lent itself to shaping the emerging states in the Arab East even for middle-class actors who did not have full sovereignty over state and society. Through collaboration with social groups who had more direct access to decision-making, such as the aristocracy or the colonial state, middle classes across the globe were able to participate in administering state and society "without taking over state power themselves", as Christof Dejung, David Motadel and Jürgen Osterhammel argued.[1] Tourism was one field for such a cooperative approach – and it allowed its proponents to reach out beyond the national level.

Whether they were striving for national sovereignty or imperial control, the advocates of tourism development in the colonial administrations, in the Yishuv and in Arab societies shared the expectation that tourism was an asset in consolidating the respective territories as they had emerged after World War I. In the Arab East, tourism development was therefore a valuable resource particularly for those actors aiming to forge territorial nation-states – the Arab middle-class nationalists as well as the Zionists. In Lebanon, Egypt and in the Yishuv, where these actors had access to political decision-making either by a sufficient degree of autonomy or by cooperation, tourism development policies were more pronounced and more consistent than in Arab Palestine and Syria.

The Egyptian *efendiyya* benefitted from the partial sovereignty of their country, which allowed them to shape domestic policies, including decisions related to tourism. As European companies largely dominated the tourism sector, subsidies in favour of tourism were more contested among Egyptian deputies than in the neighbouring countries. Advocates of tourism development engaged in strengthening tourism for political reasons though, arguing that it was an occasion to enhance the prestige of Egypt on an international stage. In addition, from the 1930s the attention of proponents of tourism shifted increasingly towards a domestic clientele, aiming to enhance the identification of the population with their country.

Attentively observing tourism development policies in Egypt and Lebanon, the advocates of tourism in the Yishuv pursued a systematic approach to tourism development. The mainly Zionist actors enjoyed wide-ranging competences with regard to shaping politics in the Yishuv and profited from cooperation with the British mandatory. As the latter's interest often consisted in reducing both manpower and financial resources to a minimum, they were able to largely define measures and contents of tourism development in the Mandate.[2] Their aim was to define itineraries and disseminate their messages among tourists during their entire journey across Mandate Palestine. The *Zionist Information Bureau for Tourists* and the Jewish Agency thus offered a wide range of services and thoroughly shaped publications and advertisement campaigns. Tourists were addressed either as witnesses to the Zionist project or as potential investors or settlers, thus as contributors to the realisation of a Zionist state.

In Lebanon, due to the existence of a regional tourism sector since Ottoman times, tourism had been on the political agenda early on, and proponents of tourism development resumed their activities right after the establishment of the mandate. A number of important families in Lebanese politics had their strongholds in the *estivage* area, which was certainly another asset in favour of tourism and a major difference to neighbouring Syria. As soon as the French "politics of cooperation" granted Lebanese deputies and associations greater room for manoeuvre in shaping cultural policies in the Mandate, they promoted tourism more systematically as a central pillar of both the national economy and national identity, and as a means of integrating the national territory.

British and French colonial officials had their own reasons to support tourism development in those cases; converging interests thus facilitated cooperation between nationalist middle-class actors and the imperial powers. As an economic sector, tourism did not compete with the manufactured goods of the imperial powers. Specifically in French colonial circles, tourism was rather perceived as a contribution to the *"mise en valeur"* of the colonial possessions. In Britain, *Thomas Cook & Sons* was an influential lobbyist in favour of tourism; and both mandatories perceived tourism as a contribution to prove their rightful administration of the mandates in both economic and educational terms.

By contrast, neither the Palestinian Arab middle classes nor its Syrian counterparts managed to shape coherent national policies of tourism development. In both mandates, the parliamentary representation of the population was severely obstructed and political participation was curtailed, relying in both cases on the cooperation between the mandate administration and Arab notables as well as the Yishuv, while largely circumventing the Arab middle classes. Whereas for Arab Palestinians, political representation was not provided for in the mandate treaty, military conflict and French tactics of rule impeded the political participation of the middle classes in Syria, and large parts of the Syrian middle classes spent the mandate period in exile.[3]

Middle-class actors in Palestine, such as George Antonius or Tawfiq Canaan, aimed to forge alliances with the better-positioned *Supreme Muslim Council* or the *Tourist Development Association of Palestine*. However, these occasional instances of collaboration did not lead to a systematic shaping of narratives and spaces of tourism. Moreover, the respective partners pursued different strategies and ambitions. While the TDA was dominated by Zionist perspectives, the tourism policies of the *Supreme Muslim Council* promoted Jerusalem as a cultural and religious symbol rather than aiming to shape visions of a Palestinian state. Kamil al-Ghazzi, pursuing a strategy of cooperation in Syria, did not even succeed in obtaining institutional support from the French mandatory. In neither of the cases, the limited room for manoeuvre of middle-class actors in the political sphere allowed for comprehensive tourism strategies at the national level – nor for any long-term strategies of state-building.

In Syria, the French High Commission shaped tourism development policies to a large extent, addressing travellers as potential supporters of the French imperial project. Advertised as a means of economic and cultural development, tourism was also conceived of as an instrument of territorial control, and the touristic and military uses of infrastructural projects overlapped. European tourists were meant to create a civilian presence in the insurgent hinterlands or at strategic outposts, and tourism advertisement, information services and organisational advice remained largely a domain of the French High Commission as well as of international entrepreneurs. As the French officials barely cooperated with the Syrian middle classes, and, unlike in North Africa, a French settler community did not exist, local support for tourism development policies remained limited.

The strategy of cooperation of the Arab and Yishuvi advocates of tourism development did not imply that divisions along the lines of coloniser/colonised, class, gender, or race were suspended in tourism. Members of the mandate administrations sometimes had serious objections against the increasing mobilities of Others.[4] They feared unintended political consequences of tourist mobilities, for example in Palestine, where British officials suspected tourism to be a way of concealing the illegal immigration of Jewish settlers. In French Syria, during the uprisings against the mandatory, French authorities feared that potential spies and foreign agents might enter the country as tourists and create unrest. Similarly, the *estivage* practices of Arab notables suspected of pan-Arab leanings aroused suspicion among the authorities, and a visa, as we have seen in the case of Al-Husayni, could be rejected on these grounds.

Restrictions of mobility also had class and gender dimensions. This did not only concern the Arab populations of the mandates and Egypt, but also citizens from the colonial metropoles as the example of the French singers and dancers showed. Imperial concerns about the reputation of the mandate power and a general anxiety about the allegedly uncontrollable mountain resorts outweighed their demands for access to imperial labour markets in times of economic crisis.

Similarly, Arab advocates of tourism development promoted tourism for lower-class citizens out of political considerations. Their approach towards the lower classes was educational, characterised by attempts to make them participate in excursions to national monuments, museums, and leisure practices that were considered adequate. As the lower classes were not considered a power base by the middle classes, their basic material interests were of secondary importance when it came to disputes about measures of tourism development. Objections against infrastructural projects like in Baalbek or concerns about rising prices were considered subordinate to the interests of a touring global bourgeoisie and the political, cultural and economic potential of their visit.

Finally, tourists had their very own perspectives on the narratives and visions presented to them. They were, after all, not just an audience, but actors themselves, and persons intending to disseminate their spatio-political visions among tourists thus depended on the willingness of the latter to consider their points of view. For nationalist actors aiming to disseminate new visions of their country it was thus difficult to modify tourist narratives. Moments of rupture were an occasion to spread new narratives. Crises and conflicts incited tourists to seek new lines of interpretation as the Syrian Revolt or the case of Mandate Palestine showed, where the conflict between immigrating Jews and the Arab Palestinian population led travellers to question established narratives of the 'Holy Land'. Yet the case of Palestine also demonstrated that the chances of actors to actually modify tourist narratives were highly unequal. In case tourists were confronted with competing narratives, they tended to follow the interpretations and explanations that seemed familiar, or trusted guides from a similar social and national background. Tourists were thus by no means a passive audience, but carefully selected information, often according to criteria of familiarity.

The access to political decision-making, obviously, did not guarantee the successful realisation of ambitions projected at tourism development. Carriages and camels remained a preferred motif for George H. Williams' photographs, the dream of an international spa resort of Tiberias did not come true, the upsurge in numbers of tourists visiting Palmyra and the Jabal al-Druze did not occur, and the Lebanese coastal towns remained on the margins of tourist maps before the war. Moreover, the economic turbulences during the interwar period menaced a number of business ventures and tourism development plans. Still, the middle-class advocates of tourism believed in its potential. They continued to set incentives for businesses, modified their plans, and addressed new target groups. In the following section I argue that they did so because of the spatial paradigm shaping their understanding of state-building, which turned tourism into an important resource.

Tourism as a transformative resource

The collective appearance of tourists as well as their desire for information and guidance allowed tour operators and associations, guidebook authors and professional tour guides to use tourism as a resource for creating meaning. These actors drafted itineraries defining the importance of sites and sights, they developed narratives about the origins and particularities of nations and attempted to shape the impressions tourists would take home and share in their countries of origin. In addition, the steadiness in tourists' movements allowed them to use it as a resource for consolidating territory. Mobilities and infrastructures related to tourism, from roads and railways to accommodation infrastructures including water and sewage systems as well as electricity cables, contributed to the integration of territory. Tourists following the same itineraries connected places and attributed sense to connections and boundaries, thereby creating space.[5] From a bird's eye view, tourism helped establishing an order of nation-states that emerged on the mental maps of visitors and local societies alike.

The transformative ambitions of the advocates of tourism development shifted from the 1920s to the 1930s, confirming Cyrus Schayegh's distinction between a spatial "Ottoman twilight" in the 1920s and an order of nation-states in the 1930s.[6] Seen through the lens of tourism, the 1920s were characterised by actors' attempts to position their states in the newly established international order, whereas the outside borders were not yet entirely defined. A united *Bilad al-Sham* was still used by some Arab guidebook authors as a framework introducing the traveller to the region, various models of a fruitful Palestinian-Yishuvi coexistence circulated, and the *Société de villégiature au Mont Liban* only started to make known Mount Lebanon as a destination to a larger, international tourist audience.

In the 1930s, the focus on tourism development shifted towards a closer integration of the national territory. This included targeting domestic audiences, a shift that might have been additionally fostered by the experience of the Great Depression and the orientation towards national economies.[7] Proponents of tourism aimed to create a national consciousness among their fellow countrymen, and they increasingly turned towards an integration of the borderlands. Both aims became manifest in new forms of local tourism aiming at both moral and physical education of the citizens.

The idea that the control of territory was vital to the sovereignty of the state was not unique to the Arab East. Charles S. Maier observed that the striving for control of bordered space ("territory", as defined by Maier) dominated political thought in the era between around 1860 and 1970. The viability of a state, he argued, was assumed to depend on the resources given in the particular territory, as well as on the capacity of the state to mobilise these resources.[8] This "territorial consciousness", as Maier termed it, explains the overarching interest of both imperial officials and nationalist middle-class actors in tourism.[9] Nationalist middle-class actors pursued tourism development policies with particular vigour, as tourism seemed to provide the control over territory that would allow them to achieve political sovereignty. This control included the classification ("mapping") of space, considered a core element of post-colonial nation-building by Anderson, yet it went beyond mere classification.[10] The integration of territory, from their point of view, was based on mobilities, infrastructures and mindsets, and tourism helped to create them.

The territorialisation characterising the transition from a post-Ottoman spatial order to the consolidation of nation-states in the Arab East can be divided into three spatial processes supported by tourism: the emergence of new centres, the integration of hinterlands and peripheries, as well as (partly) the definition of an 'outside'.

The rise of new centres

In the course of tourism development, new centres emerged on states' territory that were not necessarily defined by their administrative or economic significance. Rather, tourists or advocates of tourism development attributed historical, cultural, or aesthetic meaning to them. Among these emblematic centres were sites famous for major archaeological or cultural monuments such as Luxor, Baalbek, or Palmyra. Advocates of tourism development stipulated the objective, universal significance of the site and assumed that the grandeur of the monuments testified to the grandeur of the nation or the imperial power.[11]

On the one hand, the interests of international tourists in archaeological sites were given priority over the local interests of lower-class citizens. Michel Alouf and his supporters argued that the 'public interest' of mankind in the ruins of Baalbek, manifest in the flocks of tourists who visited the site, justified the expropriation of local inhabitants for the construction of a new road. In Luxor, the presence of tourists served as an argument for the 'embellishment' of the area around the train station and the expulsion of residents, and in Palmyra, the preservation of the ruins and their accessibility to tourists justified the displacement of local inhabitants.

On the other hand, the meaning attributed to the new centres also led to urban growth and infrastructural development. Local representatives, like Tawfiq Bishara in Luxor, used the importance of tourist centres as an argument to demand additional funds and investments in local infrastructures. At times, the successful territorial integration of a place seems to have made inhabitants reconsider their opinions regarding the new spatial order. In Baalbek, for example, the advantages its inhabitants gained from its importance as a national symbol seem to have shifted their allegiances towards the Lebanese state between the 1920s and 1936.

A certain malleability of the notion of 'tourist centre' was evident in the case of the Yishuv, where officials from the *Zionist Information Bureau for Tourists* promoted the Jezreel Valley as a particular tourist attraction. The stipulated interest derived from the promotion of the new Jewish settlements as the heartland of the projected Zionist state, while Arab Palestinian towns and villages were omitted on tourist maps. Even if tourists were not necessarily convinced by the aesthetic qualities of the settlements, most of the travellers in my sample took note of their existence and acknowledged the Yishuvi contribution to the development of Palestine.

Regarding Syria, advocates of tourism development continued to promote the urban centres of the Ottoman period as well as the estivage villages. Unlike in neighbouring states, options of fostering Syrian domestic tourism were not discussed – neither the French colonisers nor the notables, who built their influence very much on local support – had an interest in generating such larger, nationwide mobility. *Estivage* journeys of the notables as well as summer camps of the urban youth in the *estivage* resorts therefore remained the only instances of experiencing the Syrian space. 'Syria' as a new entity was thus largely defined from the perspectives of French (and British) observers, yet, without local support, its materialisation on the ground remained unstable.

Integrating peripheries

Tourism allowed its advocates to intensify state presence in regions that had been considered subordinate hinterlands or peripheries before.[12] This integration occurred in imaginaries as we have seen in Lebanon, where the reference to the Phoenician heritage initially was not related to actual policies of conservation and tourist circuits. Although the editors of tourist brochures referred to the Phoenician heritage, in terms of infrastructural accessibility and preservationist policies the port towns of Sidon, Tripoli and Tyre remained marginal until the late 1930s. Phoenicianism justified the *shape* of the nation, but only in a time of contestation it translated into actual policies of integrating these towns into the Lebanese tourist space. Increased movements were then supposed to reinforce the connections between the coastal areas and the Lebanese heartland.

In the contested French Mandate of Syria, French military officers and colonial administrators also identified tourist infrastructures and trajectories as a means of tying insurgent towns and hinterlands closer to the state. Directing tourists to contested areas was conceived of as a method of surveying and controlling territory without military means, thus reconcilable with mandate obligations. In the Jabal al-Druze and the ʿAlawite state, tourism development was fostered as a means of 'pacification' after armed conflict. It was intended to create a civilian presence across the Syrian territory, while infrastructures such as road projects served both potential tourists and the military. As it seems that, other than in the French North African settler colonies, Syria did not attract a sufficient number of French (or other) civilians, it seems likely that the strategy did not bring the intended military success though.

A similar logic explains the promotion of tourist attractions or the creation of resorts at strategically relevant outposts. Palmyra was an example, but also the *Kallia* resort at the Dead Sea, which was promoted by its managers as an escape for the Yishuvi population even during times of violent conflict. Arab Palestinian advocates of tourism, by contrast, mainly advertised Jerusalem as a tourist centre. While this demonstrated their historical rootedness in the land of Palestine, neglecting the hinterland implied that the integration of a potential Arab Palestine did not occur: they neither drafted maps of a future sovereign Arab Palestinian state nor had the means to shape infrastructural connections.

Policies of integrating nations by means of tourism development also had sociopolitical implications as they implied the creation of shared living standards across the country. Efforts of electrification and the construction of roads and water infrastructures were particularly relevant for the Lebanese mountain villages. Moreover, proponents of tourism aimed to create domestic mobilities, ambitions that were centred on the integration of the working classes and the modernisation of society. Thereby, they mirrored developments in Europe and North America, where fascist, socialist, and democratic governments fostered tourism to secure the loyalty of the population.[13]

In the Mediterranean, beach tourism turned into the symbol of modern societies in the 1930s and the youthful swimming woman became its visual icon. The excursions were associated with physical activity, thus differing from the Ottoman *estivage* habits and their focus on sociability, while generating mobilities towards other, previously peripheral regions. The promotion of beaches as sites of recreation was accompanied by the propagation of new codes of conduct, and the beach became associated with an educational agenda, as the detailed regulations on dress and behaviour demonstrated.[14] Then, from such a perspective, tourism led to a rapprochement between urban spaces and peripheral areas, both in actual movements and in cultural terms.

Defining the 'outside'

The definition of the neighbouring countries as outside Others in tourism was rare in the four case studies under scrutiny. Presumably, notions of belonging to a shared cultural space still shaped imaginaries in former *Bilad al-Sham*. Only in some of the 1920s brochures, authors claimed a Mediterranean, distinctly modern identity for Lebanon while at the same time distinguishing it from an Orientalised Syria. Egypt, by contrast, exhibited its outside Other in tourism more prominently: in the magazines of the *Tourist Development Association*, racial and civilisational boundaries were assumed to separate Arab Egyptians from Nubians as well as Sudanese, 'lifting up' the Egyptians against clichés of uncivilised Africans.

Rather than defining their countries by their Others, advocates of tourism development in all case studies carefully shaped visions of their own community, refuting stereotypes of allegedly backward Orientals and presenting their societies in terms of an international modernity. The references for such modernity differed: whereas postcards from French Mandate Syria showcased the Hamidian urban modernity, in Lebanon, Egypt and the Yishuv, modernity implied the exhibition of industrial productivity, of which leisure activities in certain defined spaces such as beaches, luxurious hotels and clubs were one facet. The showcasing of agricultural landscapes in this context was particular to the Yishuv and must be understood in Zionist discourses on restoring ancient productivity in Palestine. Arab Palestine was an exception, as Palestinian tourist narratives presented the history rather than the modernity of Palestine, thereby expressing their claims to the land.

Lebanon had a particular relationship with its outside neighbours on whom its tourism sector largely depended. Although Lebanese citizens participated in the expansion of sports courts, clubs and swimming pools as well as *estivage* and skiing practices, in contrast to Egypt and the Yishuv, domestic working-class tourism played a subordinate role, presumably due to the country's small population. Advocates of tourism development focused instead on attracting visitors from across the Arab world and beyond, pursuing less educational and rather political-diplomatic aims. In shaping Lebanon as a leisure space for the Yishuv (until 1948) and the Arab world, Lebanon not only managed to preserve the meaning of its *estivage* industry, but also maintained good relations with the neighbouring countries the recognition of which was viable to its political existence.

Continuities

Alongside the processes of transition towards an order of nation-states, alternative mental maps and spatial orders persisted. Such visions characterised the perspectives of Arab notables, for whom the mountains of Lebanon and Syria remained an important *estivage* resort that was not defined in national terms. To them, resorts such as Bludan or Ehden remained embedded in a framework of Arab sociability, allowing them to maintain relations across the newly established borders.

In Syria, such a perception of space as a network of interconnected centres continued to define the views of its inhabitants more generally. Neither the Syrian notables nor its middle classes seemed to develop a territorial idea of Syria. Efforts to foster tourism development, though generally limited, focused on the urban centres or the *estivage* resorts. Other than the French mandate power aiming to penetrate the Syrian hinterlands, the attractions suggested by Syrian advocates of tourism development were monuments situated in or associated with urban contexts. Visual representations, like on postcards, reproduced visions of Syria as a country of Ottoman modernity, characterised by infrastructures interlinking the urban centres. I assumed that this was related to the ongoing dominance of notables in Syrian politics whose power was based on local allegiances rather than national identifications. In Kamil al-Ghazzi's reflections, instances of thinking Syria in a territorial sense were discernible, yet a national forum for exchange and debate was lacking. For the same reasons, new centres of Syria, though sites like Palmyra and the Suwayda were identified and infrastructurally developed by French efforts, did not gain great prominence.

A perception of space in terms of networks also characterised the approach of Arab Palestinians in Mandate Palestine. Tawfiq Canaan's search for a distinctive Palestinian national culture in the villages seems to have been an exception. Arab middle-class nationalists such as Alexander Khoori and George Antonius continued to situate Palestine in their narratives in the larger context of *Bilad al-Sham*. The approach of Amin al-Husayni and the *Supreme Muslim Council*, who identified a political potential in tourism much earlier than the Syrian notables, similarly reflected thinking in networks and alliances. His political strategy did not include tourism development in the Palestinian peripheries and hinterlands but was based largely on the symbolic importance of Jerusalem. Its touristic attraction, al-Husayni reasoned, would allow them to place Jerusalem on the mental map of a global audience and raise global awareness of the situation of the Palestinian people. While he seems to have succeeded in this regard, such symbolic action did not bring the Arab Palestinians closer to a nation-state.

Though to different degrees, the efforts at tourism development during the 1920s and 1930s had lasting effects in the four countries under scrutiny. Deputies and state officials increasingly resorted to tourism as a resource in defining a national order. This option did not exist for Palestinian Arabs, who resorted to revolt, and the uncompromising position of the French mandatory made it a stony process for Syrians. Yet in Egypt, Lebanon and the Yishuv, nationalist middle-class actors sought to gain independence within the given national frame, and they promoted their visions in the tourist arena. Zionist advocates of tourism development side-lined the interests of tourists to the benefit of the state-building project. A comprehensive strategy of informing visitors and reaching out to potential settlers and investors resulted in a projected subordination of tourists' desires to itineraries and gazes useful for the nationalist project. Egyptian representatives consolidated the Egyptian nation on the beaches of Alexandria and used tourist contexts to define civilisational and national borders against the Sudanese, and civilisational and moral borders against Mussolini's Italy. Lebanon was a striking case, as attracting tourists from outside Lebanon remained the primary goal of proponents of tourism development throughout the whole period, and tourism allowed its representatives to fashion a place for Lebanon in both a regional and an international setting. Moreover, the tourist space came to be identified with the national space by both visitors and a large segment of the population. Tourism development initiatives spread narratives of national coherence and created powerful national symbols, while infrastructures and itineraries aimed at connecting Ottoman *estivage* centres and more recently identified tourist attractions. The incorporation of territories that had long been (and would remain) disputed became a material reality, and the expansion of infrastructures directed tourists to the proclaimed national heartland: the mountain range.

The tourist's age

When the predominantly bourgeois as well as regional tourist mobilities of the 1920s and 1930s gave way to European mass tourism in the Mediterranean in the second half of the twentieth century, the territorial hopes projected at tourism seem to have lost their relevance. Continuities in the tourism business linked mass tourism to its predecessors, as tourist movements did not entirely cease after the outbreak of World War II. The *Shepheard's Hotel* in Cairo accommodated British officers, and British soldiers were sent to Palestine for recovery. Guidebook authors pursued their efforts to shape the perceptions of these visitors even more relentlessly than before, attempting to influence their perspectives in view of the ensuing post-war order.[15]

Yet, the establishment of independent nation-states in the 1940s impacted mobilities in the region. Tourism was now clearly framed in national terms, organised as a sector of the national economy and subject to national visa regulations. Not least, the new reality of the State of Israel fundamentally altered the possibilities of crossing borders in the Arab East.[16] The struggle of Arab Palestinian actors in tourism was severely impeded, and the idea of Yishuvi tourists skiing in Lebanon was no longer imaginable. The Arab Eastern Mediterranean had crumbled into national tourist spaces by the 1950s, both in terms of imaginaries and movements.

Moreover, the independence most countries achieved set an end to international negotiations on the right to self-rule. The political significance of tourism receded into the background. Instead, the economic potential of the Mediterranean was perceived in the accommodation of a growing number of middle- and working-class members, to whom the economic boom of post-war Europe allowed the enjoyment of holidays on the beaches of the Mediterranean.[17] Packaged tours offered cheap vacations, promising the '3 S's': sand, sun, sea.[18] As they were organised by large, international tour operators, the possibilities for local actors to shape tourism, its infrastructures and narratives were limited. As a result, the boom of tourism in the 1960s and 1970s went hand in hand with a trend often described as 'deterritorialisation', thoroughly analysed (and critically nuanced) by Waleed Hazbun in his case study of Tunisia. He showed how on the one hand, the revenues from tourism became a major pillar of the national economies around the Mediterranean which strengthened the position of tourism in national politics and the coherence of state-led tourism development policies. On the other, the resorts emerging along the coastlines often formed enclaves where interaction with the local population was limited. Mobility linking resorts to other parts of the country was no longer constitutive in tourism development. Low labour costs were thus the decisive criteria for the popularity of a place, while the destinations as such seemed substitutable.[19] In the boom years of Mediterranean tourism, the Mediterranean destinations thus lost any distinctive appeal and with a further decrease in transportation costs, sun-sand-sea tourism was easily transferred from the Mediterranean to other destinations.[20]

Some instances of political uses of tourism persisted even after nation-states had come to define the global order. On the international level, tourism seems to have remained a strategy for achieving international recognition. The Spanish dictatorship, for example, fostered tourism to present the image of an open, liberal and stable country to tourists. Under different premises, in Yugoslavia under Tito, tourism was an important element of Yugoslavia's 'Third Way', presenting a friendly form of socialism to the world.[21] In Lebanon, where the business of tourism continued to be dominated largely by Lebanese entrepreneurs, tourism remained a political resource used to position the country both in a regional and an international context. Zeina Maasri argued that the 1960s marked a shift in tourism advertisement, shaping Lebanon as a distinctly modern country on a global stage and emphasising the Mediterranean over the mountain identity.[22] Yet I suggest that the coexistence of both elements, the advertising of the mountain resorts as well as the beaches, which already characterised Lebanese tourism in the interwar period, persisted well into the twentieth century. After all, the mountain resorts experienced an upsurge in popularity among Arab *estiveurs* after World War II and started to attract a growing number of Gulf tourists from the 1960s onwards.[23] Both tourism and *estivage* remained vital in Lebanon's self-positioning as a sovereign state.

On a national level, tourism remained an instrument of propagating narratives of national history and forging national identities, as in Tunisia, where state-led heritage policies promoted a secular historical narrative as an alternative to the upsurge of Islamic identifications in North Africa during the 1990s.[24] To some degree, even notions of territorial integration in tourism development persisted. Tourism remained an option for economic development in peripheral regions, while infrastructures built for tourism also connected the local population in these areas with the nation and created income and employment opportunities.[25] In Syria under Hafiz al-Asad, tourism was perceived as such a means of economically developing the national hinterlands.[26] Furthermore, Rebecca L. Stein has argued that tourism was part of Israeli practices of occupation and conquest in the occupied territories.[27]

More often than not, however, tourism seems to have been fostered for economic reasons, as it provided states with foreign currency.[28] Such reflections seem to have led some states to pursue a strategy of 're-territorialisation', as described for the late 1970s and 1980s in Tunisia by Hazbun. This shift, according to him, had emanated from the political awareness about the vulnerability of mass tourism as an economic strategy. Reacting to the crisis in tourism experienced in the mid-1970s, governments adopted a more active role in tourism development, attempting to promote national culture and heritage and hence, advertise tourist attractions specific to the national territory rather than the interchangeable Mediterranean beaches.[29] Though driven largely by the bureaucracies of the independent nation-states of the Southern Mediterranean, territorial considerations did not seem to play a role here.[30] We may conclude that the onset of globalisation, putting an end to the territorial ambitions projected at tourism development, occurred around the 1960s or 1970s in tourism.[31]

In a long history of travel and tourism, the 1920s and 1930s marked a particular stage as the organised journeys of tourists attracted the attention of political representatives and institutions. Getting back to the idea of the 1920s and 1930s as an 'experimental stage' of the twentieth century, we have seen the emergence of Mediterranean beach resorts as leisure spaces for the working classes, but also the emergence of ministries of tourism, signalling the importance tourism came to play for a number of states.[32]

At the same time, the period marked the height of tourism as a political resource, as it was expected to nourish ambitious projects of achieving sovereignty, modernising society and shaping nation-states. The historical openness of the provisional spatial order in the post-Ottoman Arab East, the international dialogue institutionalised by the mandates system, the territorial thinking of the twentieth century, and the ambitions of local middle-class actors led them to attribute major significance to tourism, which appeared to create the conditions for territorial control. Even if they did not fully achieve their aims, these proponents of tourism development drafted visions of territorially integrated nation-states that came into being at least in the Yishuv, in Egypt and in Lebanon.

Tourism thus transformed the spatio-political order of the Arab East in the 1920s and 1930s. At the same time, the 1920s and the 1930s transformed tourism. In the context of nation-building that occurred on an international stage, requiring the acknowledgement of a rather undefined global audience, and given the territorial *zeitgeist*, tourism unfolded its political power, turning from a leisurely occupation of well-off aristocrats to a practice that brought nations into being.

Notes

[1]   Dejung, Motadel and Osterhammel, 'Worlds of the Bourgeoisie', pp. 20–21.

[2]   Pappé, *A History of Modern Palestine*, pp. 93–100.

[3]   Krämer, *Geschichte Palästinas*, pp. 233–236. Khoury, *Syria and the French Mandate*, pp. 115–119, 221–227, 239, 249.

[4]   Huber, *Channelling Mobilities*, pp. 306, 318–319.

[5]   Rau, *Räume*, pp. 183–184.

[6]   Schayegh, *The Middle East and the Making of the Modern World*, pp. 8–9.

[7]   In the Egyptian Parliament, such arguments were advanced in the context of the 1925/26 economic crisis (cf. p. 98), yet in other sources I have not come across similar arguments.

[8]   Maier, *Once within Borders*, pp. 1–2, 236–243. Maier, 'Consigning the Twentieth Century', p. 815. Osterhammel, *Die Verwandlung der Welt*, pp. 130–131, 173–174. Schlögel, *Im Raume lesen wir die Zeit*, pp. 167–176, 189–210.

[9]   Maier, 'Consigning the Twentieth Century', pp. 819–820.

[10]   Anderson, *Imagined Communities*, pp. 178, 184–185.

[11] At the same time, the arguments Lebanese actors put forward in the debate about Baalbek mirrored a shift from notions of a universal cultural heritage to an understanding of cultures as diverse and distinct, as shown by Andrea Rehling: Rehling, '"Kulturen unter Artenschutz"?', even more so when the festival was established after World War II: Maasri, 'Troubled Geography', p. 137.

[12] Maier, 'Consigning the Twentieth Century to History', pp. 819–823.

[13] Zuelow, *A History of Modern Tourism*, pp. 135–145.

[14] Cf. Jacob, *Working Out Egypt*, pp. 125–155. Azaryahu and Golan, 'Contested Beachscapes'. Walton, 'Beaches, Bathing and Beauty'. The widespread rejection of bathing practices in Palestine was an exception, according to Salim Tamari: Tamari, 'The Mountain against the Sea?'.

[15] Irving, '"This is Palestine"'. Cf. also the guidebook of Rouhi Jamil: BO, Jamil, *Damascus – Palmyra – Baalbek*.

[16] Schayegh, 'The Many Worlds of 'Abud Yasin'.

[17] Hazbun showed though that Tunisia under Bourguiba initially envisioned the establishment of winter resorts for European elites, but due to a lack in demand turned towards mass tourism: Hazbun, 'Modernity on the Beach', p. 212.

[18] Hazbun, *Beaches, Ruins, Resorts*, p. 18.

[19] Hazbun, *Beaches, Ruins, Resorts*, pp. 12, 17–18. Similar patterns characterised tourism development in Spain, Portugal and Greece: Manera, Segreto and Pohl, 'The Mediterranean as a Tourist Destination', pp. 7–8.

[20] Hazbun, *Beaches, Ruins, Resorts*, p. 19.

[21] Pack, *Tourism and Dictatorship*. Crumbaugh, *Destination Dictatorship*. Grandits and Taylor, *Yugoslavia's Sunny Side*.

[22] Maasri, 'Troubled Geography', pp. 133–138.

[23] Buccianti, 'Tourisme et villégiature', pp. 26–27, 76–81.

[24] Hazbun, *Beaches, Ruins, Resorts*, pp. 69–70.

[25] Hazbun, *Beaches, Ruins, Resorts*, pp. 58–61. Berrino, 'Programmi di valorizzazione turistica', pp. 778–779.

[26] Gray, 'The Political Economy of Tourism in Syria', pp. 66–67.

[27] Stein, '#Stolen Homes'.

[28] Gray, 'The Political Economy of Tourism in Syria', pp. 58, 70. Gray, 'Economic Reform, Privatization and Tourism', pp. 93–94, 96–99.

[29] Hazbun, *Beaches, Ruins, Resorts*, pp. 21–25, 33–35, 54–55.

[30] Hazbun, *Beaches, Ruins, Resorts*, pp. 70–71.

[31] Charles Maier dated the end of the age of territoriality, on a macro-level, to around the 1970s: Maier, 'Consigning the Twentieth Century to History', p. 815.

[32] Kunkel and Meyer, 'Dimensionen des Aufbruchs'.

References

Sources

Bibliothèque Orientale, Université Saint-Joseph, Beirut
JAMIL, ROUHI, *Damascus – Palmyra – Baalbek. With a Tour Itinerary Round Syria and Lebanon*, The Green Guide Books, Damascus: Librarie Universelle, 1941, BO/68 C2/29.

Institut français du Proche-Orient, Beirut

BUCCIANTI, LILIANE, 'Tourisme et villégiature dans la montagne libanaise: Deux exemples précis. Bhamdoun-gare et Bhamdoun-village', Thèse de doctorat (Toulouse, Université de Toulouse, 1975).

Secondary Literature

ANDERSON, BENEDICT R., *Imagined Communities: Reflections on the Origin and Spread of Nationalism* (2nd edn., London, New York: Verso, 2016).

AZARYAHU, MAOZ, and GOLAN, ARNON, 'Contested Beachscapes: Planning and Debating Tel Aviv's Seashore in the 1930s', *Urban History*, 34/2 (2007), pp. 278–295.

BERRINO, ANNUNZIATA, 'Programmi di valorizzazione turistica per le regioni meridionali negli anni cinquanta del novecento', *Società e storia*, 2018, pp. 777–804.

CRUMBAUGH, JUSTIN, *Destination Dictatorship: The Spectacle of Spain's Tourist Boom and the Reinvention of Difference* (Albany, N.Y.: SUNY Press, 2009).

DEJUNG, CHRISTOF, MOTADEL, DAVID, and OSTERHAMMEL, JÜRGEN, 'Worlds of the Bourgeoisie', in Christof Dejung, David Motadel, and Jürgen Osterhammel (eds.), *The Global Bourgeoisie. The Rise of the Middle Classes in the Age of Empire* (Princeton: Princeton University Press, 2019), pp. 1–39.

GRANDITS, HANNES, and TAYLOR, KARIN, *Yugoslavia's Sunny Side: A History of Tourism in Socialism (1950s–1980s)* (Budapest, New York: Central European University Press, 2010).

GRAY, MATTHEW, 'The Political Economy of Tourism in Syria: State, Society, and Economic Liberalization', *Arab Studies Quarterly*, 19/2 (1997), pp. 57–73.

GRAY, MATTHEW, 'Economic Reform, Privatization and Tourism in Egypt', *Middle Eastern Studies*, 34/2 (1998), pp. 91–112.

HAZBUN, WALEED, *Beaches, Ruins, Resorts: The Politics of Tourism in the Arab World* (Minneapolis: University of Minnesota Press, 2008).

HAZBUN, WALEED, 'Modernity on the Beach: A Postcolonial Reading from Southern Shores', *Tourist Studies*, 9/3 (2009), pp. 203–222.

HUBER, VALESKA, *Channelling Mobilities: Migration and Globalisation in the Suez Canal Region and Beyond, 1869–1914* (Cambridge, New York: Cambridge University Press, 2013).

IRVING, SARAH, '"This is Palestine": History and Modernity in Guidebooks to Mandate Palestine', *Contemporary Levant*, 2019, pp. 1-11.

JACOB, WILSON CHACKO, *Working Out Egypt: Effendi Masculinity and Subject Formation in Colonial Modernity, 1870–1940* (Durham: Duke University Press, 2011).

KHOURY, PHILIP S., *Syria and the French Mandate: The Politics of Arab Nationalism, 1920–1945* (London: I.B. Tauris, 1987).

KRÄMER, GUDRUN, *Geschichte Palästinas: Von der osmanischen Eroberung bis zur Gründung des Staates Israel* (6th edn., München: C.H. Beck, 2015).

KUNKEL, SÖNKE, and MEYER, CHRISTOPH, 'Dimensionen des Aufbruchs: Die 1920er und 1930er Jahre in globaler Perspektive', in Sönke Kunkel and Christoph Meyer (eds.), *Aufbruch ins postkoloniale Zeitalter. Globalisierung und die außereuropäische Welt in den 1920er und 1930er Jahren* (Frankfurt a. M., New York: Campus, 2012), pp. 7–33.

MAASRI, ZEINA, 'Troubled Geography: Imagining Lebanon in 1960s Tourist Promotion', in Kjetil Fallan and Grace Lees-Maffei (eds.), *Designing Worlds. National Design Histories in an Age of Globalization* (New York: Berghahn Books, 2016), pp. 125–140.

MAIER, CHARLES S., 'Consigning the Twentieth Century to History: Alternative Narratives for the Modern Era', *American Historical Review*, 105/3 (2000), pp. 807–831.

MAIER, CHARLES S., *Once within Borders: Territories of Power, Wealth, and Belonging since 1500* (Cambridge, Mass.: Harvard University Press, 2016).

MANERA, CARLES, SEGRETO, LUCIANO, and POHL, MANFRED, 'The Mediterranean as a Tourist Destination: Past, Present, and Future of the First Mass Tourism Resort Area', in Luciano Segreto, Carles Manera, and Manfred Pohl (eds.), *Europe at the Seaside. The Economic History of Mass Tourism in the Mediterranean* (New York: Berghahn Books, 2009), pp. 1–10.

OSTERHAMMEL, JÜRGEN, *Die Verwandlung der Welt: Eine Geschichte des 19. Jahrhunderts* (München: C.H. Beck, 2009).

PAPPÉ, ILAN, *A History of Modern Palestine: One Land, Two Peoples* (2nd edn., Cambridge, New York: Cambridge University Press, 2006).

PACK, SASHA D., *Tourism and Dictatorship: Europe's Peaceful Invasion of Franco's Spain* (Basingstoke: Palgrave Macmillan, 2011).

RAU, SUSANNE, *Räume: Konzepte, Wahrnehmungen, Nutzungen* (Frankfurt a. M., New York: Campus, 2013).

REHLING, ANDREA, '"Kulturen unter Artenschutz"?: Vom Schutz der Kulturschätze als Gemeinsames Erbe der Menschheit zur Erhaltung kultureller Vielfalt', *Jahrbuch für Europäische Geschichte*, 15 (2014), pp. 109–137.

SCHAYEGH, CYRUS, 'The Many Worlds of 'Abud Yasin: or: What Narcotics Trafficking in the Interwar Middle East Can Tell Us about Territorialization', *The American Historical Review*, 116/2 (2011), pp. 273–306.

SCHAYEGH, CYRUS, *The Middle East and the Making of the Modern World* (Cambridge, Mass.: Harvard University Press, 2017).

SCHLÖGEL, KARL, *Im Raume lesen wir die Zeit: Über Zivilisationsgeschichte und Geopolitik* (Frankfurt a.M.: Fischer, 2011).

STEIN, REBECCA L., '#Stolen Homes: Israeli Tourism and/as Military Occupation in Historical Perspective', *American Quarterly*, 68/3 (2016), pp. 545–555.

TAMARI, SALIM, 'The Mountain against the Sea?: Cultural Wars of the Eastern Mediterranean', in Salim Tamari (ed.), *Mountain Against the Sea. Essays on Palestinian Society and Culture* (Berkeley: University of California Press, 2009), pp. 22–35.

WALTON, JOHN K., 'Beaches, Bathing and Beauty: Health and Bodily Exposure at the British Seaside from the 18th to the 20th Century', *Revue Française de Civilisation Britannique*, 14/2 (2007), pp. 117–134.

ZUELOW, ERIC G. E., *A History of Modern Tourism* (London, New York: Palgrave Macmillan, 2016).

Notes on persons, associations and enterprises

Abaza, Muhammad Fikri: (1897–1979) Egyptian writer, lawyer, politician and editor of the weekly magazine *al-Musawwar* between 1926 and 1961. During the 1919 Revolution, he published articles in *al-Ahram* against the British protectorate.
He joined the National Party and was elected to the Chamber of Deputies in 1926. He refused to adopt a position as minister on several occasions, as long as British troops occupied Egypt.

ʿAbd al-Nur, Amin: Notable from Beirut, who was granted the Ottoman concession for the exploitation of the Tiberias Baths in 1912.

Abu Shahla, Habib: (1902–1957) The first Minister of Tourism and *villégiature* in Lebanon, 1937–38.

Achouʿ, Emile: Lebanese banker from Beirut, who supported a Lebanese state independent from Syria. Achouʿ was a member of the *Commission du tourisme et de la villégiature* (1929) and member of the *Comité d'Etudes économiques* in 1930.

Agronsky, Gershon: (1893 1959) Journalist and editor of the *Palestine Post*, born in Ukraine and resident of the US from 1906 onwards. Agronsky enlisted in the Jewish Legion in 1918 and remained in Palestine after the demobilisation.
From 1921–1924 he worked in New York as editor of the *Jewish Telegraphic Agency*. Agronsky served as head of the Press Bureau of Zionist Organization from 1920–1921 and 1924–1947. In 1932 he founded the *Palestine Post*.
As director of information of the Zionist Executive in Jerusalem in an honorary capacity, he met local and foreign press representatives and distinguished visitors. He represented the Zionist Executive on the committee in charge of the *Zionist Information Bureau* and on the Joint Press Committee.

Al-Ahdab, Khayr al-Din: (1894–1941) Lebanese journalist and politician from Tripoli, editor of the Arab nationalist newspaper *al-Ahl al-Jadid*. Initially he adopted pan-Arabist positions, but shifted to more moderate positions in the 1930s. From 1934 he was a deputy in the Chamber; in 1937 he was appointed prime minister and established the first Ministry of Tourism in Lebanon. Al-Ahdab was member of the *Commission du tourisme et de la villégiature (1929)*.

al-Atasi, Hashim: (1875–1960) Syrian nationalist leader from a landowning family in Homs. He had served in the Ottoman bureaucracy and rejected the French Mandate. In 1936 he negotiated the Franco-Syrian Treaty and was elected President of the Republic at the end of the year.

Alouf, Michel: Conservator of the ruins of Baalbek from January 1931 onwards, member of the *Service des Antiquités* and of the *Institut Français d'Archéologie*. Author the guidebook *Histoire de Baalbek par un de ses habitants*.
Agent of the *Société de Villégiature au Mont Liban* (VML) for Damascus in 1925 and agent of the SET in Baalbek. His sons owned the Hotel Palmyra in Baalbek.

Antonius, George: (1892–1942) Author and intellectual, born in Alexandria to a family of Lebanese origin, died in Jerusalem. Antonius served in several British administrations in Egypt and Palestine. He became famous for his work *The Arab Awakening: A History of the Arab National Movement*. Beirut: Khayats, 1938.
Having earned a degree in engineering from King's College, London, in 1913, Antonius was an official in the Public Works Department in Alexandria until the outbreak of World War I. After the end of the war, he worked for the British administration in Palestine in several departments until 1930, when Antonius resigned from the mandate government. In New York, he served as a member of the Institute of Current World Affairs, inspired and funded by Charles Crane. In 1931 he became unofficial adviser for local affairs to the British High Commissioner in Palestine.
Around the same time, he started to work on *The Arab Awakening*, completed in 1938. In 1939, Antonius was appointed secretary of the Palestinian delegation to the London Round Table conference.

Arab National Office for Public Enlightenment: Founded in September 1934 in Damascus. A major driving force behind the office was the Syrian nationalist deputy Fakhri al-Barudi. Among other tasks, it aimed at disseminating information and publications for tourists, addressing mainly visitors from France or England.

Arslan, Fu'ad: Druze emir from Choueifat, a member of the committee preparing the Lebanese constitution.
He strongly supported investments in the tourism business, e.g. subsidies for transport companies in November 1922 and support for the tourism industry in December 1924.
He suggested promoting winter tourism and advocated a coherent policy of exhibiting antiquities in the early 1920s.

Ayyub, Alphonse: Director of the Lebanese *Services économiques* from February 1922 onwards.

ʿAzzam, ʿAbd al-Rahman: (1893–1976) Egyptian politician, born in the province of Giza. He studied medicine in London, but left London to serve in the Ottoman army during the Balkan war as well as in Libya against the Italian invasion.
Member of Parliament for the Wafd Party until 1932; from the mid-1930s he broke with the Wafd Party and supported King Faruq. He lobbied for pan-Arab cooperation.
In May 1945, ʿAzzam became the first Secretary General of the Arab League. After Nasser took power, ʿAzzam left Egypt and did not return until 1972.

Baehler, Charles: (1868–1937/9) A Swiss-born trained accountant, who arrived in Cairo in 1889. He started to work for the *Shepheard's Hotel* and soon became its manager. As chairman of the *Egyptian Hotels Co. Ltd.* and co-founder of the *Upper Egyptian Hotels Co.*, he became one of the most powerful hotel owners in Egypt.

Barsky, George: Jewish hotelier and millionaire.
He leased the *Allenby Hotel* (later *Fast Hotel*) in the early 1920s and the *Palace Hotel* in early 1930.

Al-Barudi, Fakhri: (1889–1966) Syrian politician born in Damascus, Ottoman military education. After the war a close ally of King Faysal, he became one of the most prominent members of the Syrian National Bloc.
Member of Parliament in 1928, 1932, 1936 and 1943. In 1929, he sponsored the 'Umayyad scouts'. He co-founded the *Arab National Office for Public Enlightenment/Bureau national arabe de recherches et d'informations* in 1934, which lobbied for the development of tourism to Syria.

Bériel, Philippe: French colonial officer from Lyon. He worked for the *Services Economiques* of the colonial administration in Tunisia, before being sent to Beirut where he served as advisor on economic affairs to the High Commissioner.
As president of the *Club Alpin Français, Section du Levant*, Bériel promoted the establishment of skiing tourism in Lebanon.

Berthelot, Philippe: (1866–1934) General Secretary at the Foreign Office in Paris since 1920; retired in 1933. He contributed significantly to the *politique de la coopération* the French administration adopted in the late 1920s.

Bishara, Tawfiq Andraous: Deputy in the Egyptian Chamber, from the constituency of Luxor.

Bowman, Humphrey: (1879–1965) British colonial administrator.
Educated at Eton and Oxford, he served as advisor to the Egyptian Ministry of Education between 1903 and 1925, and inspector for the Sudanese Education Department before WW I. After the war, he was temporary director of education in Iraq 1918–20 and director of education in Palestine 1920–1936.

Bloch, Werner: Head of the *Zionist Information Bureau for Tourists* in the late 1930s.

Brodetsky, Selig: (1888–1954) Mathematician and Zionist leader. Born in Olwiopol (Russia) and professor of Applied Mathematics at the University of Leeds 1924-1948. Member of the executive of the *Zionist Organization* and of the Jewish Agency for Palestine from 1928.

Bureau syrien et libanais d'information et de tourisme: The office, based in Paris and directed by Dr Georges Samné, had been founded jointly by the High Commission in Beirut, the *Société des Grands Hôtels du Levant*, the *Messageries Maritimes*, several banks, transport companies and other industrial companies, as well as individuals. It was associated with the *Office des Etats du Levant* in Paris.

Canaan, Tawfiq: (1882–1964) Palestinian physician and ethnographer of rural Palestine.
He claimed historical continuities between pre-Islamic society in Palestine and the contemporary Arab life in the countryside. In the guidebook of the *Tourist Development Association of Palestine*, he published an article on 'Mohammedan Saints and Sanctuaries in Palestine'.

Chaikin, Benjamin: (1885–1945) Architect born in London, resident in Palestine since 1920. Among other buildings he designed the *Nathan Strauss Health Centres* in Jerusalem and Tel Aviv as well as the Municipal Building in Haifa. Together with Emil Vogt, he planned the *King David Hotel* in Jerusalem.

Chiha, Michel: (1891–1954) A successful banker from a merchant family in Beirut. Appointed deputy in 1925, he is considered a major contributor to the Lebanese Constitution. In 1934, he founded the Newspaper *Le Jour.*
Chiha had good relations with Muslims and Christians in Beirut and across Lebanon, but also with the French authorities. Early on, he ideated Lebanon as a territorial state. An economic liberalist, Chiha aimed at establishing Lebanon as a centre for trade and commerce and advocated a Mediterranean orientation for the country. He supported the development of tourism.

Chimy, Muhammad Beghat: Representative of the Egyptian State Railways. Member of the administrative committee of the *Tourist Development Association of Egypt.*

Chouha Frères: Postcard editors from Aleppo.

Comité du tourisme: Established in 1923, the *Comité du tourisme* aimed to improve the conditions for tourists in Syria and Lebanon. Funded by both mandate states, it created incentives for the construction of a modern grand hotel in Beirut as well as several summer villas in the mountains.
It was under patronage of the High Commission in Beirut and consisted of "high-ranking personalities from the Syrian political and economic spheres", according to the French officials. The *Comité du tourisme* covered the entire area of the French Mandate and seems to have been the predecessor of the *Commission du tourisme des Etats sous Mandat* (or it had been renamed). It published in particular the brochure *Syria and Lebanon: Holidays off the beaten track*. It was not related to the *Commission du tourisme et de la villégiature* established by the Lebanese Government in 1929.

Commission (consultative) du tourisme et de la villégiature: The *Commission du tourisme*, established on 24 December 1929, was institutionally linked to the Under-Secretary of State of Economic Affairs and presided over by Under-Secretary of State Gabriel Menassa. From April 1930 onwards, its president was the Minister of Finances.
Among its members were: Emile Achouˤ, Khayr al-Din al-Ahdab, Amin Gemayel, Constantin Joannidès, Antoine Kettaneh, Norman Nairn, Jacques Tabet, Gebran Tuéni, Georges Vayssié, ˤAbdallah Zehil, and others.
The *Commission du tourisme* aimed to support the hotel industry, encourage the development of policies favourable to tourism development as well as the publication of visual material and brochures to attract tourists.
It was probably the predecessor of the *Société d'encouragement au tourisme* founded in 1935.

Corm, Charles: (1894–1963) Francophone writer and intellectual, born in Beirut. His most famous work *La montagne inspirée* (*The Sacred Mountain*) is considered a major artistic contribution to Lebanese nation-building.

Corm founded and directed the *Revue Phénicienne*, he idealised the alleged Phoenician roots of Lebanon and idealised Mount Lebanon as a core region of Greater Lebanon. This idea did not necessarily exclude Syria from the outset: he ran the pro-French organisation *Association Nationale de la Jeunesse Syrienne* (ANJS), proclaiming a federal Greater Syria under French guidance.
As a businessman, he made a fortune as the sole agent for *Ford* motorcars in the French Mandate in the 1920s.

D'Andurain, Pierre: French owner of the *Hotel Zenobia* in Palmyra.

Debbas, Charles: (1885–1935) Lebanese president from September 1926 until January 1934. Formerly member of the Arab Syrian Congress (convened in Paris in June 1913, advocating a decentralisation of the Ottoman Empire).

Doss, Tawfiq: Egyptian politician, member of the Liberal Constitutionalist Party. Minister of Agriculture, Minister of Communications and president of the committee to prepare the International Congress of Tourism in Cairo 1933.

Eddé, Emile: (1886–1949) Lebanese politician and French-educated lawyer, born in Beirut. Eddé was an ardent Lebanese nationalist and a supporter of the French mandate in Lebanon. He advocated the idea of Lebanon as a Mediterranean country. His anti-Arab policies rendered him extremely unpopular among Lebanese Muslims and Greek Orthodox. In 1936 he became Lebanese president, despite a weak majority in parliament. French support stabilised his presidency.

Eliachar, Eliahu: (1899–1981) Born in Jerusalem, he received his education at the *Université Saint-Joseph* in Beirut. Between 1922 and 1934 he was in government service for the British mandate as an officer in charge of the Trade Section of the Dept. of Customs, Excise and Trade. He was editor of the *Commercial Bulletin* of the Government and initiated the first census of industries in Palestine in 1928.

Egyptian Hotels Ltd.: A large hotel corporation, its chief shareholder was Charles Baehler. In 1925, the company took over the properties of the *Grand Hotels D'Egypte* and dominated the Egyptian sector of grand hotels.

Elizabetha Haven of Rest: Sanatorium in Tiberias, founded by Salomon Feingold and Miss Margarete Palmer, inaugrated in January 1929.

Fakhuri, Samuel: Notable from Beirut. He was granted a concession to exploit the Tiberias baths in 1912.

Feingold, Solomon: Owner of the *Elizabetha Haven of Rest* in Tiberias.

Gebeili, Nicolas and Elie: Owners of the *Grand Hôtel D'Aley* in Lebanon.

Gesundheit, Jacob: Director of the *Hamei Tiberia Company*.

Ghanem, Shukri: (1869–1921) Francophone Lebanese poet living in Paris, editor of the Paris-based paper *Correspondance d'Orient*. Ghanem had been a member of the Arab Syrian Congress (which convened in Paris in June 1913, demanding a decentralisation of the Ottoman Empire). He headed the *Comité Central Syrien*, supported and financed by France, which lobbied for a Greater Syria under French guidance. Like other Phoenicianists, he initially supported a non-Arab Greater Syria, later shifting towards a Lebanese nationalist position.

Ghattasse, Chéhadé: Syrian entrepreneur

Al-Ghazzi, Kamil: (1853–1933) Historian and chronicler from Aleppo. He was a founding member of the *Société Archéologique d'Alep* and editor of its bilingual review.

Gibran, Gibran Khalil: (1882–1931) Lebanese poet, who emigrated with his family to the US in 1895. Gibran was considered a preeminent representative of Lebanese poetry, mountain romanticism, and Phoenicianist nationalism.

Gobril, Rizkallah: Lebanese entrepreneur and owner of a travel agency in Beirut: *Messageries automobiles – Itinéraires Rizkallah*. He advocated Lebanese separatism. In 1928/29 he published a guide on Syria and Lebanon: *Visitez le Levant*.

Grandguillot, G.: French Commercial attaché for the Near East and director of the tourist office of France in Egypt.

Grossmann, Richard: (1873–1916) Swiss-German member of the Protestant Templar Colony in Haifa. In 1896, he founded the *Hotel Tiberias*.

Grossmann, Fritz: (1908–1938) Son of Richard Grossmann, born in November 1908 in Tiberias. He became owner and manager of the *Hotel Tiberias* after the death of his father.

Haardt, Georges-Marie: (1884–1932) A Belgian-French businessman and director of the *Citroën* company. Haardt organised the so-called *"Croisière noire"* as well as the *"Croisière jaune"*, expeditions by car across Africa and Asia.

Harb, Talʿat: (1867–1941) Egyptian businessman and economist. He founded *Banque Misr* in April 1920 as a purely Egyptian bank as well as numerous affiliated companies, among them *Misr Air* and *Misr Studios*, Egypt's first large cinema company, inaugurated in 1934.

Harrison, Helena: (c. 1895–1986) British botanist. In 1925, Harrison travelled to Egypt and Palestine, where she visited her brother Austen St Barbe Harrison, an architect and town planner in the Public Works Department, Palestine.

Haydar, Subhi: Lebanese politician from an influential landowning family of the Beqaa, one of the major Shiite leaders in the area around Baalbek. He served as minister in the late 1920s; in the mid-1930s he was director of education and arts.

Herrmann, Hugo: (1887–1940) Journalist, editor and Zionist activist. Born in Mährisch-Triebau (Moravská Třebově), from 1909 to 1912 he was secretary of the *Zionist Organization* of Bohemia. From 1913 until the outbreak of WW I, he was editor in chief of the *Jüdische Rundschau*, the newspaper of the German Zionist Organization. For several years, he directed the *Palestine Foundation Fund* in Bohemia. From 1925 onwards, he wrote several books and brochures on Palestine and directed the *Palestine Picture and Photo-block Agency* in Brünn/Brno which provided newspapers with photographs of Palestine and the Zionist work. In 1934 he emigrated to Jerusalem. In the same year, he established a new map of Palestine published for the *Palestine Foundation Fund.*

Hotel Baron: The only first-class hotel in Aleppo in 1931.

Hôtel Saint-Georges: Modernist hotel in Beirut, directed by the *Société des Grands Hôtels du Levant.* The building was designed by the architects Antoine Tabet, Jacques Poirrier, André Lotte and Georges Bordes. The hotel was inaugurated in 1932.

Al-Husayni, Ismaʿil: (d. 1945) Landowning notable from Palestine. He was the head of the Jerusalem branch of the Husayni family, and landowner in inner Palestine. He was a conservative pro-Ottoman loyalist who shunned away from nationalism. Interested in archaeology, he founded the first Palestinian museum in Ottoman times and he was involved in the organisation of the reception of Kaiser Wilhelm II in Jerusalem.
He cooperated with Jewish business partners and sought an understanding with the Zionist movement in the mandate period. The officials of the British mandate administration respected him and cooperated with him. Al-Husayni himself had a pro-British posture and disliked the idea of Greater Syria. He was a board member of the *Kallia Company* on the Dead Sea.

Iskandar, Naguib: Deputy in the Egyptian Chamber, from Shubra in the Cairo governorate. He favoured a strengthening of European tourism to Egypt as it enhanced the national prestige of Egypt.

Jamal, A.: Manager of the *Jamal Brothers* travel agency, Jerusalem and one of the Directors of the *Tourist Development Association of Palestine.*

Jouvenel, Henry de: (1876–1935) French journalist and politician, the first civilian French High Commissioner in Beirut (1925/26). Under his aegis, a drafting committee was formed in order to draft a constitution for Lebanon.

Joannidès, Constantin: (*1884) Agent of the *Messageries Maritimes* in Alexandria and from 1928 onwards in Beirut. He was a French national, yet spoke Greek and Turkish as well. In 1929/30 he became a member of the *Commission du tourisme e de la villégiature.*

Karim, Ibrahim Fahmi: Egyptian deputy, Minister of Public Works, Minister of Communications.

Keith-Roach, Edward: (1885–1954) British civil servant in Palestine. Keith-Roach held various government positions in Palestine between 1919 and 1943: deputy district commissioner in Jerusalem District, 1926–31; district commissioner in Haifa 1931–37 and Jerusalem 1937–1943. He co-authored the *Handbook of Palestine*, 1922 and 1930.

Kettaneh, Antoine: Lebanese entrepreneur in the transportation business. Kettaneh was the owner of the automobile transport company *Kettaneh Frères*, which offered transportation services on the route between Beirut and Baghdad. He might have been the founder/owner of the Hotel Zenobia in Palmyra, which was named Hotel Kettaneh before. He was member of the Lebanese *Commission du tourisme et de la villégiature*.

Khadder, George Assad: (*1902) Born in Jerusalem, he studied at the American University of Beirut. Khadder was secretary of the Arab Chamber of Commerce, Jerusalem and secretary of the *Tourist Development Association of Palestine*.

Khaouam, Joseph and Elias: The Khaouam brothers owned several famous hotels in Homs, Hama, Damascus, Beirut and Baalbek.

Khoori, Alexander: Also Iskandar Khouri, Alexandre Cury, Alec R. Cury.
Travel guide and author, born in Palestine. According to him, he was formerly attached to the Egyptian Department of Antiquities. He published numerous guidebooks for travellers on Luxor, Alexandria, Cairo, Jerusalem and Baghdad and was the author of the language guide *Arabic without a Teacher*. His guidebooks were translated into French by Henri Munier, the librarian of the *Egyptian Museum* in Cairo.

Al-Khoury, Fouad: Lebanese architect. In the *Revue phénicienne*, he published an article on the hotel industry in Lebanon, 1919.

King David Hotel: Hotel in Jerusalem, initiated by the Mosseri Family, founders of the *Palestine Hotels Company*. Planned by the Swiss architect Emil Vogt. The construction works began in July 1929, the inauguration was celebrated in January 1931. The manager at the time was Joseph A. Seiler from Switzerland.

Le Gras, Charles: French traveller, author of *Le Lac méditerranéen, Pèlerinage à Jérusalem et Troisière en Méditteranée orientale du 5 avril au (9 mai 1927), Notes de voyage.*

Levy, Harry: Manager of the *Kallia Hotel.*

Levy, Moss: (1892–1967) British Zionist businessman, who settled in Jerusalem in 1920. Experienced in tourism.

Löwenstein, Dr Fritz: (1892–1964) Born in Berlin. Secretary of the *Zionist Organization,* Germany, editor of the *Jüdische Rundschau* 1915–1923. In 1923, he emigrated to Palestine, where he became director of the Jewish Agency Information Bureau until 1931. Author of the guidebook *Das jüdische Palästina* (1927) and *Guide to New Palestine*. Director of the *Zionist Information Bureau for Tourists* since 1929.

Maestracci, Noël: French colonial officer in North Africa and the French mandate. He published the guide *Contemporary Syria: Everything you need to know about the territories under French Mandate,* as well as *Le Maroc contemporain.*

Maklouf, Ibrahim: Lebanese editor. Together with his brother Emile he founded the Lebanese-Phoenicianist journal *La Revue du Liban et de l'Orient Méditerranéen* in Paris in 1928, which circulated in France and in Lebanon. Maklouf published numerous articles on Phoenician archaeology, history and civilisation and in 1933, the brothers founded an organisation for Lebanese emigrants in Paris: *Nova Phenicia*. They organised an event on the occasion of Lamartine's centenary. In 1937/38, Maklouf edited the brochure *Le Liban – Pays d'art et de tourisme* for the SET.
When the brothers moved back to Beirut in 1939, they changed the name of their review to *La Revue du Liban et de l'Orient Arabe*.

Martel, Damien de: (1878–1940) French diplomat and High Commissioner of France in Lebanon and Syria between 1933 and 1939. During his office, he launched a vast economic programme and Palmyra was classified as a "historical monument".

Mascle, Joseph: (1877–1956) Military chaplain in Suwayda in the Djebel Druze. Author of the tour guide *Le Djebel Druze* (1936).

Matson, G. Eric: (1888–1977) Photographer. Born in Sweden, he emigrated with his family to Palestine in 1896. In 1934 he took over the *American Colony Photo Department* (founded in 1898) together with his wife Edith, and the department was renamed the *Matson Photo Service*. In 1946 they fled to the United States.

Matson, G. Olaf: Guidebook author and brother of Eric Matson. He was the author of the *American Colony Guidebook to Palestine*.

Mena House: A hotel near the Great Pyramids in Egypt, founded by a wealthy British couple in the 1880s. In 1904 the owners sold it to the hotel company of George Nungovich, which was taken over by Charles Baehler's *Egyptian Hotels Ltd.* in 1925.

Menassa, Gabriel: He became Lebanese Under Secretary of State of Economic Affairs in November 1929, responsible for tourism, *villégiature* and emigration. Menassa belonged to the circle of New Phoenicians around Michel Chiha, advocating liberal economic policies. Carla Eddé considered him in this position a "hidden minister". Menassa drafted a plan for a "reconstruction of the Lebanese economy and a reform of the state". He presided over the *Commission du tourisme et de la villégiature*.

Mosseri family: Jewish family originating from Cairo and Alexandria, founders of the *King David Hotel*. They owned a bank that held *Egyptian Hotels Ltd.* In 1921 they founded the *Palestine Hotels Company*, registered in Jerusalem in 1929.

Naccache, Albert: Lebanese civil engineer from an influential Maronite family, owner of the *Société de Kadisha*, which ran an electrical power plant in the Qadisha valley. Trained in engineering at Lausanne, he worked in Leningrad and for the Turkish Government. During the mandate, he became director of the services of public works in 1920.
In the *Revue phénicienne*, he published an article on *Notre avenir économique* (Our economic future), arguing in favour of a Greater Lebanon and lobbying for investments in the estivage

business. Despite claims for independence from European capital, he later cooperated with Zionist investors.

Al-Nahas, Mustafa: (1879–1965) Egyptian judge and leader of the Wafd Party, cabinet minister, and five-time premier. He was the successor of Saʿd Zaghlul as leader of the Wafd Party. Communications minister in 1924. Vice-president of the Chamber and president the following year. Prime minister in 1928, 1930, 1936–1937, 1942–1944, 1950–1952.
Al-Nahas headed the delegation in the negotiations with Britain that led to the Anglo-Egyptian Treaty in 1936. His alliance with the British in World War II damaged his reputation as a nationalist leader.

Nahhas, Antoine: Modernist architect; among his major projects was the Lebanese National Museum.

Nassif, Sulayman: Entrepreneur who was granted the concession for the exploitation of the Tiberias Baths together with Amin Rizq by the British mandatory. After the concession was transferred to the Zionist *Hamei Tiberia*, he continued to exploit the baths of al-Hamma.

Nimr, Faris: (1856–1951) Lebanese editor and intellectual. From a Greek Orthodox family in Wadi al-Taym, he studied at the Syrian Protestant College (later American University of Beirut). As editor of *al-Muqtataf*, he was one of the pioneers of Arabic journalism. With the co-founder of *al-Muqtataf*, Ya'qub Sarruf, he moved with the journal to Cairo in 1884 (following a quarrel over Darwin's theory). In 1889 they founded a daily newspaper, *al-Muqattam*. The editorials of the paper advocated individualism and a liberal economic system. Faris Nimr was a member of the committee to prepare the International Congress of Tourism in Cairo.

Nungovich, George: (d. 1908) A Greek-Cypriot hotel impresario in Egypt, whose hotels rivalled Baehler's grand hotels. His hotel company (*Nungovich Hotels Company*, later *Grand Hotels d'Egypte*) was founded in 1899, and taken over by Baehler's *Egyptian Hotels Ltd.* in 1925.

Office économique et touristique des Etats de Syrie, du Grand-Liban et des Alaouites: Established under High Commissioner Henry de Jouvenel on May 17, 1926. Its main office was in Cairo. By distributing information and advertisements on tourism to the French mandate, the *Office économique* hoped to attract tourists and *estiveurs*. The commercial attaché at the French Legation in Cairo had the administrative and financial control, but the costs for running the office were shared between the mandate states.

Office du tourisme et de la villégiature au Grand Liban: It was created on 17 March 1925 under the patronage of the Lebanese Government. The director of the *Services Economiques* presided over the office. It controlled the conditions of hygiene in hotels, organised advertisement campaigns and rented out cars to tourists. In February 1930, the office signed an agreement of collaboration with the *Société de Villégiature au Mont Liban* (VML).

Palace Hotel: In 1928, the *Supreme Muslim Council* decided to have the *Palace Hotel* constructed, a grand hotel for Jerusalem. It was leased to George Barsky in February 1929, thus before its inauguration in December 1929.

Palestine and Egypt Lloyd Ltd.: Travel agency, successor of the *Palestine Express* travel agency in 1928.

Palestine Express Travel Agency: Travel agency created in the early 1920s by the Jewish Agency and the *Jewish National Fund* under the auspices of the *Anglo-Palestine Bank.*

Hotel Palmyra: Grand hotel in Baalbek. It was established in 1880, rebuilt and newly furnished in 1924. Previously owned by P. Mimikaki, during the interwar period the Alouf brothers managed the hotel, the sons of Michel Alouf.

Perera, Guy U.: An industrialist and agent of the shipping company *Triestino Lloyd.* Perera was a member of the *Touring Club of Egypt.* In 1933/34 he became a member of the administrative committee of the *Tourist Development Association of Egypt.* He edited the illustrated brochure *A trip to Egypt/Un voyage en Egypte* with A. Otto.

Philippar, Georges: (1877/1883–1959) French manager. Since 1918 he was general director of the shipping company *Messageries Maritimes*; from 1925–1948 its president. He had a pronounced interest in tourism, published travelogues (e.g. *En Méditerranée* and *Voyage en Egypte)* and served as member of the French *Conseil Supérieur du Tourisme.* Around 1920 he became board member of the *Compagnie française du tourisme.* Philippar was well connected in French colonial circles, for example serving as Vice-President of the *Institut colonial français* and since December 1934 as board member of the illustrated magazine *Le Monde colonial illustré.* In 1929, he published a brochure on *Le tourisme en Syrie.*

Piérard, Louis: Belgian journalist and deputy of the Belgian Socialist Party. He travelled to the French Mandate of Syria in February 1927.

Qatinqa, Barukh: (1887–1945) Born in Russia, Jewish civil engineer and building contractor, who emigrated to Palestine in 1908. During World War I, he served as chief engineer of the *Hedjaz railway* for the Ottoman army. As a contractor, he was involved in the construction of several buildings, among them the YMCA building in Jerusalem and the *Palace Hotel.*

Radwan, ʿAli Ibrahim: Egyptian deputy from el-Tallein in the Mudiriyya of Sharqiyeh in the Eastern Delta. In the Chamber, he advocated a reduction for railway tickets towards the summer stations.

Rechter, Zeev: (1899–1960) Architect. Rechter was born in Russia, studied architecture and engineering in Italy and France, and went to Palestine in 1919. He had a considerable influence on housing construction in Israel and in particular in Tel Aviv, and is considered one of the leading modernist architects of Israel. He was the architect of the *Kallia Hotel.*

Rizq, Amin: Lebanese entrepreneur from Broummana/Lebanon who was granted the concession for the exploitation of the Tiberias Baths together with Sulayman Nassif.

Rockefeller, John D.: (1839–1937) American oil magnate and philanthropist. His *Rockefeller Foundation* donated money to allow for the construction of the *Palestine Archaeological Museum*.

Roe, Donovan: British naval officer. During his service in the 1930s, he undertook touristic excursions in the Eastern Mediterranean.

Rosenblatt, Bernard A.: (1886–1969) Lawyer from New York. Born in Grodok (Poland/Russia), he lived in the United States since 1891. In 1915 he organised the *American Zionist Commonwealth*, a land development and settlement company. Rosenblatt was vice president and a member of the executive committee of the *Zionist Organization of America* from 1920 to 1946, and a member of the Zionist Executive of Palestine between 1923 and 1927. He actively supported business enterprises in Palestine during the mandate period. President of the *Jewish National Fund* and *Keren Hayesod of America*.

Rowan-Hamilton, Norah: (d. 1945) Born Norah Logan Phillips, she married the British Lieutenant Archibald James Rowan Hamilton (1877–1915) in 1908. Rowan-Hamilton published several travelogues: *Through Wonderful India and Beyond* (1915), *Under the Red Star* (1930) and *Both sides of the Jordan: A Woman's Adventures in the Near East* (1928). She travelled with a delegation of the *British Archaeological Society* to the Arab East.

Sabri, Husayn: (*1897) Businessman, president and director of numerous companies. Born in Cairo, he worked and lived in Alexandria, where he was for example vice-president of the shipping company *The Khedivial Mail Line*, administrator of the *Cairo Electric Railways & Industrial Co*. Sabry also served as governor of Alexandria.

Sacher, Miriam: (1892–1973) British Zionist. She was one of the early founders of WIZO (*Women's International Zionist Organisation*) and played a leading role in the Federation of Women Zionists of Great Britain and Ireland.
Her husband, Harry Sacher (1881–1971), was a Zionist lawyer and politician from London, and member of the Executive of the *Zionist Organization* between 1927–1931.
Miriam Sacher was board member of the *Kallia company*.

Sadiq, Ahmad: Governor of the Municipality of Alexandria, member of the *Tourist Development Association of Egypt* and director of the *Office du Tourisme de l'Etat Egyptien* from 1935 onwards.

Salama, Edwin: Lawyer and administrator of societies. Salama was an agent of the *American Express* agency and one of the directors of the *Tourist Development Association of Palestine*.

Salameh, Dimitri: Manager of *Thomas Cook & Son* in Palestine, one of the directors of the *Tourist Development Association of Palestine*.

Samné, Georges: Born in Egypt into a Greek Catholic family from Damascus. Samné lived in France since the turn of the century. He supported a non-Arab Greater Syria, and opposed the idea of an independent Lebanon.

In May 1918, he authored a memorandum to the Quai d'Orsay, in which he made a case for French aid and intervention in Greater Syria after the war. Despite his pro-French inclinations, he considered the Ottoman system better than French rule in Algeria and the British system in Egypt.
Samné was co-founder and Director of the *Bureau syrien et libanais d'informations et de tourisme (Information and Tourist Office for Syria and Lebanon)* and director of the review *Correspondance d'Orient*.

Samuel, Herbert: (1870–1963) British High Commissioner of Palestine 1920–1925 and president of the *Tourist Development Association of Palestine*.

Sarrafian Brothers: Abraham (c. 1875–1926), Boghos (1876–1934) and Samuel (1884–1941) Sarrafian. Photographers and postcard editors of Armenian origin. Originally from Diyarbakir, they settled in Beirut in 1897 as a consequence of the massacres in 1895. The Sarrafian Brothers were the biggest postcard editors in the Arab East.

Al-Sawda, Yusuf: Born in Bikfaya, educated at the *Université Saint-Joseph* in Beirut. In 1908 he emigrated to Alexandria, where he worked for the *Tribunaux mixtes* and served as secretary of the *Alliance Libanaise*.
Author of *Fi Sabil Lubnan* on the Phoenician origins of the Lebanese; strong advocate of the Phoenician identity of the Lebanese.

Al-Sayyid, Husayn Ahmad: Egyptian overseer of the works in the expedition of Lord Carnavaron and Howard Carter.

Seiler, Joseph A.: (1896–1948) Scion of one of the most famous hotel dynasties in Switzerland, the Seiler family who established tourism in Zermatt and were among the pioneers of tourism in Switzerland. Joseph A. Seiler was manager of the *King David Hotel* in Jerusalem until 1937.

Shepheard, Samuel: (1816–1866) Director of the *Shepheard's Hotel*. Shepheard arrived in Egypt in the 1840s, took over the hotel from its British owner in 1846. In 1860, he sold the hotel (to Philip Zech, a Bavarian hotelier) and left Egypt. In the 1890s the building was torn down and replaced by a new structure designed by the architect George Somers Clarke Jr. It was inaugurated in 1893. Charles Baehler was employed by the heirs of Philip Zech, until Baehler bought out the shares in 1905 and became its owner. The *Shepheard's* attracted in particular British travellers in Egypt until its destruction in 1952.

Sidqi, Ismaʿil: (1875–1950) Lawyer, cabinet minister, twice premier. Born in Alexandria to a wealthy bourgeois family.
Sidqi was a member of the Wafd in 1918 and interned in Malta with Saʿd Zaghlul. Later, he broke with the Wafd and helped to establish the *Constitutional Liberal Party* in 1922. As interior minister in 1924–25 he worked closely with King Fuʾad. Sidqi founded the *Shaʿb Party* (People's Party) to support his campaign for prime minister under the 1930 constitution. When he took power in 1930, he headed a pro-palace government, dissolved the parliament and ruled by decree. The new constitution he introduced enlarged the powers of the king. Sidqi served on the Egyptian delegation that negotiated the 1936 Anglo-Egyptian Treaty.

Throughout his career, one of his major concerns was the development of the Egyptian industry. In 1917, he created a *Committee for Commerce and Industy* to foster the Egyptian industry. President of the *Egyptian Federation of Industries* in 1930.

Sidqi, Mahmud: Governor of Cairo.

Smith, Edwin W.: Former Protestant missionary in South Africa with profound ethnographic interests. He travelled to Palestine in late 1929.

Société Archéologique d'Alep: Founded in January 1930, directed by Kamil al-Ghazzi. Its members published the *Revue archéologique* from May 1931 onwards.

Société d'encouragement au tourisme au Liban (SET): Lebanese interest group based in Beirut. It was founded in 1935 and issued numerous publications to assist tourists in planning their stay, such as hotel guides or information brochures on sights in Lebanon. The SET seems to have been the successor organisation of the *Commission du tourisme et de la villégiature* (1929). It was founded in 1935, and declared of public interest in 1936. It was not an official government institution, but it received subsidies from the Lebanese state and regularly cooperated with political bodies.

Société des Grands Hôtels du Levant: Enterprise that constructed hotels in Beirut, Damascus, Aleppo, the most famous of which is the *Hôtel Saint-Georges* in Beirut. The *Société des Grands Hôtels* promoted the foundation of the *Bureau syrien et libanais d'information et de tourisme*.

Société de Villégiature au Mont Liban (VML): The VML offered the services of a travel agency from 1913. It was not operational during the war and resumed its activities only in the summer of 1922.
Its main office was in Cairo, but it had branches in Beirut, Haifa, Qantara, Alexandria, Port-Said, Jerusalem, Jaffa, Damascus, Paris and Marseille and agents in numerous Lebanese mountain villages. The VML published guidebooks on the region of *Bilad al-Sham* and Egypt, with a focus on Lebanon. Its agents offered organised tours and reliable guides to visitors. Moreover, its agents offered villas, pensions and rooms for rent in the major *estivage* centres, sold tickets for transportation or organised package tours. For summer guests in the major destinations, they offered postal services and delivered newspapers from neighbouring Arab countries or the private mail of summer guests.
The dense network of agents all over Lebanon and in Syria addressed mainly a regional clientele, in particular from Egypt. Among the letters of recommendation, American travellers, and dignitaries in particular from Cairo recommended the VML's services.

Steele-Maitland, Mary: Wife of Sir Arthur Ramsay Steele-Maitland (1876–1935), a conservative politician who served as under-secretary for the colonies during World War I and as Minister of Labour in 1924-25. He remained MP until his death in March 1935. In winter 1931/32, the couple travelled to the Eastern Mediterranean.

Strekalovsky, Nicholas: Jewish painter and illustrator, who had immigrated to Palestine from Eastern Europe. He designed the cover illustration of the Lebanese *estivage* guide *al-Istiyaf fi Lubnan* (1926) and posters for the Egyptian *Misr Airlines* in the 1930s.

Tabet, Antoine: (1907–1964) French-educated modernist architect of the *Hôtel Saint-Georges*, 1930, and of *Almaza Beer Factory*, 1934.

Tabet, Ayoub: Lebanese politician, a nationalist from Bhamdoun and the right hand of Emile Eddé.
He had been a member of the Arab Syrian Congress (convened in Paris in June 1913, demanding a decentralisation of the Ottoman Empire). He spent the war years in New York, where he lobbied for a non-Arab unified Greater Syria in the League for the Liberation of Syria and Lebanon, over which he presided. Other members were Amin al-Rihani, Gibran Khalil Gibran and Mikhail Nuʿaym.
During the Mmndate, Tabet was a member of the *Conseil Représentatif*, minister, and director of the Lebanese *Services Economiques*. At the end of the mandate, Catroux established him as president and premier.
Tabet supported the strengthening of *estivage*, which he considered a contribution to the improvement of the economic situation of Lebanon. Moreover, from his point of view, Baalbek, the Lebanese historical heritage and the National Museum had to be promoted among international visitors as they fostered the prestige of Lebanon.

Tabet, Jacques: (1885–1956) Influential francophile Phoenicianist, former publisher of *La Syrie*. He supported the idea of a Greater Syria, considered as the successor state of ancient Phoenicia. Tabet co-founded the *Amis du Musée*, an association for fund-raising and lobbying for a large Lebanese National Museum. Later, he became the first treasurer and director of the Lebanese National Museum. Tabet was a member of the *Commission du tourisme et de la villégiature 1929/30*.

Taqla, Gabriel: (1891–1943) Egyptian journalist. His family had roots in the village Kfar Shima near Beirut. Taqla was the owner of the newspaper *al-Ahram* (founded in 1875 by his father and uncle) and a member of the committee to prepare the International Congress of Tourism in Cairo.

Touring Club de Syrie et du Liban: Founded in 1919, its head office was located in the offices of the journal *La Syrie* in Beirut. Its president Georges Vayssié was both president of the *Touring Club* and editor of *La Syrie*. It gave advice on tourism in the Mandate, and assisted travellers in getting the necessary official permissions. Moreover, the *Touring Club* distributed maps and guidebooks. Affiliated with the *International Automobile Association* and international touring clubs, it counselled the Lebanese Government in developing tourism. Two representatives of the *Touring Club* were members of the *Office du tourisme et de la villégiature au Grand Liban*, 1925.
Among its 471 members were Emile Achouʿ, ʿAbdallah Beyhum, Constantin Joannidès, Pharaon, Sursock, Tabet and Habib Trad.

Tourist Development Association of Egypt: The TDA was founded in 1912, though regular activities were discernible only in the 1930s. It was involved in the organisation of the International Congress of Tourism in Cairo, published guidebooks and brochures.
The TDA was under the patronage of the Egyptian king and presided over by the general manager of the Egyptian State Railways, Telegraphs and Telephones. Egyptian dignitaries were among the members as well as numerous representatives from European travel companies and leisure organisations.

Tourist Development Association of Palestine: The creation of the TDA was promoted notably by the tour operator *Palestine Lloyd Ltd.* and High Commissioner Herbert Samuel served as president of the association. Upon official registration of the *Tourist Development Association of Palestine* in 1932, its founders stated that its purpose consisted in making Palestine and Trans-Jordan known abroad in order to attract a greater number of tourists and visitors.
The TDA gathered both Arab Palestinian and Zionist members and included representatives of the big companies involved in tourism to Palestine.

Tuéni, Gebran: (1890–1947) Lebanese editor and minister from Beirut. Tuéni was a journalist in Egypt and returned to Beirut in 1923. He was co-founder of the newspaper *al-Ahrar* in 1924 and founded the Lebanese daily *al-Nahar* in 1933.
In 1931, Tuéni became Minister of Education. He was a member of the *Commission du tourisme et de la villégiature* 1929/30.

Tulloch, Thomas G.: British engineer, who was granted the concession for the extraction of minerals from the Dead Sea together with Moise A. Novomeyski. He was one of the directors of the *Palestine Potash Company* and manager of the *Kallia Resort* on the Dead Sea.

Upper Egypt Hotels Ltd.: The hotel corporation owned among other hotels the *Winter Palace Hotel* and the *Luxor Hotel*, both in Luxor, as well as the *Cataract Hotel* in Aswan. *Thomas Cook & Son* held shares in it; the consortium was headed by Charles Baehler.

Vacaresco, Hélène: (1866–1947) Romanian diplomat and writer.
Born in Bukarest, she lived mainly in France, and wrote in French, e.g. *Nuits d'Orient*, 1889. From 1919–1940 she served as a Romanian delegate to the League of Nations. She was co-founder and deputy of the *Institut international de coopération intellectuelle* (1924), the predecessor of UNESCO.

Vacca, Virginia: Italian Arabist and journalist, she reported notably on Syrian economic development.

Vallotton-Warnery, Henry: Member of the Swiss National Council from Geneva and a member of the *Swiss Touring Club*. He undertook a journey to Syria in 1926.

Vayssié, Georges: Vayssié was an editor and journalist and very active in tourism development. He was general director of the *Havas* press agency in the French Mandates and owner of the paper *La Syrie*.

Vayssié had been involved in the founding of the *Automobile Club of Egypt*; he was a member of the *Touring Club* in Lebanon and served as a member of the *Comité du Tourisme* established by the High Commission in 1923 as well as of the *Commission consultative du tourisme et de la villégiature* 1929/30.

Vilnay, Zev: (1900–1988) Born in Kishinev he emigrated to Palestine with his parents. He authored *Steimatzky's guide to Palestine*.

Williams, George H.: (1893–1975) British businessman, entrepreneur in textiles and amateur photographer, who travelled to the Arab East in 1925.

Witasse, Pierre de: (1878–1956) French representative at the French Legation in Cairo. President of the French *Comité des correspondants du tourisme*.

Yusuf, Muhammad Amin: Egyptian deputy and inspector general of the Department of Commerce and Industry. He was member on the administrative committee of the *Tourist Development Association of Egypt*.

Zaccour, Michel: (*1896) Editor of the Lebanese paper *al-Maʿrid*. Lebanese deputy, Minister of the Interior in 1937.

Zehil, ʿAbdallah: General Agent of the *Compagnie française de naviagtion à vapeur Cyprien Fabre (Fabre Line*. Zehil was the author of the guidebook *Petit guide historique* (1929) and a member of the *Commission du tourisme et de la villégiature* (1929).

Zionist Information Bureau for Tourists: The ZIB was established by the Jewish Agency, *Keren Hayesod*, as well as the *Jewish National Fund* (JNF, *Keren Kayemet Le-Yisrael*) in the early 1920s in order to establish organised tourism to Palestine. According to a letter to the editor by Mr Grunhut, its manager, it was founded in 1923 (Cohen-Hattab and Shoval dated its foundation to 1925).
An additional branch was established in Tel Aviv in 1935. The ZIB published guidebooks such as the ones written by Fritz Löwenstein, *Das jüdische Palästina*, and the *Guide to New Palestine*. Its directors were Grunhut, Fritz Löwenstein, and, from 1938, Werner Bloch.

Index of persons, associations and enterprises

Printed and bound by CPI Group (UK) Ltd, Croydon, CR0 4YY

30/06/2026

14910892-0002